Protein Electron Transfer

Protein Electron Transfer

Edited by
Derek S. Bendall
Department of Biochemistry, University of Cambridge,
Cambridge, UK

CRC Press
Taylor & Francis Group
Boca Raton London New York

CRC Press is an imprint of the
Taylor & Francis Group, an **informa** business

First published 1996 by Bios Scientific Publishers Limited

Published 2019 by CRC Press
Taylor & Francis Group
6000 Broken Sound Parkway NW, Suite 300
Boca Raton, FL 33487-2742

© 1996 by Taylor & Francis Group, LLC
CRC Press is an imprint of Taylor & Francis Group, an Informa business

First issued in paperback 2019

No claim to original U.S. Government works

ISBN-13: 978-0-367-44870-7 (pbk)
ISBN-13: 978-1-85996-040-0 (hbk)

Visit the Taylor & Francis Web site at
http://www.taylorandfrancis.com

and the CRC Press Web site at
http://www.crcpress.com

Copyright held by Taylor & Francis Group, LLC for all chapters except Chapter 10 on pp. 249–272 which is © The Nobel Foundation 1993 or *Les Prix Nobel* 1992.

A CIP catalogue record for this book is available from the British Library.

Typeset by Chandos Electronic Publishing, Stanton Harcourt, UK.

Front cover: Crystal structure of complex between yeast cytochrome c peroxidase and cytochrome c (2pcc) drawn using MOLSCRIPT. Kraulis PJ (1991) *J. Appl. Crystallogr.* **24**: 946–950.

Contents

Contributors

Bendall, D.S. Department of Biochemistry and Cambridge Centre for Molecular Recognition, University of Cambridge, Tennis Court Road, Cambridge CB2 1QW, UK

Beratan, D.N. Department of Chemistry, University of Pittsburgh, Pittsburgh, PA 15260, USA

Durley, R.C.E. Department of Biochemistry and Molecular Biophysics, Washington University School of Medicine, St Louis, MO 63110, USA

Dutton, P.L. The Johnson Research Foundation and Department of Biochemistry and Biophysics, University of Pennsylvania, Philadelphia, PA 19104, USA

Farver, O. Department of Analytical and Pharmaceutical Chemistry, The Royal Danish School of Pharmacy, 2 Universitetsparken, DK-2100 Copenhagen, Denmark

Marcus, R.A. Noyes Laboratory of Chemical Physics, 127-72, California Institute of Technology, Pasadena, CA 91125, USA

Mathews, F.S. Department of Biochemistry and Molecular Biophysics, Washington University School of Medicine, St Louis, MO 63110, USA

Moore, G.R. School of Chemical Sciences, University of East Anglia, Norwich NR4 7TJ, UK

Moser, C.C. The Johnson Research Foundation and Department of Biochemistry and Biophysics, University of Pennsylvania, Philadelphia, PA 19104, USA

Northrup, S.H. Department of Chemistry, Tennessee Technological University, Cookeville, TN 38505, USA

Onuchic, J.N. Department of Physics, University of California, San Diego, La Jolla, CA 92093, USA

Parson, W.W. Department of Biochemistry, Box 357350, University of Washington, Seattle, WA 98195, USA

Rich, P.R. The Glynn Research Foundation, Glynn, Bodmin, Cornwall PL30 4AU, UK

Abbreviations

AO	ascorbate oxidase
Az	azurin
BChl	bacteriochlorophyll
BD	Brownian dynamics
BPh	bacteriophaeophytin
CCP	cytochrome c peroxidase
CDNP	carboxydinitrophenyl
CH	horse heart cytochrome c
CT	charge transfer
CY	yeast iso-1-cytochrome c
ENDOR	electron–nuclear double resonance
EPR	electron paramagnetic resonance
ESE	electron self-exchange
ESR	electron spin resonance
ET	electron transfer
FAD	flavin–adenine dinucleotide
FC	Franck–Condon factor
FCB2	L-lactate:ferricytochrome c oxidoreductase
FCSD	flavocytochrome c sulphide dehydrogenase
FMN	flavin mononucleotide
GR	glutathione reductase
HOMO	highest occupied molecular orbital
IR	infrared
LRET	long-range electron transfer
LUMO	lowest unoccupied molecular orbital
MADH	methylamine dehydrogenase
MCD	magnetic circular dichroism
MPD	2-methyl-2,4-pentanediol
NMR	nuclear magnetic resonance
NQR	NADH-ubiquinone oxidoreductase
Pc	plastocyanin
PCMH	p-cresol methylhydroxylase
PDLP	protein dipoles/Langevin dipoles
PS	photosystem
Q	quinone
SCF	self-consistent field
SOD	superoxide dismutase
T1	type 1 copper
T2	type 2 copper
T3	type 3 copper
TTQ	tryptophan tryptophylquinone
UQ	ubiquinone

Some useful constants

Elementary charge	e	1.602177×10^{-19} C
Boltzmann constant	k_B	1.380658×10^{-23} J K^{-1}
Avogadro constant	N_A	6.022137×10^{23} mol^{-1}
Planck constant	h	6.626076×10^{-34} J s
Permittivity of vacuum	ε_o	8.854188×10^{-12} C^2 J^{-1} m^{-1}
Gas constant	R	8.314510 J K^{-1} mol^{-1}

Equivalence table for energy units

eV	cm^{-1}	kJ mol^{-1}	kcal mol^{-1}
1	8065.5	96.485	23.061
1.23984×10^{-4}	1	1.19627×10^{-2}	2.8591×10^{-3}
1.03643×10^{-2}	83.594	1	0.23901
4.3364×10^{-2}	349.76	4.184	1

Thermal energy

At $T = 298.15$ K:
$RT = 2.4790$ kJ mol^{-1} = 0.59250 kcal mol^{-1}
$k_B T = 25.693$ meV; $k_B T/hc = 207.23$ cm^{-1}

Preface

This book is an attempt to meld together theory and experiment in a fundamental aspect of protein biology. Electron transfer reactions are essential for life. Not only do they provide the means for harnessing solar energy and transforming the chemical energy of foodstuffs into a readily utilizable form, they also extend into a range of metabolic processes that support the life of the cell. They are the simplest of chemical reactions in that they do not normally involve the making or breaking of bonds, but on the other hand proteins are necessary to control the rates and this imposes a degree of complexity. Chemical reactions are strongly localized in space, but electron transfer between the metal centres on which an electron might reside in a protein can take place over distances many times larger than the length of a chemical bond. Sometimes, as in photosynthetic reaction centres, the reaction does not show the usual Arrhenius type of temperature dependence. For example, in *Rhodobacter sphaeroides*, the rate constant for electron transfer from the excited special pair to bacteriophaeophytin *decreases* as the temperature *increases*. Even more remarkable is the fact that site-directed mutagenesis of a single residue (Tyr-M210) can drastically modify this unusual temperature response. How is such behaviour to be explained?

The basic questions, then, are: what are the physical factors that determine the rate constants for protein electron transfer? How do protein structure, and the structure of the redox centre itself, influence these factors? The proteins concerned may either be organized into a larger complex or interact by diffusion and collision. In the former case the protein acts as a scaffold on which to hang two or more redox groups. In the latter, the proteins will have to seek out an appropriate configuration of a short-lived collision complex, and the nature of the forces concerned poses an additional question. Physical theory can now provide an understanding of electron transfer at a level which is sufficiently explicit and detailed to guide experimentation by those attempting to understand the structural basis of the properties of particular systems.

The subject is not the organization of electron transport chains, but what determines the underlying rate constants. The unusual feature of this book for biochemists is that it involves the explicit application of quantum mechanics to biological behaviour. This may appear daunting at first sight, but the spirit of the book is that this need not be so. Quantum-mechanical tunnelling is shown to have predictive power in biology, and an ability to explain otherwise puzzling phenomena. The chapters are arranged in such a way that different aspects of the theory of electron transfer are introduced first in a manner which places the emphasis on concrete physical understanding rather than mathematical rigour. The more mathematical aspects of the subject are reserved for two appendices. The aim has been to take theory and experiment together to develop a framework of ideas that can be applied to a variety of systems. This is the concern of the first four chapters. The fifth discusses the crystal structures of electron transfer proteins and complexes; it forms an essential bridge between the general and the particular because true understanding at the molecular level is dependent on detailed structural information. In the next four chapters the most thoroughly investigated systems, grouped under the main classes of redox protein, are discussed as tests of the theory. First and foremost is the photosynthetic reaction centre of purple bacteria; other chapters discuss copper proteins and haem proteins before the final topic of linkage between electron transfer and proton translocation in complexes such as cytochrome oxidase is tackled.

If the problems of biological electron transfer can now be perceived with a degree of clarity, it is due as much to the participation of chemists and physicists as of biologists. The complexities of biological systems for long clouded the issues. The old arguments about the nature of biological oxidations were not completely resolved by Keilin's discovery of cytochromes and his concept of a respiratory chain. For a long time it was unclear whether the cytochrome system was a chain of electron carriers or of hydrogen carriers, although the chemiosmostic hypothesis took advantage of this uncertainty. The basic question of how a series of protein carriers could operate in a chain was not resolved until it was established that electron transfer could take place at reasonable rates over a distance of 10–20 Å through a protein. Chemists were able to avoid these difficulties by at first tackling the simpler cases of small metal complexes, both experimentally and theoretically. The outstanding contribution to the theory of electron transfer in such chemical systems was made by R.A. Marcus, starting in the late 1950s. Later it appeared that biological systems could also be brought into the same theoretical framework and the review on electron transfers in chemistry and biology by Marcus and Sutin, published in 1985, represents a milestone in the development of the subject. It is appropriate that the last chapter of the book is Marcus's own account of how the subject of electron transfer developed and is still developing, and it opens several windows on to areas of interest which are beyond the scope of this book.

In the 1990s it seems that research on electron transfer in proteins has entered its exponential phase of growth. This book is a collaborative effort, made possible by the willingness of a group of leading figures in the subject to write about their own areas of expertise. It has been planned primarily with the needs of biochemists in mind, in the hope that it will aid those who wish to understand the structural and physical basis of particular electron transfer systems and plan their own experiments; it should also be of value for graduate students and advanced undergraduates.

D.S. Bendall *(Cambridge, UK)*

Outline of theory of protein electron transfer

Christopher C. Moser and P. Leslie Dutton

1.1 Introduction

The chemiosmotic hypothesis describes the central feature of biological energy transduction as the creation of a transmembrane proton gradient from redox potential energy of photosynthesis and respiration (Mitchell, 1961). The gradient is built up by a series of guided electron transfers within and between membranous proteins that selectively move electrons and protons without compromising the insulation of the membrane. Energy transduction relies on the integration of three mechanisms: (i) diffusional motion of protons and electron carriers between specific sites on the major redox complexes; (ii) coupling of electron and proton motion at sites designed to control the bond making and breaking chemistry; and (iii) long-range electron transfer through the insulating medium between sites.

This last mechanism relies on electron tunnelling, a phenomenon that was clearly shown to be important in biology by early experiments of DeVault and Chance (1966). Light-activated electron transfer in the photosynthetic bacterium *Chromatium vinosum* was shown to have a relatively constant millisecond rate over a wide cryogenic temperature range down to liquid helium temperatures, the hallmark of a long distance electron tunnelling reaction. Since this time, biological electron transfer has benefited from advances in the physical theory of tunnelling and, in return, biology has provided a fine experimental field for testing theory (DeVault, 1980).

Biological systems appear astonishingly complex. Even small protein units have several thousands of atoms integrated into convoluted structures. Attempts to build a complete theoretical description of the behaviour of these systems are forced immediately into drastic simplifications (Friesner, 1994; Newton, 1991). For the small percentage of electron transfer complexes where we fortunately have crystals in which X-ray diffraction provides some approximation of one possible geometry, no amount of available computing ability can support *ab initio* calculations of all the wave functions involved. Instead various

Protein Electron Transfer, Edited by D.S. Bendall
© 1996 BIOS Scientific Publishers Ltd, Oxford

approximations and functionals are needed to begin a simulation. In this sort of environment, simple models and formalisms have an important role, especially if such pictures integrate a large number of observations and provide a much needed set of guidelines with which to test electron transfer theory, stimulate predictive insight into biological function and plan further experiments.

1.2 Fundamental tunnelling theory

1.2.1 Simple electron tunnelling through a barrier

The simplest picture of electron tunnelling involves the localization of an electron on a donor in a narrow potential well. Surrounding the donor is an electrically insulating region which represents a barrier to the electron. Classically the electron does not have enough energy to surmount the barrier yet, in the quantum mechanical view, it is possible to find the electron with some small but finite probability inside the barrier region. That is, the wave function describing the electron penetrates the insulating barrier. Indeed, the wave function penetrates far enough to reach the potential well of an acceptor, and thus there is a finite probability that an electron localized on the donor will be found at the acceptor. In the simple case of narrow donor and acceptor wells of equal depth, and a uniform intervening barrier of uniform height V above the wells, the penetration of the electron will fall off exponentially with distance (R) and will be given by Equation 1.1 (Gamow, 1928):

$$H_{AB}^2 \propto \exp\left(-2R\sqrt{2mV}/\hbar\right) = \exp(-\beta R) \qquad (1.1)$$

where H_{AB}^2 represents the square of the electronic coupling of the donor and acceptor states which is related to the square of the integrated overlap of the donor and acceptor electronic wave functions, m is the mass of the electron and $\hbar$ is Planck's constant/2π. In a fortunate coincidence of dimensional units, the exponential coefficient (β) in units of Å^{-1} will be approximately the square root of the barrier height in eV.

There is no temperature term in this expression. This is a 'deep-well' picture, in which the barriers sensed by the electron are measured in electron volts, much larger than physiological thermal energies of about 25 meV. This tunnelling phenomenon contrasts with a classical transition state view in which thermal energy is required to cross over, as opposed to pass through, a barrier, thereby revealing an Arrhenius activation energy. This absence of a significant activation energy barrier at cryogenic temperatures, together with the relatively slow millisecond reaction rate, suggested long distance electron tunnelling to DeVault and Chance (1966).

1.2.2 Fermi's golden rule

To relate the tunnelling theory to observation, we must be able to derive an electron transfer rate. However, in the simple quantum mechanical view just described, an electron will penetrate the barrier to the acceptor well but then return through the barrier back to the donor in a resonant process that depends on the distance between and the relative energy of the wells. In any real biological system there will be a range of acceptor wells of slightly different depths (i.e. there is a density of acceptor states) which break up the resonance and lead to a time dependence which can define an electron transfer rate. The corresponding mathematical description of this electron transfer rate (k_{et}; Equation 1.2) is the fundamental equation of non-adiabatic electron transfer theory, Fermi's golden rule:

$$k_{et} = \frac{2\pi}{\hbar} H_{AB}^2 FC \qquad (1.2)$$

FC refers to the Franck–Condon weighted density of states. As in spectroscopy, the Franck–Condon term refers to an electron jumping from one quantum state to another in a time too short for the nuclei to respond; a large FC term corresponds to a favourable geometric overlap of the nuclear wave functions of the reactant and product. Thus Equation 1.2 also serves to separate the dependence of the electron transfer rate into electronic and nuclear terms. This equation is applicable only to cases in which the donor and acceptor are relatively well separated and distinct, in other words long distance electron transfer. It is critical to remember that, in cases where the donor and acceptor are close enough to interact strongly, electron transfer is adiabatic and this simple description has to be abandoned; prediction of electron transfer rates becomes much more difficult.

1.2.3 *Marcus theory and nuclear geometries*

A particularly straightforward interpretation of the Frank–Condon term has been provided by Marcus (Marcus, 1956; Marcus and Sutin, 1985). In this view (*Figure 1.1*), the nuclei of the reactant and its immediate environment are represented by a single simple harmonic oscillator potential along a reaction coordinate. The minimum of this potential represents the equilibrium geometry of the reactant. The product is represented by a similar harmonic potential with a minimum that is displaced along the reaction coordinate to the equilibrium geometry of the product, and lowered in potential by ΔG, the free energy of the electron transfer reaction. The change in geometry upon electron transfer is characterized by the reorganization energy (λ), defined as the energy that must be added to a reactant and its environment to distort its equilibrium geometry into the equilibrium geometry of the product without, however, allowing the electron to be transferred. It includes, for example, the energy required to re-arrange the solvent dipoles from a geometry that electrostatically favours the electron on the donor, to a geometry that favours the electron on the acceptor.

In this non-adiabatic view, the reactant can be viewed as moving back and forth along the parabola surface, passing through the region of intersection with the product parabola many times before the weak coupling permits electron transfer to take place. In contrast, in an adiabatic reaction, the reactant and product are so strongly coupled that motion in this region can carry the reactant directly to the product. In the adiabatic limit, the bottoms of the two roughly parabolic wells are joined into a single, more complex surface which is separated, or split, from the remnants of the upper parabolic surfaces by the large electronic coupling, $2H_{AB}$.

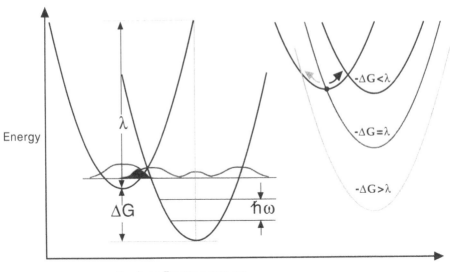

Nuclear Rearrangement

Figure 1.1. In a Marcus description, all the nuclear motion of the reactant and the surrounding environment are approximated as a single, simple harmonic oscillator potential (upper parabola on left). The equilibrium geometry corresponds to the bottom of the potential well. The product potential (lower parabola) has the same shape but is shifted to a more negative potential energy by an amount ΔG, corresponding to the free energy of the reaction. The product potential is also shifted in nuclear geometry, corresponding to the rearrangement of the equilibrium nuclear geometry upon electron transfer. The reorganization energy (λ) corresponds to the energy that must be added to the reactants to move from the equilibrium reactant geometry to the equilibrium product geometry, but remaining on the reactant surface, that is without transferring an electron. In a quantized view of the simple harmonic oscillator potentials, only certain energy levels are permitted, shown as horizontal lines spaced by the quantum energy of the oscillator $\hbar\omega$. This energy corresponds to the characteristic frequency of the oscillator. In the quantized view, nuclear tunnelling from reactant to product depends upon the overlap of the harmonic oscillator wave functions. In this figure, we show schematically the square of the wave functions (corresponding to the probability density of the wave functions) as shaded regions. On the right, the activation energy of the reaction, corresponding to the height of the intersection of the reaction and product parabolas above the bottom of the reactant well, depends upon the free energy of the reaction relative to the reorganization energy of the reaction. When $-\Delta G = \lambda$, there is no activation energy. When $-\Delta G < \lambda$ (the normal region) and when $-\Delta G > \lambda$ (the Marcus inverted region) the activation energy is non-zero and the reaction rate is slower.

In the simpler non-adiabatic case, the height of the intersection of the reactant and product parabolas relative to the bottom of the reactant potential, which is analogous to the activation energy ($\Delta E^{\ddagger}$) of the transition state of a reaction in a traditional (adiabatic) chemical reaction, is given by Equation 1.3:

$$\Delta E^{\ddagger} = (\Delta G + \lambda)^2/4\lambda \qquad (1.3)$$

This activation energy predicts that the rate of the electron transfer reaction has a simple Gaussian dependence on the free energy:

$$k_{et} \propto FC = (4\pi\lambda k_B T)^{-1/2} e^{-(\Delta G + \lambda)^2/4\lambda k_B T} \tag{1.4}$$

where k_B is the Boltzmann constant. The strongest temperature dependence is found in the Gaussian exponent; however, when this exponent goes to zero (an activationless reaction) there is a small, inverse temperature dependence in the normalizing pre-factor which becomes noticeable.

Beginning from small free energies, increasing the driving force of a reaction speeds the reaction until an optimum free energy is reached. At this point the free energy matches the reorganization energy ($-\Delta G = \lambda$, *Figure 1.1*), and the re-action is activationless. Increasing the driving force still more actually decreases the reaction rate, defining the Marcus 'inverted region' (*Figure 1.1*). Several electron transfer systems have demonstrated the existence of this inverted region (see, for example, Closs *et al.*, 1986; Miller *et al.*, 1984).

In this classical Marcus description, the Gaussian describing the rate dependence on free energy becomes infinitely narrow as the temperature drops towards absolute zero (*Figure 1.2*). Indeed, it is frequently assumed that an essentially activationless electron transfer reaction means that the free energy must match the reorganization energy. This is not true if quantum corrections are made to the classical Marcus expression. A semi-classical expression (Hopfield, 1974; Equation 1.5) retains the Gaussian form with a mean of λ, but replaces the variance of $2\lambda k_B T$ with $\lambda \hbar\omega \coth(\hbar\omega/2k_B T)$.

$$k_{et} \propto FC = [2\pi\lambda\hbar\omega\coth(\hbar\omega/2k_B T)]^{-1/2} e^{-(\Delta G + \lambda)^2/2\lambda\hbar\omega\coth(\hbar\omega/2k_B T)} \tag{1.5}$$

$\hbar\omega$ corresponds to the energy of the characteristic frequency of nuclear motion (reorganization) coupled to electron transfer. If the temperature is large relative

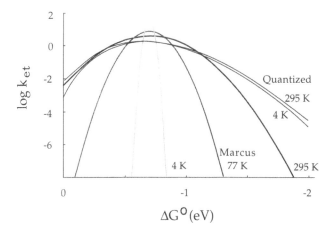

Figure 1.2. Temperature sensitivity of the electron transfer rate as a function of free energy for quantum and classical Marcus oscillators. Thick lines show the theoretical Gaussian Marcus rate relationship at a series of temperatures (4, 77 and 295 K). Thin lines show the rate relation for quantized harmonic oscillators with a characteristic frequency of 70 meV. In the classical picture, only free energies close to the reorganization energy (here 0.7 eV) are relatively temperature insensitive. In the quantum picture, as long as the characteristic frequency is significantly larger than the physiological Boltzman thermal energy, even free energies relatively far from the reorganization energy can be relatively temperature insensitive.

to $\hbar\omega/2k_B$, then the variance is the Marcus classical $2\lambda k_B T$. On the other hand, as the temperature falls below $\hbar\omega/2k_B$, the variance stabilizes at $\lambda\hbar\omega$ (*Figure 1.2*) and the rate becomes nearly temperature independent at any free energy.

This characteristic frequency is effectively an average of the manifold of all the vibrations coupled to electron transfer, weighted by the contribution of each vibrational motion to the overall reorganization energy. For example, if there are two frequencies ($\hbar\omega_1$ and $\hbar\omega_2$) coupled to electron transfer with reorganization energies λ_1 and λ_2, then the overall free energy dependence of the rate is the convolution of the two Gaussians with means λ_1 and λ_2. The result is a third Gaussian with a mean $\lambda = \lambda_1 + \lambda_2$, and a summed variance. If both vibrations are greater than $2k_B T$, the new variance is $\lambda_1\hbar\omega_1 + \lambda_2\hbar\omega_2$; thus a single characteristic frequency would be a λ-weighted average of the two frequencies. However, if vibration 2 were much smaller than $2k_B T$ the variance would be $\lambda_1\hbar\omega_1 + 2\lambda_2 k_B T$.

A fully quantized harmonic oscillator description is given in Equation 1.6 (see, for example, Jortner, 1976; Levich and Dogonadze, 1959).

$$k_{et} \propto FC = \frac{1}{\hbar\omega}\exp\left[-S(2n+1)\left(\frac{n+1}{n}\right)^{P/2} I_P[2S\sqrt{n(n+1)}]\right]$$

$$S = \frac{\lambda}{\hbar\omega}\; ; P = \frac{-\Delta G^\circ}{\hbar\omega}\; ; n = \frac{1}{\exp[\hbar\omega/k_B T] - 1}$$

(1.6)

where S and P are the reorganization energy and the free energy normalized by the characteristic frequency, and n reflects the average vibrational level populated at temperature T. I_P is the modified Bessel function of order P. The free energy dependence of this relationship is not exactly Gaussian; rates in the inverted region fall off less quickly with increasing free energy. This becomes most obvious if the characteristic frequency is relatively large compared with the thermal energy (*Figure 1.3*). Once again, this characteristic frequency reflects the spectrum of vibrations coupled to electron transfer (Ulstrup and Jortner, 1975).

The most significant difference in moving from a classical Marcus description to a quantized model is that the electron transfer rate becomes mostly temperature independent when the temperature is below $\hbar\omega/2k_B$, even when the free energy does not match the reorganization energy (*Figure 1.2*). The temperature independence can be understood in this way. In the fully quantized picture, the Franck–Condon factors correspond to the integrated overlap of the vibrational wave functions at each energy level. Thus if the $\hbar\omega$ is small compared with the thermal energy, many vibrational levels will be populated and the *FC* overlap will have a noticeable temperature dependence. If $\hbar\omega$ is large, then essentially only the bottom vibrational level will be occupied and the overlap, and hence the reaction rate, will be mostly temperature independent, even when $-\Delta G \neq \lambda$. Another way to view this case is as follows: when thermal energies are well below the characteristic frequency coupled to electron transfer, there is insufficient energy for the nuclei to reach the crossing point of the reactant and product parabolas, and electron transfer not only involves electron tunnelling, but also nuclear tunnelling.

While we have been emphasizing the effect of changes in free energy on the electron transfer rate with a given reorganization energy, it is also useful to

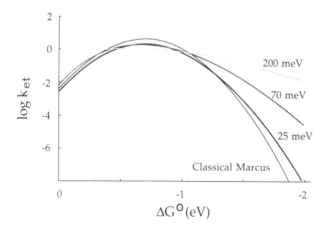

Figure 1.3. The electron transfer rate versus free energy relationships at three characteristic frequencies (200, 70 and 25 meV) for quantized harmonic oscillators are compared with the classical, Gaussian Marcus relation. All theoretical curves are at room temperature ($k_B T = 25$ meV). As the characteristic frequency coupled to electron transfer becomes large compared with temperature, the quantized curve becomes broader and less Gaussian, becoming shallower in the inverted region. It is also important to recognize that if $-\Delta G$ is much larger than λ, it may be possible for electron transfer into an excited state of the product, making the free energy dependence in the inverted region even more shallow.

consider the corresponding effect of changes in reorganization energy for a given free energy (*Figure 1.4*). Since reorganization energy reflects energy changes associated with nuclear motions upon the charge rearrangement of electron transfer, electron transfers that take place in relatively unpolarizable environments will induce relatively minor rearrangements of nuclei, leading to small reorganization energies typically around 0.2–0.7 eV (Warshel *et al.*, 1989). Electron transfers that involve redox centres near a relatively easily polarized medium, such as water, will induced more dramatic dipole and charge rearrangements resulting in much larger reorganization energies, for example from 1.0 to 1.4 eV.

Cofactor size and polarizability is an important partner to the environmental contribution to the reorganization energy. Relatively large redox centres like chlorins which distribute the transferred electron density over a 10 Å diameter typically induce smaller changes upon electron transfer, in both the cofactor itself and in the immediate environment. Thus, large cofactors such as chlorophylls, and especially dimers as in the photosynthetic reaction centre, tend to have relatively small reorganization energies. Small redox centres of just a few atoms with relatively localized electrons typically undergo larger nuclear rearrangements upon electron transfer, together with significant changes in the immediate environment, leading to larger reorganization energies.

Analogously, electron transfer over relatively small distances compared with the size of the redox centres causes relatively minor rearrangements in the equilibrium geometries of the various nuclei, and is associated with relatively small reorganization energies. It also appears that ultrafast electron transfers,

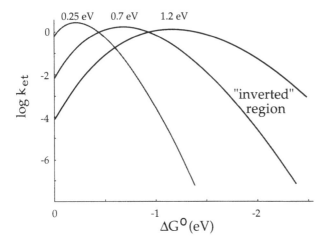

Figure 1.4. The effect of reorganization energy on the rate versus free energy relationship. Reorganization energy corresponds to the free energy at the maximal rate. Three examples are shown. The lowest reorganization energy of around 0.25 eV appears in special non-polar environments with large redox cofactors; 0.7 eV appears typical of many reactions occurring in a protein interior; 1.2 eV seems typical of reactions occurring at the polar surface of a protein.

occurring in picoseconds, may result in significantly smaller reorganizations than slower electron transfer reactions.

1.3   Tunnelling in proteins

1.3.1 *Characteristic free energy dependence of biological electron tunnelling*

An approximately Marcus-like free energy dependence has been seen in intraprotein electron transfer, with photosynthetic reaction centres providing many examples (Bixon *et al.*, 1995; Dutton and Moser, 1994; Giangiacomo and Dutton, 1989; Gunner and Dutton, 1989; Gunner *et al.*, 1986; Jia *et al.*, 1993; Lin *et al.*, 1994a,b; Moser *et al.*, 1992; Warncke and Dutton, 1993). Indeed, the temperature and free energy dependence of quinone-substituted photosynthetic reaction centres suggests that a quantized picture is needed with a characteristic frequency of about 70 meV or three times the room temperature Boltzman energy (Gunner and Dutton, 1989; Gunner *et al.*, 1986; Moser *et al.*, 1992). It appears that one characteristic frequency is sufficient to describe the width of the approximately Gaussian dependence of photosynthetic intraprotein electron transfer. However, not all these reactions have been explored over a large enough free energy or temperature range to resolve any differences in the vibrational spectrum of the various reactions. At room temperature, the theoretical effect of changing the characteristic frequency from 70 meV to a much smaller value will be relatively minor. Hence, the information available at present suggests that the characteristic frequency is a parameter that has not been naturally selected specifically for the control of biological electron transfer rates. It might be that a single apparent characteristic frequency of around 70 meV is a simple consequence of a seemingly random selection of vibrations coupled to electron transfer from the relatively similar broad spectrum of vibrations associated with most protein materials, and analogous to the similar heat capacities found for different proteins.

On the other hand, there appear to be clear differences in the measured values of the reorganization energies of the different reactions. These range from high values of 1.23 eV (Labahn et al., 1994) for Q_B^- to bacteriochlorophyll $(BChl)_2^+$ (presumably the result of a polar watery environment of the Q_B site), to 0.7 eV, typical for electron transfer in the protein interior, and much smaller values (<0.1 eV) for the very early reactions (Jia et al., 1993). It seems reasonable that just as ΔG can be naturally selected to control electron transfer, so can λ.

1.3.2 Characteristic distance dependence of biological electron tunnelling

According to Fermi's golden rule (Equation 1.2), when it is possible to estimate and isolate the optimum FC factor, then it is possible to examine the mostly electronic coupling term H_{AB}^2. The relationship between the free energy optimized rate and the edge-to-edge distance between cofactors in the bacterial photosynthetic reaction centres is shown in *Figure 1.5* (Moser et al., 1992). These reactions include the physiologically productive charge separation reactions that are responsible for the creation of a transmembrane electrochemical proton gradient, as well as physiologically unproductive charge recombination reactions, which must be suppressed *in vivo* because they would lead to the death of the organism. There is a remarkably linear relationship between the rate and the distance over a range of about 12 orders of magnitude of rate and 20 Å of distance. Apparently, a single β value of 1.4 Å^{-1} comes close to providing an estimate of the electron transfer rate at any distance.

An analogous free energy optimized rate versus distance plot for chemically synthesized systems that have a direct rigid covalent link between donor and acceptor show a distinctly different β value of about 0.7 Å^{-1} (Moser et al., 1992; see also Newton, 1991). Thus, it seems that the propagation of the electronic wave function through a protein medium is intermediate between that of covalently linked systems and that anticipated for a vacuum [~2.8 Å^{-1} if the depth of the well is ~8 eV, corresponding to the typical ionization energy of chlorins and other biological cofactors (Adler et al., 1978; Gutman and Lyons, 1967; Meot-Ner et al., 1973)]. Indeed, the β value for intraprotein electron transfer in reaction centres is similar to the value of 1.2 Å^{-1} found for a frozen organic solvent (2-methyl-tetrahydrofuran; Miller et al., 1984). Thus, it appears that the barrier experienced by an electron tunnelling between redox centres in a protein is similar to that encountered in a glassy solvent, composed of regions which are covalent-like in which the barrier is on the order of the spacing between bonding and anti-bonding orbitals, roughly 0.5 eV, and regions which are vacuum-like, in which the barrier approaches many electron volts.

An apparent single β value for these various reactions, including physiologically productive and unproductive reactions, suggests that, for the photo-

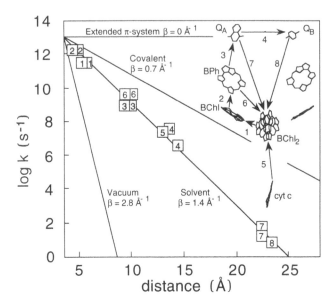

Figure 1.5. Effect of distance on electron transfer rate. Intraprotein electron transfer within the reaction centre and many other proteins cluster about a single line defining an effective β of about 1.4 Å^{-1}, similar to that found in organic solvents. This β value is intermediate between that found in redox pairs linked by rigid organic spacers (~0.7 Å^{-1}) and that expected for a vacuum barrier based on typical cofactor ionization energies of around 8 eV (~2.8 Å^{-1}). For simplicity, the line with slope β = 0.7 Å^{-1} is shown with the same intercept of 10^{13} s^{-1} at van der Waals contact, although for unknown reasons the intercept for rigid, covalently linked systems appears to be closer to 10^{12} s^{-1} (Moser *et al.*, 1992). cyt *c*, cytochrome *c*; BPh, bacteriophaeophytin; BChl, bacteriochlorophyll.

synthetic reactions at least, there has not been any natural selection to adjust the structure of the intervening protein medium to enhance electron propagation for productive reactions and inhibit the unproductive reactions. For example, there does not appear to be any evidence that covalent-like links with a β of 0.7 Å^{-1} have been used consistently between redox centres engaged in productive electron transfer, although the covalent-like link between quinones in reaction centres represents an interesting case which we will discuss below (see Section 1.3.4). Perhaps the changes in protein structure needed to add or delete a series of such links are difficult to accomplish by evolutionary mutagenesis. Such changes could affect protein assembly or stability dramatically in a manner that affects overall function more profoundly than the connectivity changes themselves (Moser *et al.*, 1995). Since distance has such a dominant effect on electron transfer rate, the most effective way to select electron transfer may be to adjust the position of the residues that lead to cofactor binding such that physiologically productive partners are closer to one another than partners that would engage in unproductive reactions.

1.3.3 *A simple three-parameter predictor of rate*

A review of the data on native intraprotein electron transfer reactions (with an emphasis on the photosynthetic reaction centres) suggests that we can often predict electron transfer rates with moderate accuracy using a simple equation with only three parameters. Apparently, the free energy dependence of the rate is at least roughly Gaussian, with a free energy optimum (reorganization energy) that is best uncovered by experiment, but can often be estimated by assessing the polarity of the cofactor environment. It also appears that the width of the Gaussian is at least roughly proportional to the square root of the reorganization energy (as described by Marcus), but that the classical Marcus temperature dependence must be replaced with a quantum view with a characteristic frequency coupled to electron transfer that has the effect of slightly broadening the Gaussian. It does not appear to be necessary to vary the characteristic frequency in order to accommodate the data currently available. Thus Franck–Condon terms can be estimated usefully using only two parameters, the reorganization energy (λ) and the free energy (ΔG). Furthermore, the electronic coupling appears to be moderately well described with a single value for β and the extrapolated rate at van der Waals contact, leaving only the edge-to-edge distance (R) between cofactors as a parameter. Thus, Equation 1.7 should be useful for order of magnitude estimates of native intraprotein electron transfer rates that are not rate limited by non-tunnelling adiabatic reactions:

$$\log k_{et} = 15 - 0.6R - 3.1\,(\Delta G + \lambda)^2/\lambda \qquad (1.7)$$

where R is in Å, and ΔG and λ are in eV (Moser and Dutton, 1992).

1.3.4 *Variable distance dependence*

The remarkable fit of a single β value to the reaction centre tunnelling data (*Figure 1.4*) seems to suggest that the protein presents a single uniform barrier of about 2 eV to the tunnelling electron. However, as indicated by Gamow (1928), the β value of Equation 1.1 can be understood in terms of the average barrier height along the tunnelling length. In an attractive simple view, the protein medium acts like a mix of relatively low, covalent-like barriers (~0.5 eV) in regions near atomic nuclei and along chemical bonds, and relatively high, vacuum-like barriers (~8 eV) relatively far from the nuclei. Indeed, if the percentage of low barrier region corresponds to the approximately two-thirds of the protein volume within the van der Waals radii of the nuclei, and the high barrier percentage to the rest, then the average β would be close to the 1.4 Å^{-1} observed.

Nevertheless, even in this simple first approximation, some variation in β due to the natural heterogeneity of protein structure is expected, if only for statistical reasons. Some regions of the protein should have somewhat greater than average connectivity, and hence lower β, while other regions will have less connectivity and higher β. For example, the crystal structure of the *Rhodopseudomonas viridis* reaction centre (Michel *et al.*, 1986) shows a direct series of bonds from Q_A to histidine to an iron atom to histidine to Q_B. Some crystal structures of *Rhodobacter sphaeroides* (Allen *et al.*, 1987) show a similar sequence, although the crystal structure of Ermler *et al.* (1994) shows that Q_B

is not bonded to the histidine and is further away from Q_A. In most of the structures, the connection between quinones appears much like that of covalent linked systems, suggesting that the β for Q_A to Q_B electron transfer should be smaller than 1.4 Å^{-1}. Indeed, from a simple calculation based on a covalent–vacuum barrier mix [described below (Page *et al.*, in preparation)] the optimum electron transfer rate would be 3×10^{10} s^{-1}. This is much faster that the 6×10^7 s^{-1} expected for a β of 1.4 Å^{-1} and suggested by early experiments probing the free energy dependence of the reaction at small ΔG, far from λ. The resolution to this puzzle could come from an extended study of the free energy relationship which may either define a faster optimal rate, or perhaps uncover an adiabatic component, such as motion of Q_B within the binding site or associated protonation–deprotonation reactions, that introduces non-electron transfer dynamics that effectively limit the observed rate. In contrast, similar calculations done on other regions of the protein, such as that between BChl$_2$ and bacteriophaeophytin (BPh), seem to have somewhat less than average electronic connectivity, and suggest a rate slower than measured. Here the issue may be resolved by the discovery of water or membrane fatty acid tails not resolved and included in the crystal structure; phospholipid molecules have been recognized in the *R. viridis* four-haem subunit.

The experimental evidence for variations in β within a protein is most clear in a series of experiments in which artificial redox centres have been placed at selected locations on the surface of water-soluble redox proteins (Casimiro *et al.*, 1993a,b; Farver and Pecht, 1992; Karpishin *et al.*, 1994; Langen *et al.*, 1995; Pan *et al.*, 1988; Wuttke *et al.*, 1992a,b). The electron transfer rate for a given distance can vary noticeably (Farid *et al.*, 1993; Karpishin *et al.*, 1994; Willie *et al.*, 1992; Wuttke *et al.*, 1992a,b). Intentional positioning of a ruthenated histidine redox centre at a surface position from which the most likely structure suggests a relatively poor connectivity from the interior haem of cytochrome *c* has shown a rate that is about 40 times slower than that expected using a β of 1.4 Å^{-1} (Farid *et al.*, 1993; Wuttke *et al.*, 1992a,b). Also, a series of ruthenated mutants in the β-sheet protein azurin has shown that the best β value for transfer along the length of the relatively well connected sheet is closer to 1.1 Å^{-1} (Langen *et al.*, 1995). Moreover, a cytochrome b_5 has been ruthenated (Willie *et al.*, 1992) to create a redox pair that looks as well connected as the rigid covalently linked synthetic systems, and falls on the same $\beta = 0.7$ Å^{-1} line as these systems.

There are a number of ways in which possible variations in β can be assessed (for review, see Closs *et al.*, 1986; Friesner, 1994; Newton, 1991). In the simple one-dimensional view just described, a line can be drawn between an atom on the edge of the donor and an atom on the edge of an acceptor, and a β value can be calculated based on the fraction of the line that passes within the van der Waals radius of intervening atoms in a protein crystal structure. This type of β calculation tends to have a large sampling error as the line grazes certain atoms and penetrates others. The β value tends to be very dependent on the particular choice of edge atoms. To some extent, the sampling can be improved by extending this one-dimensional analysis to include an averaging over all pairs of donor and acceptor atoms (Page *et al.*, in preparation).

Other approaches extend the potential barrier analysis to three dimensions, with a corresponding increase in complexity and parameters. For example, the approach of Kuki *et al.* (Gruschus and Kuki, 1993; Kuki and Wolynes, 1987) represents the tunnelling potential by a relatively complicated function centred on each atom, and uses a Monte-Carlo scattering method to calculate Feynman path integrals. In this way, the entire region between and around the donor and acceptor is taken into account. In this view the tunnelling electron is significantly affected by an ellipsoid that includes the donor and acceptor and the intervening region which can be 5 or 10 Å in diameter, depending on the donor–acceptor spacing. Nevertheless, β values calculated in this way are typically found to be near 1.4 Å^{-1}.

A different approach focuses on the coupling of molecular orbitals of the donor, acceptor and intervening medium, rather than on the potential barriers. These methods concentrate on superexchange, in which mixing or coupling of orbitals places some electron density from the donor into nearby medium orbitals of higher energy, followed by coupling of medium orbitals to other nearby medium orbitals, and so on until the final medium orbital is coupled with the acceptor. This method also works for hole transfer. In one phenomenological approach of this style, the protein is topologically reduced to a series of through-bond (either covalent or hydrogen bond) and through-space couplings, with a β value for each of the three coupling types calibrated from rates of ruthenium-modified electron transfer proteins (Beratan *et al.*, 1991, 1992a,b; Betts *et al.*, 1992; Onuchic *et al.*, 1992). This algorithm is sensitive to the structural connectivity of the protein and clearly shows a better correlation of rate with structure than an approach which completely ignores variations in connectivity. The topological nature of the approach inevitably leads to a particular path which represents the sequence of bonds and gaps which contribute the most to the electronic coupling between redox centres. However, it is important not to misinterpret this approach by suggesting that proteins have selectively constructed a particular pathway between redox centres to effect electron transfer, or that modifications of that topology by site-directed mutagenesis will lead to significantly different electron transfer rates. Indeed, site-directed mutants most frequently have little effect on electron transfer rates. Instead, it must be remembered that the entire region between redox centres has an effect on the relative ease with which an electron can tunnel through the medium, and that this volume can be relatively large in dynamic biological systems with relatively large redox centres.

There are a number of other orbital-based approaches that begin with *ab initio* electronic structure calculations, and allow couplings between orbitals to take any value (see, for example, Evenson and Karplus, 1993; Newton, 1991; Siddarth and Marcus, 1993). In principle, these approaches should be able to take account of constructive and destructive interference between different sets of orbitals. In order to handle the relatively large protein systems, these approaches introduce essential assumptions and simplifications.

Regardless of the analytical approach, there is little doubt that a highly connected structural medium between donor and acceptor will speed up electron transfer, while an unusually unconnected structure will impede electron transfer. This is clear despite the experimental difficulties of dissecting the effects of the medium on the rate upon mutagenic and chemical modification of the protein, from the effects of small changes in distance, free energy and reorganization energy. The issue for which there is doubt, and which needs exploration beyond the handful of examples available, is whether any biological system has incorporated through natural selection a specific structural motif designed to promote (or retard) electron transfer. At the present time there is no evidence to suggest that such motifs exist.

1.4   Redox protein engineering principles

1.4.1 *The dominance of distance*

The overwhelming importance of distance on the rate of electron transfer provides the basic engineering principle that makes a Mitchell chemiosmotic-type mechanism of bioenergetics possible. It is possible to lead the tunnelling electron in any direction through an insulating medium simply by positioning the closest redox centres at a zero or exothermic ΔG in the proper direction. This could be in a direction mostly perpendicular to the membrane for the electrogenic membrane charging electron transfers, or towards protonation–deprotonation coupling sites for interfacing with electroneutral coupled electron–proton transmembrane transfer. In order to accomplish intraprotein electron transfer on a submillisecond time scale (the approximate time scale of the slower diffusive adiabatic reactions) at modest free energies and typical reorganization energies, distances between the edges of cofactors should be less than 17 Å. This means that electrogenic electron transfer entirely across the approximately 35 Å low dielectric region of the membrane will have to be done by relay, with at least one intermediate redox centre embedded in the low dielectric interior. To put this another way, redox centres separated by the 35 Å dielectric interior will be safely insulated from one another on a time scale of decades, while redox partners separated by the full 50 Å thickness of the membrane would require, in principle, billions of years to tunnel.

This 17 Å distance is also relevant from another perspective. An electron on a redox centre that is poised for participation in a relatively slow adiabatic reaction (e.g. at a site involving the reduction and protonation of quinone or oxidation and deprotonation of lactate and succinate) must be protected from exogenous redox acceptors that diffuse through the aqueous and membrane phases and can come close enough to capture the electron through tunnelling. This can be assured by surrounding the redox centre with approximately 17 Å of protein bulk that will act as an effective insulator on this time scale. For example, the highly reactive radical species of semiquinone in the bacterial reaction centres are insulated from the aqueous phase by about 20 Å of protein. However, for weaker oxidants and reductants, less insulation may be sufficient because, in the absence of some sort of binding association between the adventitious redox centre and the peptide, the slowness of a second order diffusional reaction will often make the time of unguided electron transfer longer than milliseconds.

1.4.2 *Distance and binding in interprotein electron transfer*

A similar concern is faced by redox centres at the interface between two proteins in interprotein electron transfer. On the one hand, it is important to have the relevant pair of redox centres close to the edge of the protein so that the distance of electron transfer is less than 17 Å and electron tunnelling will be rapid on the time scale of the binding interaction that brings the redox centres into proximity. If this difference were evenly split between the proteins, as a general rule a redox centre participating in interprotein election transfer should be within about 9 Å of the surface. On the other hand, the redox centre should be sheltered by sufficient protein medium to prevent accidental unproductive electron transfer.

An excellent example is provided by the relatively small redox protein cytochrome c, in which the haem is positioned asymmetrically. The haem edge is within 5 Å of the surface on one side of the protein, yet the approximately 30 Å diameter of the protein means that most of the rest of the surface of cytochrome c will be effectively insulated from adventitious electron transfer. Cytochrome c complexes in which both the distance and free energy dependence of electron transfer have been measured (Cheung *et al.*, 1986; Lin *et al.*, 1994b; Tiede *et al.*, 1993; Whitford *et al.*, 1991) generally support the view that the same β of 1.4 Å^{-1} can be useful for estimating both intra- and interprotein electron transfer rates.

Although the asymmetric placement of redox centres in proteins provides the insulation that mostly excludes adventitious electron transfer, it can also prevent the interaction with desirable redox partners unless there is an attractive binding site that brings the redox centres into relatively close proximity (McLendon, 1988). Generally, this is accomplished by the decoration of the binding sites with complementary electrical charges, creating an attractive potential that draws the physiological redox partners together and enables submillisecond electron transfer. For example, Tiede *et al.* (1993) have examined cytochromes c with a wide variation of patterns of charged amino acids near the haem edge as they interact with the photosynthetic reaction centre. Even though all of these cytochromes are thermodynamically capable of re-reducing the light-oxidized bacteriochlorophyll dimer of the reaction centre, those with non-complementary binding site charge distributions show no electron transfer. So far, only relatively few examples of such electron transfer protein pairs have been examined. More are needed, but the picture that cytochrome c and the reaction centre project is one of large areas on each protein surface through which the electron can pass when the two reactants are bound in any of several positions (Aquino *et al.*, 1995; Moser *et al.*, 1995). Indeed, kinetics and binding measurements suggest the highest affinity binding geometry need not correspond to the fastest rate (Moser and Dutton, 1988; Tiede *et al.*, 1993). Cytochrome c appears to move into a more weakly bound state that presents a much shorter distance and hence faster rate for electron transfer. X-ray structures of co-crystals of redox protein partners need to be approached with caution [see, for example, the dispute over the electron transfer geometry of the cytochrome c–cytochrome c peroxidase complex (Northrup *et al.*, 1988; Pelletier and Kraut, 1992; Zhou and Hoffman, 1994)].

1.4.3 *Distance and stability themes in intraprotein charge separation*

The strong dependence of electron transfer rate on distance dictates the basic redox centre architecture surrounding the critical energy-coupling centres of bioenergetic membranes: the primary charge separating centres of photosynthesis and respiration.

Light-induced charge separation. In photosynthetic systems, the excited state of chlorophylls or bacteriochlorophyll represents a very strong reductant. However, typical excited states lose energy by fluorescence in less than 1 ns; charge separation must be faster for efficient energy trapping. According to Equation 1.7, efficient subnanosecond charge separation requires that the redox pair initiating charge separation must be separated by less than 7 Å if the reorganization energy is small (~0.3 eV) or by less than 3 Å if the reorganization energy is large (~1 eV). In the bacterial photosynthetic reaction centres illustrated in *Figure 1.4*, the closest redox centre to the excited bacteriochlorophyll special pair is placed at an edge-to-edge distance of about 4.5 Å. It appears that the chlorins in photosystem I are similarly closely spaced.

Apparently, the reorganization energies of these ultrafast steps are naturally selected to be small. This permits the charge separation to be fast enough while also allowing the redox centres to be well separated and to avoid the strong electronic couplings that result in the electron being effectively shared between the two redox centres. This low reorganization energy is possible partly because the charge separation reactions take place in the relatively hydrophobic protein interior and because the relatively large chlorin redox centres effectively distribute the electric charge of the tunnelling electron over many Å^2.

Once a charge-separated pair is created, there is always the possibility for charge recombination. Recombination to the ground state converts the energy of the incident photon into wasteful heat. In light-activated charge separation, the sum of the ΔG for the initial charge separation and the ΔG for the charge recombination from this charge-separated state is close to the energy of the absorbed quantum which created the excited state. In reaction centres of the photosynthetic bacterium *R. sphaeroides*, the total ΔG is about 1.4 eV. According to the Marcus relation of Equation 1.4, the rate of an electron transfer depends on exp-$(\Delta G + \lambda)^2$. Thus, λ must be less than half the total ΔG or charge separation will be followed by an even faster charge recombination. The small λs that seem to be associated with the initial charge separation of photosynthesis allow large free energy recombinations to fall into the slow Marcus inverted region while charge separation can take place at modest free energies at close to the optimal rate for that distance.

Another means of assuring stable charge separation in the face of charge recombination is to place a third redox centre closer to the second redox centre, so the second step of charge separation is faster and more efficient than charge recombination. Charge recombination from this widely separated state is over a much larger distance than the initial charge separation and thus considerably stabilized.

Charge separation from quinone coupling sites. Distance also plays a crucial role in the stabilization of the redox potential-driven charge separation at the special quinone coupling sites of the respiratory complexes III (Ding *et al.*,

1992) and possibly complex I (Dutton *et al.*, in preparation). It appears that a two-step electron transfer, analogous to that just described in photosynthetic systems, operates to separate charges and preserve the redox energy of these respiratory systems by preventing short circuit electron transfers that would result in the production of heat.

In complex III, reducing equivalents are delivered to the (double) Q binding site designated Q_o in the form of reduced ubiquinone (ubihydroquinone, QH_2) that diffuses in from the membranous Q pool. Oxidizing equivalents are present in the nearby [2Fe-2S] cluster and cytochromes c_1 and c (or c_2). This redox-potential energy difference between cytochrome c and QH_2 at the Q_o site represents the driving force for the transmembrane proton and electron transport performed by this complex. The [2Fe-2S] cluster (oxidized by the isopotential cytochromes c and c_1)extracts a single electron from the Q, forming an unstable semiquinone intermediate state. Deprotonation of this semiquinone forms a strong reductant which now donates an electron to another nearby acceptor, the low potential cytochrome b of complex III (cytochrome b_L). At this stage the two electrons originally present on the QH_2 have been separated and launched onto two separate chains of redox centres, one at relatively high potential, one at low potential. There is a danger, however, of electron transfer between the chains at the point of contact with the Q site, for example between reduced low potential cytochrome b_L and high potential [2Fe-2S] cluster reoxidized by cytochrome c_1 (or cytochrome c_1 or c themselves). Indeed, this electron transfer is analogous to charge recombination in the photosynthetic reaction centre. The [2Fe-2S] cluster and cytochrome b_L are in relatively close proximity to the Q_o site. They may, for example, share histidine ligands with bound quinones, in which case the edge-to-edge separation may be as close as 18 Å, with an electron transfer rate predicted by Equation 1.7 to be around 10^3 s^{-1}. A short circuit electron transfer can be avoided by a relatively rapid charge separation away from the Q site by means of cytochrome b_L to cytochrome b_H electron transfer. According to Equation 1.7, a transmembrane cytochrome b_L to cytochrome b_H electron transfer with a centre to centre distance of 13 Å can take place in 1.2 μs, fast enough to prevent the short circuiting reaction (Ding *et al.*, 1995).

An analogous situation may also be operating in complex I. However, in this case, rather than QH_2 donating electrons to high and low potential chains, via a semiquinone intermediate, it is proposed that Q can accept electrons from high and low potential donors, via a semiquinone intermediate. It appears that the short circuit electron transfer between the high and low potential arms is once again prevented by rapid transmembrane electron transfer (this time between quinone redox centres) establishing the critical insulating distance and preventing uncoupling.

1.4.4 *The role of adiabaticity*

We have emphasized a non-adiabatic theory of electron transfer which allows us to begin to understand single electron transfer reactions between many of the redox centres found in bioenergetic membranes. Although there are a great many of these non-adiabatic reactions, the vast majority of oxidoreductases (the largest class of enzymes in the IUPAC categorization) involve at least one adiabatic reaction that requires the making and breaking of bonds. Indeed, it is these adiabatic reactions that set the ultimate design constraints on non-adiabatic electron transfer systems. The most obvious case is the diffusion of redox substrates to the enzyme active sites, substrate binding, and the conversion of $n = 2$ electron redox couples into a pair of $n = 1$ single electron transfers that can participate in long distance non-adiabatic electron transfer. These diffusional reactions are often quite slow, sometimes taking up to a millisecond. Thus, without some means of pulse activating these enzymes, it may be very difficult to reveal any underlying non-adiabatic long distance electron transfer rates. Even after the diffusional processes, adiabatic reactions involving bond breaking may dominate electron transfer rates. The proton binding and release reactions that are central to the energy coupling of quinone redox sites can easily become rate limiting and obscure the underlying electron transfer rates. Thus, while the parameters of the simple and general non-adiabatic electron transfer we have presented make it possible to use reaction rates to provide insight into the structure of electron transfer proteins, there must first be a reasonable assurance that the rates in question are not limited by adiabatic processes.

In spite of the time limits imposed on overall electron-transfer throughput rates in respiratory and photosynthetic systems by adiabatic reactions, the rapid reaction rates provided by long-range non-adiabatic electron transfer perform an essential role in respiratory and photosynthetic systems by stitching together adiabatic reaction sites placed on opposite sides of the chemiosmotic membrane and assuring that electron transfer takes place in a sequence that maintains the efficiency of redox energy transduction. Indeed, the relatively simple rules that permit quick estimates of the magnitude of intraprotein electron transfer rates encourage us to believe that we can successfully engineer and construct, *de novo*, unique electron transfer proteins that will perform in ways we find useful (Robertson *et al.*, 1994). Ninety years ago, Emil Fischer predicted that our understanding of protein action would take us to this point:

> "If we wish to catch up with Nature, we shall need to use the method as she does, and I can foresee a time in which physiological chemistry will not only make greater use of natural enzymes but will actually resort to creating synthetic ones." (Fischer, 1905).

While traditional non-redox enzyme actions may still present severe design challenges, we suggest that the principles of intraprotein electron transfer now permit the rational design of long-range electron transfer proteins and a foray into the construction of dehydrogenases.

References

Adler AD, Longo FR, Kampas F. (1978) Solid state phenomena in porphyrins and related materials. In: *The Porphyrins*, Vol. 5 (ed. D Dolphin), Academic Press, New York, pp. 483–492.

Allen JP, Feher G, Yeates TO, Komiya H, Rees DC. (1987) Structure of the reaction center from *Rhodobacter sphaeroides* R-26: the cofactors. *Proc. Natl Acad. Sci.* USA **84:** 5730–5734.

Aquino AJA, Beroza P, Beratan DN, Onuchic JN. (1995) Docking and electron transfer between cytochrome c_2 and the photosynthetic reaction center. *Chem. Phys.* **197:** 277–288.

Beratan DN, Betts JN, Onuchic JN. (1991) Protein electron transfer rates set by the bridging secondary and tertiary structure. *Science* **252**: 1285–1288.

Beratan DN, Betts JN, Onuchic JN. (1992a) Tunneling pathway and redox-state-dependent electronic couplings at nearly fixed distance in electron-transfer proteins. *J. Phys. Chem.* **96**: 2852–2855.

Beratan DN, Onuchic JN, Winkler JR, Gray HB. (1992b) Electron-tunneling pathways in proteins. *Science* **258**: 1740–1741.

Betts JN, Beratan DN, Onuchic JN. (1992) Mapping electron tunneling pathways: an algorithm that finds the 'minimum length'/maximum coupling pathway between electron donors and acceptors in proteins. *J. Am. Chem. Soc.* **114**: 4043–4046.

Bixon M, Jortner J, Michel-Beyerle ME. (1995) A kinetic-analysis of the primary charge separation in bacterial photosynthesis: energy gaps and static heterogeneity. *Chem. Phys.* **197**: 389–404.

Casimiro DR, Richards JH, Winkler JR, Gray HB. (1993a) Electron-transfer in ruthenium-modified cytochromes-*c*: sigma-tunneling pathways through aromatic residues. *J. Phys. Chem.* **97**: 13073–13077.

Casimiro DR, Wong LL, Colon JL, Zewert TE, Richards JH, Chang IJ, Winkler JR, Gray HB. (1993b) Electron-transfer in ruthenium zinc porphyrin derivatives of recombinant human myoglobins: analysis of tunneling pathways in myoglobin and cytochrome-*c*. *J. Am. Chem. Soc.* **115**: 1485–1489.

Cheung E, Taylor K, Kornblatt JA, English AM, McLendon G, Miller JR. (1986) Direct measurements of intramolecular electron transfer rates between cytochrome *c* and cytochrome *c* peroxidase: effects of exothermicity and primary sequence on rate. *Proc. Natl Acad. Sci. USA* **83**: 1330–1333.

Closs GL, Calcaterra LT, Green NJ, Miller JR, Penfield KW. (1986) Distance, stereoelectronic effects, and the Marcus inverted region in intramolecular electron-transfer in organic radical-anions. *J. Phys. Chem.* **90**: 3673–3683.

DeVault D. (1980) Quantum mechanical tunnelling in biological systems. *Q. Rev. Biophys.* **13**: 387–564.

DeVault D, Chance B. (1966) Studies of photosynthesis using a pulsed laser. I. Temperature dependence of cytochrome oxidation rate in *Chromatium*. Evidence for tunneling. *Biophys. J.* **6**: 825–847.

Ding H, Robertson DE, Daldal F, Dutton PL. (1992) Cytochrome bc_1 complex [2Fe-2S] cluster and its interaction with ubiquinone and ubihydroquinone at the Qo site: a double-occupancy Qo site model. *Biochemistry* **31**: 3144–3158.

Ding H, Moser CC, Robertson DE, Tokito MK, Daldal F, Dutton PL. (1995) A ubiquinone special pair central to the primary energy conversion reactions of cytochrome bc_1 complex. *Biochemistry* **31**: 3144–3158.

Dutton PL, Moser CC. (1994) Quantum biomechanics of long-range electron transfer in protein: hydrogen bonds and reorganization energies [comment]. *Proc. Natl Acad. Sci. USA* **91**: 10247–10250.

Ermler U, Fritzsch G, Buchanan SK, Michel H. (1994) Structure of the photosynthetic reaction-center from *Rhodobacter-sphaeroides* at 2.65-angstrom resolution: cofactors and protein–cofactor interactions. *Structure* **2**: 925–936.

Evenson JW, Karplus M. (1993) Effective coupling in biological electron-transfer: exponential or complex distance dependence. *Science* **262**: 1247–1249.

Farid RS, Moser CC, Dutton PL. (1993) Electron transfer in proteins. *Curr. Opin. Struct. Biol.* **3**: 225–233.

Farver O, Pecht I. (1992) Long range intramolecular electron transfer in azurins. *J. Am. Chem. Soc.* **114**: 5764–5767.

Fischer E. (1905) Synthesen in der Purin- und Zuckergruppe. In: *Les Prix Nobel en 1902* (eds PT Cleve, C-B Hasselberg, K-A-H Morner, C-D Wirsen, C-G Santesson). P.-A. Norstedt & Fils, Stockholm.

Friesner RA. (1994) Comparison of theory and experiment for electron transfers in proteins: where's the beef? *Structure* **2**: 339–343.

Gamow G. (1928) Zur Quanten theories des Atom kernes. *Z. Phys.* **51**: 204–212.

Giangiacomo KM, Dutton PL. (1989) In photosynthetic reaction centers, the free-enegy difference for electron-transfer between quinones bound at the primary and secondary quinone-binding sites governs the observed secondary site specificity. *Proc. Natl Acad. Sci. USA* **86**: 2658–2662.

Gruschus JM, Kuki A. (1993) New Hamiltonian model for long-range electronic superexchange in complex molecular structures. *J. Phys. Chem.* **97**: 5581–5593.

Gunner MR, Dutton PL. (1989) Temperature and $\Delta G°$ dependence of the electron-transfer from Bph.- to Q_A in reaction center protein from *Rhodobacter sphaeroides* with different quinones as Q_A. *J. Am. Chem. Soc.* **111**: 3400–3412.

Gunner MR, Robertson DE, Dutton PL. (1986) Kinetic studies on the reaction center protein from *Rps. sphaeroides*: temperature and free energy dependence of electron transfer between various quinones in the Q_A site and oxidized bacteriochlorophyll dimer. *J. Phys. Chem.* **90**: 3783–3795.

Gutman F, Lyons LE. (1967) *Organic Semiconductors*. Wiley, New York.

Hopfield JJ. (1974) Electron transfer between biological molecules by thermally activated tunneling. *Proc. Natl Acad. Sci. USA* **71**: 3640–3644.

Jia YW, Dimagno TJ, Chan CK, Wang ZY, Du M, Hanson DK, Schiffer M, Norris JR, Fleming GR, Popov MS. (1993) Primary charge separation in mutant reaction centers of *Rhodobacter capsulatus*. *J. Phys. Chem.* **97**: 13180–13191.

Jortner J. (1976) Temperature dependent activation energy for electron transfer between biological molecules. *J. Chem. Phys.* **64**: 4860–4867.

Karpishin TB, Grinstaff MW, Komarpanicucci S, McLendon G, Gray HB. (1994) Electron-transfer in cytochrome-*c* depends upon the structure of the intervening medium. *Structure* **2**: 415–422.

Kuki A, Wolynes PG. (1987) Electron tunneling paths in proteins. *Science* **236**: 1647–1652.

Labahn A, Paddock ML, McPherson PH, Okamura MY, Feher G. (1994) Direct charge recombination from $D^+Q_AQ_B^-$ to DQ_AQ_B in bacterial reaction centers from *Rhodobacter sphaeroides*. *J. Phys. Chem.* **98**: 3417–3423.

Langen R, Chang IJ, Germanas JP, Richards JH, Winkler JR, Gray HB. (1995) Electron tunneling in proteins: coupling through a beta strand. *Science* **268**: 1733–1735.

Levich VG, Dogonadze RR. (1959) Teiriya bezizluchatelnikh electronnikh perekhodov mezhdu ionami v rastvorakh. *Dokl. Akad. Nauk. SSSR* **124**: 123–126.

Lin X, Murchison HA, Nagarajan V, Parson WW, Allen JP, Williams JC. (1994a) Specific alteration of the oxidation potential of the electron-donor in reaction centers from *Rhodobacter sphaeroides*. *Proc. Natl Acad. Sci. USA* **91**: 10265–10269.

Lin X, Williams JC, Allen JP, Mathis P. (1994b) Relationship between rate and free-energy difference for electron-transfer from cytochrome c_2 to the reaction-center in *Rhodobacter sphaeroides*. *Biochemistry* **33**: 13517–13523.

Marcus RA. (1956) On the theory of oxidation–reduction reactions involving electron transfer: I. *J. Chem. Phys.* **24**: 966–978.

Marcus RA, Sutin N. (1985) Electron transfers in chemistry and biology. *Biochim. Biophys. Acta* **811**: 265–322.

McLendon G. (1988) Long-distance electron transfer in proteins and model systems. *Acc. Chem. Res.* **21**: 160–167.

Meot-Ner M, Green JH, Adler AD. (1973) Electron-impact mass spectrometry of porphyrin systems. *Ann. N. Y. Acad. Sci.* **206**: 641–648.

Michel H, Deisenhofer J, Epp O. (1986) Pigment protein interactions in the photosynthetic reaction center from *Rhodopseudomonas viridis*. *EMBO J.* **5**: 2445–2451.

Miller JR, Beitz JV, Huddleston RK. (1984) Effect of free energy on rates of electron transfer between molecules. *J. Am. Chem. Soc.* **106**: 5057–5068.

Mitchell P. (1961) Coupling of phosphorylation to electron and hydrogen transfer by a chemi-osmotic type of mechanism. *Nature* **191**: 144–148.

Moser CC, Dutton PL. (1988) Cytochrome *c* and c_2 binding dynamics and electron transfer with photosynthetic reaction center protein and other integral membrane redox proteins. *Biochemistry* **27**: 2450–2461.

Moser CC, Dutton PL. (1992) Engineering protein structure for electron transfer function in photosynthetic reaction centers. *Biochim. Biophys. Acta* **1101**: 171–176.

Moser CC, Keske JM, Warncke K, Farid RS, Dutton PL. (1992) The nature of biological electron transfer. *Nature* **355**: 796–802.

Moser CC, Page CC, Farid R, Dutton PL. (1995) Biological electron transfer. *J. Bioenerg. Biomembr.* **27**: 263–274.

Newton MD. (1991) Quantum chemical probes of electron-transfer kinetics: the nature of donor–acceptor interactions. *Chem. Rev.* **91**: 767–792.

Northrup SH, Boles JO, Reynolds JC. (1988) Brownian dynamics of cytochrome c and cytochrome c peroxidase association. *Science* **241:** 67–70.

Onuchic JN, Beratan DN, Winkler JR, Gray HB. (1992) Pathway analysis of protein electron-transfer reactions. *Annu. Rev. Biophys. Biomol. Struct.* **21:** 349–377.

Pan LP, Durham B, Wolinska J, Millett F. (1988) Preparation and characterization of singly labeled ruthenium polypyridine cytochrome c derivatives. *Biochemistry* **27:** 7180–7184.

Pelletier H, Kraut J. (1992) Crystal structure of a complex between electron transfer partners, cytochrome c peroxidase and cytochrome c. *Science* **258:** 1748–1755.

Robertson DE, Farid RS, Moser CC, Urbauer JL, Mulholland SE, Pidikiti R, Lear JD, Wand AJ, Degrado WF, Dutton PL. (1994) Design and synthesis of multi-heme proteins. *Nature* **368:** 425–431.

Siddarth P, Marcus RA. (1993) Electron-transfer reactions in proteins: an artificial intelligence approach to electronic coupling. *J. Phys. Chem.* **97:** 2400–2405.

Tiede DM, Vashishta AC, Gunner MR. (1993) Electron-transfer kinetics and electrostatic properties of the *Rhodobacter sphaeroides* reaction center and soluble c-cytochromes. *Biochemistry* **32:** 4515–4531.

Ulstrup J, Jortner J. (1975) The effect of intramolecular quantum modes on free energy relationships for electron transfer reactions. *J. Chem. Phys.* **63:** 4358–4368.

Warncke K, Dutton PL. (1993) Influence of Q_A site redox cofactor structure on equilibrium binding, *in situ* electrochemistry, and electron-transfer performance in the photosynthetic reaction center protein. *Biochemistry* **32:** 4769–4779.

Warshel A, Chu ZT, Parson WW. (1989) Dispersed polaron simulations of electron transfer in photosynthetic reaction centers. *Science* **246:** 112–116.

Whitford D, Gao Y, Pielak GJ, Williams RJ, McLendon GL, Sherman F. (1991) The role of the internal hydrogen bond network in first-order protein electron transfer between *Saccharomyces cerevisiae* iso-1-cytochrome c and bovine microsomal cytochrome b_5. *Eur. J. Biochem.* **200:** 359–367.

Willie A, Stayton PS, Sligar SG, Durham B, Millett F. (1992) Genetic engineering of redox donor sites: measurement of intracomplex electron transfer between ruthenium-65-cytochrome b_5 and cytochrome c. *Biochemistry* **31:** 7237–7242.

Wuttke DS, Bjerrum MJ, Winkler JR, Gray HB. (1992a) Electron-tunneling pathways in cytochrome c. *Science* **256:** 1007–1009.

Wuttke DS, Bjerrum MJ, Chang I-J, Winkler JR, Gray HB. (1992b) Electron tunneling in ruthenium-modified cytochrome c. *Biochim. Biophys. Acta* **1101:** 168–170.

Zhou JS, Hoffman BM. (1994) Stern–Volmer in reverse: 2/1 stoichiometry of the cytochrome c–cytochrome c peroxidase electron-transfer complex. *Science* **265:** 1693–1696.

The protein bridge between redox centres

David N. Beratan and José Nelson Onuchic

2.1 Introduction

The unifying theme of the electron transfer (ET) processes discussed in this book is motion of a single electron from one centre of localization (chlorophyll special pair, haem, blue copper centre, flavin, tyrosyl radical, etc.) to another one that is separated in distance by several angstroms (Bertini *et al.*, 1994; Lippard and Berg, 1994). Furthermore, the centres (in most cases) are insulated from one another. The electrons that are being shuttled, whether photoexcited or not, propagate from donor to acceptor without forming real long-lived intermediate (reduced or oxidized) protein states (Hopfield, 1974), that is the protein is not acting as a conductor or a photoconductor. As we shall see, the ET reactions of interest involve long-range tunnelling between localized donor and acceptor states that fall (energetically) within the energy gap between the frontier orbitals of an otherwise insulating material.

2.1.1 *What tunnels?*

The DeVault–Chance experiment first demonstrated, and later experiments confirmed, that long distance electron transfer reactions are viable from room temperatures to cryogenic temperatures (DeVault, 1984). Insensitivity to temperatures is the signature of activationless processes and/or tunnelling. Two types of motion are associated with electron transfer: motion of the electron itself from donor to acceptor and nuclear motion associated with readjustment of protein and solvent nuclei to the electron redistribution of the oxidation–reduction event. The nuclear rearrangement involves high frequency inner sphere modes (covalent bonds of the donor and acceptor) as well as outer sphere modes (protein and solvent) that adjust to the charge redistribution. While most of this chapter deals with single electron transfer processes, much of biology relies upon coupled multielectron and proton-coupled electron transfer. Our understanding of these coupled processes is less well developed than that of the single electron process.

Protein Electron Transfer, Edited by D.S. Bendall
© 1996 BIOS Scientific Publishers Ltd, Oxford

In most electron transfer reactions of interest, electrons flow between redox active cofactors in protein: blue copper centres, haems, iron–sulphur clusters, chlorophylls, etc. Aromatic amino acid side chains occasionally act as electron donors or acceptors, usually in the presence of a highly reactive redox partner. Oxidized or reduced amino acid intermediates do not form in most ET reactions. As such, true intermediate oxidized–reduced states of the protein do not exist. The energy needed to form such intermediates far exceeds thermal energies, $k_B T$. Since the donor and acceptor are not in direct contact, the electron must tunnel to move from donor to acceptor. Thus, in the long distance ET reactions of interest in this book, the dynamics are dominated by electron tunnelling; the nature of the nuclear motion (classical or quantum) will depend upon the specific system. To describe the electron tunnelling further, we need to consider the strength and the nature of the electronic interactions between donor and acceptor imbedded in the protein as well as the vibronic coupling.

2.1.2 *Strong versus weak electronic coupling regimes*

Most biological ET reactions involve donors and acceptors that interact weakly, and the reactions are non-adiabatic. As such, the ET reaction rate depends upon both an electronic interaction and a term associated with the nuclear barrier (DeVault, 1984; Marcus and Sutin 1985). In contrast, ET between donor and acceptor species in direct contact and strong communication with one another is controlled by nuclear motion, as electronic interactions are so strong that they cease to control the reaction rate. This may be the relevant regime for exchange of electrons between proteins and other small mobile reactants (e.g. inorganic complexes), but seems to be of limited physiological importance. Most biological ET reactions seem to involve reactant cofactors imbedded within proteins. The protein provides steric constraints that prohibit the reactants from contacting one another directly, leading to the weak interaction non-adiabatic regime.

Rates of non-adiabatic ET can be written as:

$$k_{et} = \frac{2\pi}{\hbar} |T_{DA}|^2 \ (FC) \tag{2.1}$$

The reaction free energy dependence is contained in the Franck–Condon factor (*FC*). Effects associated with nuclear tunnelling and with thermally activated barrier crossing are contained in this factor. T_{DA} describes the donor–acceptor interaction associated with electron tunnelling.

Physical systems are most readily described in terms of their Hamiltonian, that is the sum of their kinetic and potential energies. For quantum systems, the classical quantities are replaced by operators. The small energy of interest, T_{DA}, does not appear explicitly in the Hamiltonian of the system. Methods to manipulate the Hamiltonian in order to compute T_{DA} are described in Section 2.3 and in Appendix A.

The non-adiabatic electron transfer rate in Equation 2.1 is valid in the weak interaction regime, that is the matrix element T_{DA} is very small. We now discuss what we mean by small. Appendix A shows that the electron can only tunnel between the donor and acceptor sites when their energies are resonant. That happens at particular reaction coordinate geometries that modulate the electronic energies of the donor and acceptor. The question becomes how long does this degeneracy last compared with the resonant donor–acceptor electron hopping frequency $(T_{DA}/\hbar)$ once this resonance is achieved? If the system stays resonant for times short compared with this electronic oscillation time, the non-adiabatic limit is valid. In the opposite limit, the rate is adiabatic. This description is very intuitively appealing, but is not entirely complete. A quantum description of these issues should be consulted for more details (Onuchic and Wolynes, 1988).

In the case of an overdamped harmonic reaction coordinate (see Appendix A), the resonance time can be calculated analytically. Resonance occurs exactly at the crossing point of the donor and acceptor surfaces (Marcus parabolas). The resonance time is therefore the time that the system takes to drift away from this crossing point:

$$t_{drift} = \frac{2T_{DA}}{\lambda \omega_c} \qquad (2.2)$$

λ is the reorganization energy and $\omega_c = \omega^2/\eta$ is the overdamped frequency. The complete expression for the electron transfer rate is then

$$\frac{2\pi}{\hbar}|T_{DA}|^2 \frac{1}{1 + 2\pi|T_{DA}|^2/\hbar\omega_c\lambda}(FC) \qquad (2.3)$$

Clearly, when $t_{drift} \ll \hbar/T_{DA}$, the non-adiabatic limit is valid.

This overdamped limit is particularly interesting because it is mathematically equivalent to a Debye-like model for polar solvents. The reaction coordinate in this case is the total polarization coupled to the transfer electron. The polarization dynamics of a Debye solvent is equivalent to that of an overdamped harmonic oscillator. The coupling of this solvent to the electron transfer rate is obtained by replacing ω_c by τ_L^{-1} where $\tau_L = \tau_D/\varepsilon_o$ (Calef and Wolynes, 1983; Garg et al., 1983; Zusman, 1994).

2.1.3 Tunnelling electrons

What is the meaning of T_{DA} in the ET rate equation? In the hydrogen molecule ion, the energy splitting between the σ and σ^* orbitals can be associated with an oscillation frequency, $\omega = \Delta E/2\hbar$. That is, if an electron is started out on the left-hand atomic orbital, its probability of being found on the left orbital at a later time oscillates as $\cos^2\omega t$. In the more general case of donor and acceptor separated from one another by a bridge consisting of many thousands of orbitals, we need to compute the frequency of exchange between donor and acceptor. Here too, the frequency that we call $T_{DA}/\hbar$ is associated with a splitting, but the splitting arises from the donor–acceptor interaction provided by the presence of the intervening and surrounding protein orbitals. The formal procedure to calculate this mediated splitting, or coupling, lies at the heart of the protein ET problem (some details of this procedure are discussed in Appendix A). Defining the effective two-state problem and estimating the associated splittings has occupied numerous groups over the last few years. Simple approximations to this splitting are estimated with our Pathway methods; numerical methods that deal explicitly with multiple interfering paths utilize Green function or path integral techniques. At the present time, it appears that the remaining bottleneck is not our ability to perform the splitting computation itself for thousands or orbitals, but rather to build a meaningful effective Hamiltonian for the protein ET problem.

2.1.4 *Basics of vibronically coupled electron tunnelling*

Electrons tunnel in biological systems because thermal energies ($k_B T = 1/40$ eV at room temperature) are much smaller than the energies needed to promote an electron from the donor onto the bridge or from the bridge onto the acceptor. Nuclei, however, may tunnel or move classically, depending upon the nuclear barrier to reaction. Formulation of the non-adiabatic ET rate usually relies upon a Born–Oppenheimer/Franck–Condon description. These approximations are discussed in Appendix A. Usually, a frozen nuclear geometry is assumed for the protein in which the electron tunnelling interaction between donor and acceptor is computed. What does this mean?

The donor and acceptor localized states must be brought into degeneracy or near degeneracy in order for the electron to tunnel (DeVault, 1984; Hopfield, 1974; Marcus and Sutin, 1985). Electronic energy (binding energy, in eV for example) can be associated directly with a redox potential (Bard and Faulkner, 1980). So the problem of electron tunnelling in proteins is one of electron tunnelling in nuclear geometries around the 'activated complex' structure. This geometry differs from the equilibrium geometry in that nuclei around the donor and acceptor have begun to adjust to accommodate the change in charge associated with oxidation of D and reduction of A. If modes with similar frequencies and couplings exist on donor and acceptor, a useful approximation to the tunnelling energy is derived from the average redox potentials of D/D^+ and A/A^- (Onuchic *et al.*, 1986). In some cases, the donor–acceptor interaction may be hypersensitive to nuclear coordinates and the Condon approximation may not be appropriate. A near degeneracy with a bridge state or an *extremely* long distance transfer event may complicate the Condon approximation (Onuchic *et al.*, 1986).

The electronic portion of the protein ET problem, then, is one of an electron with tunnelling energy E_{tun} penetrating a three-dimensional structured barrier. The critical difference between electron tunnelling in proteins and conductivity in metals is that the electron does not form true intermediate states of the 'bridge', that is reduced bridge states do not persist for any detectable period of time. However, slight admixtures of the bridge states enter the effective donor and acceptor states (see Appendix A). This slight 'contamination' of the donor and acceptor orbitals by the bridge orbitals facilitates electron tunnelling.

2.1.5 *Tunnelling energetics*

Two types of virtual (i.e. energetically forbidden) intermediates facilitate tunnelling. These are species with extra electron(s) on the protein or electron(s) removed from the bridge [hole(s) injected] (Curry *et al.*, 1995). The energy of a particular virtual intermediate is defined as the energy needed to oxidize or reduce the bridge (at the frozen geometry prior to electron removal or insertion). The energy of a given virtual intermediate defines the contribution of that intermediate to the tunnelling process. This claim can be understood with the standard perturbation theory-based argument of physical chemistry – states close in energy mix more strongly than those distant in energy. It is possible that electron and hole mediation propensity differ in distinct regions of a protein, are sensitive to side chain, structure and hydrogen bonding motif, are tunnelling energy dependent, etc. Bonding orbitals and lone pairs of electrons as well as anti-bonding orbitals of the protein are tunnelling mediators. Because of the large energy gap between these sets (defined by the binding energy and the electron affinity of the protein), the overall shape of the tunnelling energy dependence forms a 'smiley face'; coupling is strong for energies near the filled or empty states, but weaker and (weakly energy dependent) in between (Beratan and Hopfield, 1984). Details of these figures (the energy gap and asymmetry of the curves) depend upon the nature of the electronic structure model used. Considerable effort is now aimed at comparing the shape and asymmetry of these plots at different levels of electronic structure theory.

2.1.6 *How does exponential decay arise from delocalized bridge states?*

Much of physical organic chemistry is based upon orbital symmetry arguments and much of ET theory can be summarized with a few orbital interaction concepts (e.g. Fleming, 1976). However, in stark contrast to many examples in physical organic chemistry, frontier orbital mixing arguments fail. For example, consider a localized donor interacting with the frontier orbital of an alkane bridge. The HOMO (highest occupied molecular orbital) [or LUMO (lowest unoccupied molecular orbital)] of the bridge is completely delocalized, with approximately equal electron density on each CH_2 unit of the chain. Thus, if the chain has 10 repeat units, about 1/10 of the electron density is located on each CH_2 of the HOMO or LUMO. Frontier orbital analysis would mix one (or both) of these states with the donor, forming a molecular orbital $\Psi \approx \phi_D$ $+\gamma\phi_{HOMO}$. γ is expected to be small, but since ϕ_{HOMO} is delocalized, the donor electron would be predicted to leak equally onto all sites of the bridge. Clearly a simple square barrier approximation of the alkane would suggest exponential decay of amplitude away from the donor. What is wrong with the frontier orbital analysis?

What is missing in the frontier orbital analysis is the fact that the energies of the orbitals below the HOMO (the HOMO-1, HOMO-2, ... orbitals) are all about as close in energy to the tunnelling energy (E_{tun} – the energy of the donor and acceptor states in the electron transfer active geometry that lies in the HOMO–LUMO energy gap of the bridge) as is the HOMO itself. Thus, all of these states mediate the coupling of the donor electron through the bridge. Neglecting even one of the bridge orbitals in calculating the donor propagation can lead to non-physical results. The subtle cancellation of oscillatory bridge wave functions leads to (approximately) exponential decay of coupling with distance. The need to include *all* states to reproduce the decay, as well as the difficulty of conducting quantum calculations on large many-body systems, has favoured simple models that build in the essential physics of the tunnelling problem. The development of quantum chemical methods to calculate electronic coupling in modest to large chemical and biochemical systems is the goal of many current efforts.

2.1.7 Through-bond and through-space decay length scales

There are two qualitatively different types of electronic interactions in chemical systems: through-bond interactions and through-space interactions. In ET reactions, we need to understand how through-bond versus through-space interactions influence tunnelling mediation. The strength of bond-mediated coupling depends upon how rapidly a localized donor localized state decays. What determines how rapidly this state decays? The distance scale is determined principally by the atomic orbital hybridization of the atoms constituting the bonds, the energies of these orbitals and the tunnelling energy. For example, the very simplest analysis of carbon–carbon bonds interacting in a polyethylene chain suggests that the decay should be proportional to the 2s–2p ionization potential difference for carbon divided by the tunnelling energy (Beratan and Hopfield, 1984). Here, the key ratio is one of bond–bond interaction strength in units of the tunnelling energy. A typical coupling decay per bond is found to be 0.6 in the experiments of Closs and Miller, for example (Closs and Miller, 1986). In contrast, the through-space distance decay is determined mainly by the tunnelling energy. In the extreme through-space limit (in one dimension, 1D), this decay scales exponentially with a decay factor proportional to the square root of the tunnelling energy. Through-space decay of R Å is approximately $0.6\exp[(-1.7)(R-1.4)]$ for an 8 eV binding (tunnelling) energy electron (Onuchic and Beratan, 1990). This expression interpolates between the decay per bond parameters (known from experiments) and the 1D decay expression for vacuum tunnelling.

2.2 Tunnelling pathways in proteins

Proteins consist of a sequence of peptide backbone interactions augmented by non-bonded interactions arising from protein folding. In an attempt to mesh the relatively well understood through-bond decay process with 'short cuts' to tunnelling provided by hydrogen bonds and van der Waals contacts in the folded structure, we built the Pathway model to estimate T_{DA}.

We have already presented enough information to estimate the strength of long-range interactions in proteins! Since electron tunnelling involves a combination of through-bond and through-space coupling, we need to assemble a combination of the two types of decay factors. Thinking of these through-bond and through-space factors as highly renormalized or rescaled parameters (more on this later) we write (Beratan et al., 1987, 1992b; Onuchic and Beratan, 1990).

$$T_{DA} = A\prod_i \varepsilon_C(i)\prod_j \varepsilon_H(j)\prod_k \varepsilon_S(k) \qquad (2.4)$$

Here, A is a pre-factor that depends on details of the interaction between donor or acceptor with the first or last bond of the tunnelling pathway, and each ε is a decay per unit, described by the experiments or simple estimates above. The hydrogen bonds can be analysed as a whole (as was done originally) or broken into a combination of through-bond and through-space steps (as was done later) (Beratan *et al.*, 1990; Onuchic and Beratan, 1990; Regan *et al.*, 1993).

Within the pathway strategy, any protein structure [from X-ray or nuclear magnetic resonance (NMR) experiments] defines a network of pathway decay parameters. That is, every pair of atoms is connected through a covalent bond, a hydrogen bond or through space. A list of simple decay factors corresponding to each connection type can be written using the simple parameterization above. Simple analysis gives:

$$\varepsilon_C = 0.6 \tag{2.5a}$$
$$\varepsilon_H = 0.6^2 \, e^{-1.7(R - 2.8)} \tag{2.5b}$$
$$\varepsilon_S = \tfrac{1}{2}0.6 \, e^{-1.7(R - 1.4)} \tag{2.5c}$$

The hydrogen bond decay is from heteroatom to heteroatom. The factor of 1/2 in the through-space factor is associated with orbital angular overlap efffects. Within this prescription, estimating the coupling is simply a matter of listing the through-bond or through-space decay for every pair of atoms in the protein and then finding the combination of these interactions, beginning at the donor and ending at the acceptor, that maximizes the product of these penalty factors. This problem is no longer one of electronic structure theory; it is one of graph theory (Betts *et al.*, 1992). We have used numerous graph theoretical search strategies (breadth-first and depth-first searches, for example) to find the very strongest path or families of pathways in proteins. This computation is rapid even in very large proteins; software exists to perform these searches on workstations and personal computers.

2.2.1 *What is the physical meaning of the dominant pathway?*

The pathway decay parameters are taken from experiments and from simple theoretical estimates (Beratan *et al.*, 1991). The estimated pathway coupling between donor and acceptor is the simple product $\varepsilon_1 \varepsilon_2 ... \varepsilon_N$. Also, the 'physical pathway' is comprised of the atoms associated with each of these decay factors. However, the physical pathway is an 'effective' object because the decay parameters take into account the surrounding bonded and non-bonded interactions in an average sense. Since our through-bond decay factors are estimated from experiment, and the experimental couplings incorporate bonded and non-bonded interactions to infinite order, the dominant physical pathway is really at the centre of a set of physical pathways that enter the electronic coupling. We are developing new methods currently to dissect the contributions of the strongest physical pathway and surrounding paths to the electronic coupling. This analysis has led to the idea of fattened pathways (Regan *et al.*, 1993) (*Figure 2.1*), pathway tubes (Regan *et al.*, 1995), contact importance factors (Skourtis *et al.*, 1994) and pathway interference effects (Skourtis *et al.*, 1995). It is important to realize that snipping one contact on the strongest pathway will not shut down electron transfer in general. Pathways arise in families, and *the strongest pathway is characteristic of those paths that dominate the electronic coupling*. Recent studies of fattened pathways and pathway tubes have shown more precisely what the shape of the relevant region

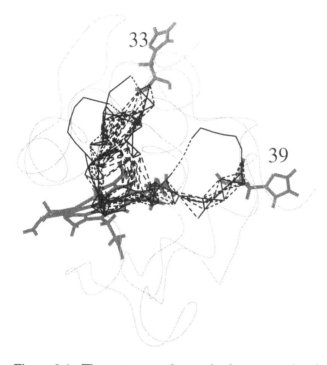

Figure 2.1. The strongest pathways clearly represent bonds at the centre of a tube of paths that contribute to the donor–acceptor interaction mediated by the protein. Reprinted with permission from Regan *et al.* (1993) Protein electron transport: single versus multiple pathways. *J. Phys. Chem.* **97:** 13083–13088. Copyright 1993 American Chemical Society.

surrounding the best path is in a given protein. To move beyond the simple pathway estimate of the coupling, all possible sequences for the propagation of wave function amplitude between atomic orbitals of the bridge that begin with an orbital coupled to the donor and end with one coupled to the acceptor must be added together, bearing in mind that some contributions may have a negative sign. Such summations can be computed using perturbation theory or Dyson expansion strategies, or by using techniques that explicitly sum all bridge pathways to infinite order (Goldman, 1991; Gruschus and Kuki, 1993; Onuchic *et al.*, 1991; Ratner, 1990; Regan *et al.*, 1993).

2.2.2 *When does a simple product approximate the tunnelling matrix element?*

The product expression for T_{DA} looks like a perturbation theory (McConnell, 1961) result, but it is not. Product forms for electronic coupling are in fact much more general. For example (daGama, 1990), any linear chain of orbitals with nearest-neighbour coupling interactions can be shown to mediate electronic coupling in a manner such that the coupling appears as a product. Linear chains of orbitals with dangling side chains also yield a product expression for T_{DA} (Equation 2.6).

$$T_{DA} \propto \prod_{i=1}^{N-1} \frac{\beta_{i,i+1}}{E_{tun} - \alpha_i - \Delta_i} \qquad (2.6)$$

In this equation, α_i is a site energy and Δ_i is a site energy shift that arises from all possible 'scattering pathways' of electron amplitude throughout both the main chain and the dangling side chains (daGama, 1990). E_{tun} is the tunnelling energy and $\beta_{i,i+1}$ is the interaction between the (i) and $(i+1)$ bridge units. In the extreme perturbation theory limit, a product expression arises in general as well. Once loops appear (provided by bonded or non-bonded interactions) in the bridging structure, a simple product expression need no longer be valid. Complications may arise when simple Hamiltonians are built neglecting dangling side chains, loops, etc. and the complicating aspects can be reintroduced and their role probed. The results, as one might expect, are protein structure dependent (Regan *et al.*, 1995).

2.2.3 *Key role for hydrogen bonds predicted*

From model compound ET data, we know that proteins would be unable to transfer electrons if the electrons were able only to propagate down the covalent peptide backbone. The reason for this is that points close in terms of distance are often far apart in terms of amino acid sequence. As such, any realistic model of protein electronic interactions confronts the issue of what creates viable 'short-cuts' for electron propagation. We proposed hydrogen bonds as tunnels for the electrons on the grounds of simple energy and overlap analysis. The hydrogen bond, in this framework, can be thought of as an interaction that creates very short through-space gaps between otherwise unconnected segments of the protein. Since the electron must propagate at least a small distance through-space to avoid a circuitous purely peptide backbone path, the ubiquitous hydrogen bond would seem to provide copious short-cuts. Since the average electron lone pair to hydrogen atom distance in a hydrogen bond is typically much smaller than van der Waals contact distances between other non-bonded atoms, we predicted a very special mediating role for hydrogen bonds.

The key role for hydrogen bond contacts in ET proteins was first demonstrated by Therien and Gray in ruthenium-modified systems (Therien *et al.*, 1991). By fixing an ET probe at a specific surface site with a well-defined protein medium in between it and the native redox centre, the predictions of the pathway model could be compared directly with experiments. Agreement was excellent. It was, perhaps, surprising that the decay of coupling across a hydrogen bond was predicted to be about the same as that across two covalent bonds. This prediction itself was confirmed in small molecule experiments by the groups of Sessler (Harriman *et al.*, 1992), Nocera (Turró *et al.*, 1992) and Therien (de Rege *et al.*, 1995). This feature of hydrogen bonds is remarkable, and further studies of the energetic, structural and dynamic aspects of the hydrogen bond in the context of ET is needed.

2.3 Predictions of the tunnelling pathway model

What are the key predictions of the tunnelling pathway model? They are basically fivefold (Beratan *et al.*, 1991, 1992a).

(i) Coupling can be approximated as a product of decay factors characteristic of the chemistry of the medium intervening between donor and acceptor.

(ii) A key role for covalent and hydrogen bonds as tunnelling mediators and a less important role for van der Waals interactions is anticipated.

(iii) The average scale of coupling decay with distance arises from the balance of through-bond and through-space decay. Coupling down a straight chain (as in a β-strand) is expected to be about the same as that found in simple covalent model compounds with similar donor–acceptor energetics. In systems with somewhat more circuitous tunnelling routes (i.e. most ET systems), the decay of coupling with distance is expected to be somewhat faster and to depend upon the secondary and tertiary structure of the folded protein.

(iv) Decay of coupling throughout proteins is anisotropic. Coupling is expected to drop with distance, but considerable scatter about a single exponential line is anticipated (*Figure 2.2*). This anisotropy gives rise to hot and cold spots for electron transfer. Subsets of coupled sites that fall on a single exponential line can always be found within a given protein.

(v) The average distance decay is tied to protein secondary structure. All of these predictions have been scrutinized in the context of ruthenium-modified proteins (Onuchic *et al.*, 1992; Winkler and Gray, 1992) and the photosynthetic reaction centre (Curry *et al.*, 1995).

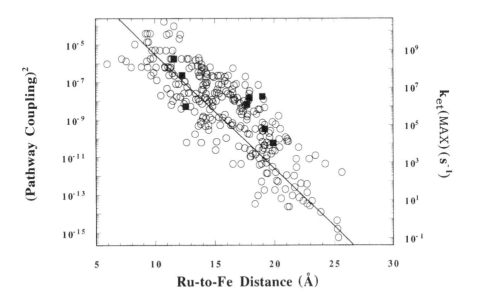

Figure 2.2. The pathway model captures the anisotropy of the protein as a tunnelling barrier. This scatter plot shows pathway couplings (open circles) to surface side chain heavy atoms in cytochrome *c* plotted versus the shortest haem to surface atom distance. Experimental rates are also shown (filled squares), mirroring the predicted degree of scatter. The solid line represents the best exponential fit to the pathway data.

2.3.1 *Tests of pathway predictions*

The essential role of hydrogen bonding in protein ET was demonstrated by Gray (Onuchic *et al.*, 1992; Therien *et al.*, 1991). The anisotropy of rates in proteins was shown in numerous Ru cytochrome *c* experiments (*Figure 2.3*). Derivatives with distances differing by up to 7 Å were shown to have the same rate. Other derivatives with essentially the same distance were shown to have rates differing by three or four orders of magnitude. Through-backbone coupling cannot account for these rates at all, nor can a model that assumes that purely through-space contacts control ET. If the through-space contacts dominated, much of the structure dependence of the rates would simply be washed out.

Just as early experiments showed the importance of hydrogen bonding, later experiments confirmed the cost associated with through-space contacts. The Ru(His72) cytochrome *c* data point is critical in this regard. In this derivative, the transfer distance is only 8.4 Å (edge-to-edge) but the maximal rate (computed for an activationless process) is a surprisingly slow 9×10^5 s^{-1} (Wuttke *et al.*, 1992). The pathway analysis clearly shows a through-space gap along the dominant donor–acceptor pathway (or pathway family). Other modified protein systems show rate anomalies or secondary structure dependencies that can only be understood in the context of pathway analysis (Karpishin *et al.*, 1994; Langen *et al.*, 1995; Moreira *et al.*, 1994; Therien *et al.* 1991; Wuttke *et al.*, 1992).

2.3.2 *Pathways in the exponential distance regime*

Much of our discussion has emphasized the non-exponential decay of coupling with distance that can arise in a protein. However, subsets of couplings in a given protein may be *highly* exponential. For example, we predicted that coupling down the β-strand of azurin would be just such a case (Beratan *et al.*, 1992a). This prediction was confirmed recently in a family of Ru-azurin experiments (Langen *et al.*, 1995; Regan *et al.*, 1995). These subsets of data associated with a single strand are essentially exponential, despite the fact that scatter of coupling with distance in azurin is predicted, over the entire protein, to be as large as in cytochrome *c*.

Dutton has pointed out that the rates of ET in the photosynthetic reaction centre (Gunner, 1991), when corrected for activation energy differences, fall on an exponential distance line (Moser *et al.*, 1992). This correlation is entirely consistent with pathway analysis as well (Curry *et al.*, 1995) and arises from relatively direct pathways between the chromophores. Thus, one must realize that subsets of data may be exponential or non-exponential with distance, depending upon how the pathway coupling within that family scales with distance.

2.3.3 *Docking and intermolecular ET*

An important step in biological ET is often *inter*molecular (see Pelletier and Kraut, 1992; Stemp and Hoffman, 1993, for examples). Much theoretical work on protein ET has been confined to *intra*molecular reactions. The pathway

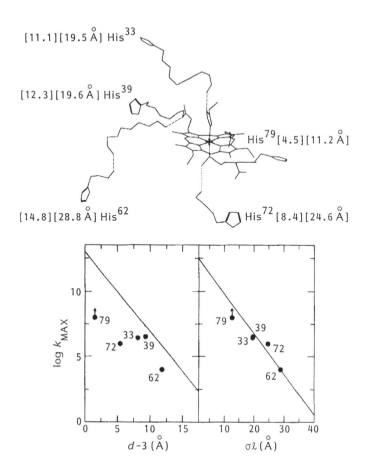

Figure 2.3. The strongest coupling pathways linking surface Ru(histidine) groups to the haem in modified cytochomes c are shown. Solid lines are covalent bonds, dashed lines hydrogen bonds and dotted lines through-space contacts. Edge-to-edge distances and tunnelling pathways lengths $[d][\sigma l]$ are shown in brackets. The tunnelling pathway length, σl, is defined as $(1.4 \text{ Å}) [\ln(\Pi_i \varepsilon_i)]/\ln(0.6)$ (Beratan et al., 1992a). The pathway length is the distance (through-bond) along an extended purely covalent chain that would give a coupling equal to that computed from $\Pi_i \varepsilon_i$ within the pathway model. Reprinted with permission from Beratan et al. (1992a) Electron tunneling pathways in proteins. *Science* **258**: 1740–1741. Copyright 1992 American Association for the Advancement of Science.

model was recently extended to describe *inter*molecular ET by combining the pathway couplings in each individual protein with an additional interface (through-space or hydrogen bond) decay factor using Equation 2.5 (Aquino et al., 1995):

$$T_{DA} \propto \Pi_i \varepsilon_D(i) \Pi_j \varepsilon_A(j) \varepsilon_{inter} \qquad (2.7)$$

Here, $\Pi_i \varepsilon_D(i)$ is the electronic coupling decay between the electron donor site and the surface of the protein containing the electron donor; $\Pi_i \varepsilon_A(j)$ is the electronic coupling decay between the electron acceptor site and the surface of the protein containing the acceptor; and ε_{inter} is the electronic coupling decay between protein surfaces (i.e. a through-space or hydrogen bond coupling between surface atoms on the two proteins). Note that in Equation 2.7, $\varepsilon_D(i)$ and $\varepsilon_A(j)$ can be any of the three types of decay (i.e. covalent, hydrogen-bonding or through-space). This factored electronic coupling separates quantities that are well defined, $\Pi_i \, \varepsilon_D(i)$ and $\Pi_i \varepsilon_A(j)$, from a quantity that is less well defined, ε_{inter}.

Surface to redox centre couplings are obtained for each protein in the usual way; maps of these couplings identify regions of the protein that efficiently couple ET between the redox site. Matching the strongly coupled regions will result in the maximal intermolecular electron transfer coupling (if ε_{inter} is not too small) and can be used as a criterion to evaluate possible docked structures (*Figure 2.4*). This criterion will supplement other energetic constraints on possible structures of functional importance in bimolecular ET.

2.3.4 *Concerns related to simple pathway analysis*

A number of basic questions enter the analysis of when pathway methods break down. The obvious chemical issue is how appropriate it is to approximate all atom types and bonds as being the same. This approximation sounds disastrous, but is justified by the fact that the tunnelling energy is several eV removed from the energies of the mediating bonds, and a distribution of bond types appears in pathways. As such, small energy differences might be less important that in other kinds of chemical reactions, and a certain amount of averaging may take place.

A large number of physical (and an even larger number of scattering) paths exist between donor and acceptor. How significant is the role played by quantum interference between pathways? The answer to this question should depend upon seconday and tertiary structure. In cytochrome c, we showed that couplings to ruthenation sites could be well approximated by adding to the best pathway the chemical bonds dangling from the path and including all scattering contributions arising from these groups. Within this set of orbitals, the coupling can be calculated in a way that includes all pathway interferences (to infinite order), and the final result is nearly identical to that found when the entire protein is included in the calculation. The case of azurin, where donor and acceptor sites are separated by many β-strands, is more complicated.

In proteins, additional complications are clearly present and are currently receiving attention. For example, fluctuations in atom–atom distances must lead to averaging of pathway couplings, especially the through-space couplings. This may be yet another reason that a simple parameter set works so well. Much of biological ET involves bimolecular processes. We are just beginning to probe the balance between docking energetics and pathway matching, and this may be another case where intra- and inter-protein fluctuations are critical. Finally, the donor and acceptor electronic structure plays an essential role in the ET process. Efforts are under way to splice improved local electronic structure results together with pathway or pathway-like calculations (I.V. Kurnikov and D.N. Beratan, in preparation). The deepest conceptual concern about the pathway approach is the neglect of explicit coupling interactions between interacting pathways.

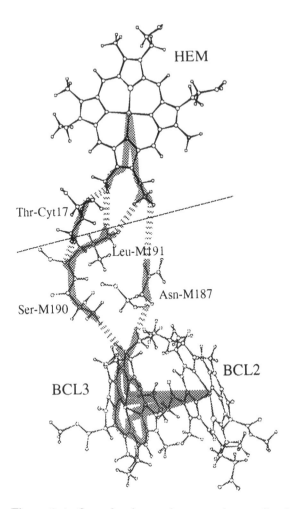

Figure 2.4. Sets of pathways that enter the coupling between the haem (HEM) of the cytochrome c_2 and the special pair of the photosynthetic reaction centre (BCL) (based upon coordinates provided by Tiede and Chang, 1988). The entrance points for the pathway tubes into the reaction centre are Leu-M191, Ser-M190 and Asn-M187. Analysis of preliminary co-crystal coordinates obtained by N. Adir *et al.* (submitted) indicate entrance points at Asn-M187, Asn-M188 and Asp-M184.

2.4   Beyond the single pathway view

2.4.1   *What is the physical meaning of a tunnelling pathway?*

As was discussed in Section 2.3, a tunnelling pathway lies at the core of a protein region that mediates the donor–acceptor interaction. The pathway model is parameterized such that it builds in the effect of surrounding mediating residues because parameters were extracted from chemical model systems with multiple paths. As such, one should not expect, necessarily, that breaking a contact in the best path will shut down (or even noticeably influence!) the rate of ET. It should be easier to strengthen a weak pathway by making protein mutations than to weaken a family of strong pathways. Thus, the best pathway is characteristic of the cluster of paths that provides donor–acceptor coupling. To quantify the role of other paths, approaches that build an explict Hamiltonian for the electronic interactions in the protein are necessary. Thus, the challenge is to build such a Hamiltonian and then to interpret it in the language of pathways.

2.4.2 Multipath models and electron/hole propagators

How can the average multipath effects in the pathway analysis be made more explicit? The donor–acceptor interaction mediated by a multiorbital bridge can in fact be described in a concise manner that includes multiple pathways and interference between pathways. Imagine solving the electronic structure problem for the isolated bridge. In the linear combination of atomic orbitals–molecular orbitals approximation, this calculation would produce a set of bridge molecular orbitals:

$$\Psi^{(i)}_{\text{protein}} = \sum_{j=1}^{M} c_j^{(i)} \phi_j \tag{2.8}$$

where j is the atomic orbital index and i is the molecular orbital index.

For simplicity, we assume that the donor interacts only with site number 1 of the bridge (with interaction energy V_{D1} and that the acceptor interacts only with site number N of the bridge (with interaction energy V_{NA}. This is a special case of Equation A.7 in Appendix A. Then the donor–acceptor interaction mediated by the bridge is:

$$T_{DA} = V_{D1} \left[\sum_i \frac{c_1^{(i)} c_N^{(i)}}{E_{\text{tun}} - E^{(i)}} \right] V_{NA} \tag{2.9}$$

Here $E^{(i)}$ is the energy of the ith molecular orbital of the protein (Larsson, 1981).

This expression accounts for all physical and scattering pathways to infinite order within the protein. The donor protein and acceptor protein interactions enter to first order. Protein molecular orbitals are weighted in Equation 2.9 based on their proximity to the tunnelling energy (denominator of equation) as well as their degree of localization near the donor (the atomic orbital coefficents appearing in the numerator). The coefficients and their products in the numerator are expected to be oscillatory in magnitude and sign for a given molecular orbital. However, *the sum* of coefficient products divided by the energy terms decays with the physical distance between sites 1 and N.

The pathways strategy provides a means of approximating the tunnelling matrix element T_{DA}. The ability of a bridge to propagate single or multiple electrons and/or holes is contained in the Green's function, G, which is closely related to Equation 2.9. The Green's function includes the energetic and orbital effects described above and all bridge pathway effects. In the molecular orbital language, the Green's function matrix element G_{1N} is the quantity in the square brackets of Equation 2.9. In this simple case, G_{1N} multiplied by D and A interactions is equal to T_{DA}. In cases where D and A interact with more than one bridge orbital, a sum over the interacting sites must be performed to calculate T_{DA}.

2.4.3 *Hamiltonian-based models of electronic coupling in proteins*

The critical features in coupling calculation are balancing through-bond versus through-space interactions and choosing appropriate values for the tunnelling energy.

Pathway-based strategies. An average decay per bond of ε in a linear chain molecule can be associated with a few molecular parameters. For example, if the chain is approximated as consisting of one orbital per bond (say a C–C σ bond), dangling bonds (perhaps CH bonds) are neglected, and overlap is ignored, the relationship between ε, the inter-bond coupling (γ) and the tunnelling energy is simply

$$\varepsilon + 1/\varepsilon = E_{tun}/\gamma \qquad (2.10)$$

Note that this equation (valid in the long chain limit) includes all high order (scattering) pathways. As such, given a value for the per bond decay constant, a Hamiltonian-based method can be parameterized (for a given tunnelling energy). Recently, this kind of model was used to analyse electronic propagation in azurin (Regan *et al.*, 1995). In these calculations, Hamiltonian elements (γ) provide interactions between neighbouring chemical bonds that share a common atom. Hydrogen bonds were treated as two interacting covalent bonds, and through-space interactions were set to zero.

Other strategies. Other approaches have used standard Hamiltonians from the literature or have built custom ones for electron- and/or hole-mediated super-exchange. The earliest work in this direction assumed an electron-mediated superexchange mechanism to construct an effective Hamiltonian; the coupling element was calculated with a path integral strategy (Kuki and Wolynes, 1987).

Most other calculations have utilized Green function strategies (see Section 2.4.2 above) and a molecular orbital approach. Custom Hamiltonians aimed at treating hole-mediated superexchange (Gruschus and Kuki, 1993) and standard Hamiltonians that include both electron and hole superexchange have been utilized. Some of these strategies first determine important protein regions with a pathway search and then compute coupling on a fragment of the protein using a Green function method built upon a simple electronic structure model. These methods are often carried out at the extended-Hückel level [see Siddarth and Marcus (1993) for example]. These calculations can be tested for size dependence to determine whether or not all relevent amino acids have been included. Some of the fragment approaches use self-consistent field methods, which generate electronic densities of states for the proteins that are more meaningful than those of one-electron methods [see Curry *et al.* (1995) for discussion of this important point].

Extended-Hückel calculations on proteins the size of cytochrome c are now routinely accessible (Regan *et al.*, 1993), and self-consistent field (SCF) methods should be approachable in the near future. SCF calculations will provide an important advance, as the extended-Hückel calculations are known to misrepresent the nature of the virtual oxidized and virtual reduced mediating states. It is fairly clear, too, that new methods for digesting the vast array of information generated in these large-scale electronic structure calculations are needed.

As more advanced methods are developed, the simplicity and obvious ease of interpretation associated with pathway analysis is lost. Hybrid strategies that compute interactions using Green's function methods (adding up high order interfering pathways) beginning with a simple pathway, fattened pathway or pathway tube provide a helpful, perhaps essential, context in which to understand electronic propagation in proteins. For example, analysis of this kind can show whether trivial interferences (arising from bonds dangling off the dominant path but not participating in paths themselves) or qualitatively distinct multiple pathway effects influence coupling in a particular protein structure. Analysis of this kind can also explain whether or not electronic propagation in specific proteins is anisotropic.

2.5   Current challenges

There are at least three aspects of bridge-mediated electronic interactions that theoretical methods are yet to come to terms with. The first is the combined electron and hole mediation of tunnelling. That is, models must be built in such a way that virtual motion of excess electrons as well as excess holes is accounted for in proper balance. Hole mediation is expected to be of particular significance in thermal ET processes, while electron mediation should be of greater importance in photoinduced ET (although hole mediation can dominate even in this regime). The mediation of tunnelling across hydrogen bonds is anticipated to have a significant hole component arising from the electron lone pair. The second critical aspect of the bridge treatment is that the theoretical methods must build in the proper qualitative energetics of the bridge virtual states. That is, methods must reproduce the energetics of virtual cation or anion states of the bridge. One-electron methods (with standard parameterizations) reproduce the N-electron properties of the bridge qualitatively, but cannot incorporate simultaneously both the electron and hole energetics using standard parameterizations [Larsson et al. (1988) proposed a parameterization to avoid this problem]. Finally, a challenge for Hamiltonian or Green's function methods is to construct the proper through-space decay of the electronic states. Basis functions commonly used in quantum chemistry are built from atomic orbitals. The decay of atomic orbital wave functions is set by atomic binding energies. Far from the nuclei (i.e. in through-space contacts), molecular orbitals must decay with an exponent determined by the molecular orbital binding energy, not the energies of the constituent atoms. It is not yet clear if standard basis set strategies will reproduce the proper through-space decay in the intermediate distance regime associated with van der Waals contacts (3–4 Å).

Whether derived from first principles or not, ET analysis requires a better understanding of through-space and hydrogen bond-mediated couplings. These interactions are surely orientation and energy dependent, yet we have limited information from model systems or theory. Of particular importance is the role of fluctuation in the through-space contacts and the possibilty of dynamically averaging these couplings and their decay factors.

Electronic structure methods based upon independent-electron models (those based upon pathway decay parameters or the extended-Hückel Hamiltonian) can be parameterized through the tunnelling energy to give reasonable estimates of coupling for closely related families of reactions. However, in these simple Hamiltonians, the bridge energetics lack qualitative agreement with experiment. As such, these methods are not expected to display the proper overall tunnelling energy dependence. Methods that reproduce experimental ionization potentials and electron affinities of the bridge (rather than optical gaps) are anticipated to give more reliable estimates of protein-mediated coupling. Large scale SCF calculations are expected, therefore, to play an increasing role in ET coupling analysis.

A current related challenge in biomolecular ET is to understand the recent reports of ET in DNA (Brun and Harriman, 1992; Meade and Kayyem, 1995; Murphy *et al.*, 1993). One particular system shows quenching, assumed to arise from electron transfer over extremely long range. Recent theoretical analysis based on an appropriate tunnelling energy scale predicts average distance decays with distances as large or larger than were found in proteins (S. Priyadarshy *et al.*, submitted). This can be understood in the context of tunnelling pathways involving 3.4 Å through-space contacts between base pairs or by analysing the through-bond pathways along the ribose phosphate backbone. Because the electron affinity and ionization potential in DNA are similar to those found in proteins (compared with a tunnelling energy of about –5 eV), the average distance dependences are predicted to be qualitatively similar.

Pathway analysis reveals some simple expectations related to secondary structure and tertiary protein motifs. For example, decay with distance is expected to be relatively soft when donor and acceptor are connected with a β-strand of protein. Propagation down an α-helix decays somewhat more rapidly, and coupling orthogonal to helices is predicted to be even more rapidly decaying. This analysis arises from the protein connectivity, and has been confirmed by experiment. A current challenge is to understand whether pathway interference in various secondary structures and tertiary motifs causes substantial deviations from the more basic pathway analysis. This should be a rich area for the creative use of theory and experiment in the future.

Acknowledgements

We are grateful to our collaborators for discussion of these ideas with us. This work is supported by the National Science Foundation (CHE-9257093 and MCB-9316186), the National Institutes of Health (GM48043-2), and the Department of Energy (DE-FG36-94G010051).

References

Aquino AJA, Beroza P, Beratan DN, Onuchic JN. (1995) Docking and electron transfer between cytochrome c_2 and the photosynthetic reaction center. *Chem. Phys.* **197:** 277–288.

Bard AJ, Faulkner LR. (1980) *Electrochemical Methods.* Wiley, New York.

Beratan DN, Hopfield JJ. (1984) Calculation of electron tunneling matrix elements in rigid systems: mixed valence dithiaspirocyclobutane molecules. *J. Am. Chem. Soc.* **106:** 1584–1594.

Beratan DN, Onuchic JN, Hopfield JJ. (1987) Electron tunneling through covalent and noncovalent pathways in proteins. *J. Chem. Phys.* **86**: 4488–4498.

Beratan DN, Onuchic JN, Betts JN, Bowler BE, Gray HB. (1990) Electron tunneling pathways in ruthenated proteins. *J. Am. Chem. Soc.* **112**: 7915–7921.

Beratan DN, Betts JN, Onuchic JN. (1991) Protein electron transfer rates set by the bridging secondary and tertiary structure. *Science* **252**: 1285–1288.

Beratan DN, Onuchic JN, Winkler JR, Gray HB. (1992a) Electron tunneling pathways in proteins. *Science* **258**: 1740–1741.

Beratan DN, Betts JN, Onuchic JN. (1992b) Tunneling pathway and redox state dependent electronic couplings at nearly fixed distance in electron transfer proteins. *J. Phys. Chem.* **96**: 2852–2855.

Bertini I, Gray HB, Lippard S, Valentine J. (1994). *Bioinorganic Chemistry.* University Science Books, Mill Valley, CA.

Betts JN, Beratan DN, Onuchic JN. (1992) Mapping electron tunneling pathways: an algorithm that finds the "minimum length"/maximum coupling pathway between electron donors and acceptors in proteins. *J. Am. Chem. Soc.* **114**: 4043–4046.

Brun AJ, Harriman A. (1992) Dynamics of electron transfer between intercalated polycyclic molecules — effect of interspersed bases. *J. Am. Chem. Soc.* **114**: 3656–3660.

Calef DF, Wolynes PG. (1983) Classical solvent dynamics and electron transfer. 1. Continuum theory. *J. Phys. Chem.* **87**: 3387–3400.

Closs G, Miller JR. (1986) Distance, stereoelectronic effects, and the Marcus inverted region in intramolecular electron transfer in organic radical anions. *J. Phys. Chem.* **90**: 3673–3683.

Curry WB, Grabe MD, Kurnikov IV, Skourtis SS, Beratan DN, Regan JJ, Aquino AJA, Beroza P, Onuchic JN. (1995) Pathways, pathway tubes, pathway docking, and propagators in electron transfer proteins. *J. Bioenerg. Biomembr.* **27**: 285–293.

daGama AAS. (1990) Through bond electron transfer interaction in proteins. *J. Theor. Biol.* **142**: 251–260.

deRege PJF, Williams SA, Therien MJ. (1995) Direct evaluation of electronic coupling mediated by hydrogen bonds: implications for biological electron transfer. *Science* **269**: 1409–1413.

DeVault D. (1984) *Quantum Mechanical Tunneling in Biological Systems,* 2nd edn. Cambridge University Press, New York.

Fleming I. (1976) *Frontier Orbitals and Organic Chemical Reactions.* Wiley, New York.

Garg A, Onuchic JN, Ambogoakar V. (1983) Effect of friction on electron transfer in biomolecules. *J. Chem. Phys.* **83**: 4491–4503.

Goldman C. (1991) Long range electron transfer in proteins – a renormalized perturbation expansion approach. *Phys. Rev. A* **43**: 4500–4509.

Gruschus JM, Kuki A. (1993) New hamiltonian model for long range electronic superexchange in complex molecular structures. *J. Phys. Chem.* **97**: 5581–5593.

Gunner M. (1991) The reaction center protein from purple bacteria: structure and function. *Curr. Top. Bioenerg.* **16**: 319–367.

Harriman A, Kubo Y, Sessler JL. (1992) Molecular recognition via base pairing: photoinduced electron transfer in hydrogen-bonded zinc porphyrin–benzoquinone conjugates. *J. Am. Chem. Soc.* **114**: 388–390.

Hopfield JJ. (1974) Electron transfer between biological molecules by thermally activated tunnelling. *Proc. Natl Acad. Sci. USA* **71**: 3640–3644.

Karpishin TB, Grinstaff M, Komar-Panicucci S, McLendon G, Gray HB. (1994) Electron transfer in cytochrome *c* depends upon the structure of the intervening medium. *Structure* **2**: 415–422.

Kuki A, Wolynes PG. (1987) Electron tunneling paths in proteins. *Science* **236**: 1647–1652.

Langen R, Chang IJ, Germanas JP, Richards JH, Winkler JR, Gray HB. (1995) Electron tunneling in proteins: coupling through a β strand. *Science* **268**: 1733–1735.

Larsson S. (1981) Electron transfer reactions in chemical and biological systems. Orbital rules for non-adiabatic transfer. *J. Am. Chem. Soc.* **103**: 4034–4040.

Larsson S, Broo A, Kallebring B, Volosov A. (1988) Long distance electron transfer. *Int. J. Quant. Chem. Quant. Biol. Symp.* **15**: 1–22.

Lippard SJ, Berg JM. (1994) *Principles of Bioinorganic Chemistry.* University Science Books, Mill Valley, CA.

McConnell H. (1961) Intramolecular charge transfer in aromatic free radicals. *J. Chem. Phys.* **35**: 508–515.

Marcus RA, Sutin N. (1985) Electron transfers in chemistry and biology. *Biochim. Biophys. Acta* **811**: 265–322.

Meade TJ, Kayyem JF. (1995) Electron transfer through DNA – site specific modification of duplex DNA with ruthenium donors and acceptors. *Angew. Chem. (Engl. Edn)* **34**: 352–354.

Moser CC, Keske JM, Warncke K, Farid RS, Dutton PL. (1992) Nature of biological electron transfer. *Nature* **355**: 796–802.

Moreira I, Sun J, Cho M, Wishart J, Isied S. (1994) Electron transfer from the heme of cytochrome *c* to two equidistant redox modified sites, histidine 33 and methionine 65 – the importance of electronic effects and peptide networks. *J. Am. Chem. Soc.* **116**: 8396–8397.

Murphy CJ, Arkin MR, Jenkins Y, Ghatlia ND, Bossmann SH, Turro NJ, Barton JK. (1993) Long range photoinduced electron transfer through a DNA helix. *Science* **262**: 1025–1029.

Onuchic JN, Beratan DN. (1990) A predictive theoretical model for electron tunneling pathways in proteins. *J. Chem. Phys.* **92**: 722–733.

Onuchic JN, Wolynes PG. (1988) Classical and quantum pictures of reaction dynamics in condensed matter – resonances, dephasing and all that. *J. Phys. Chem.* **92**: 6495–6503.

Onuchic JN, Beratan DN, Hopfield, JJ. (1986) Some aspects of electron transfer reaction dynamics. *J. Phys. Chem.* **90**: 3707–3721.

Onuchic JN, deAndrade PCP, Beratan DN. (1991) Electron tunneling pathways in proteins – a method to compute tunneling matrix elements in very large systems. *J. Chem. Phys.* **95**: 1131–1138.

Onuchic JN, Beratan DN, Winkler JR, Gray HB. (1992) Pathway analysis of protein electron transfer reactions. *Annu. Rev. Biophys. Biomol. Struct.* **21**: 349–377.

Pelletier H, Kraut J. (1992) Crystal structure of a complex between electron transfer partners, cytochrome *c* peroxidase and cytochrome *c*. *Science* **258**: 1748–1755.

Ratner MA. (1990) Bridge assisted electron transfer – effective electronic coupling. *J. Phys. Chem.* **94**: 4877–4883.

Regan JJ, Risser SM, Beratan DN, Onuchic JN. (1993) Protein electron transport: single versus multiple pathways. *J. Phys. Chem.* **97**: 13083–13088.

Regan JJ, DiBilio AJ, Langen R, Skov LK, Winkler JR, Gray HB Onuchic JN. (1995) Electron tunneling in azurin: the coupling across a β-sheet. *Chem. Biol.* **2**: 489–496.

Siddarth P, Marcus RA. (1993) Correlation between theory and experiment in electron transfer reactions in proteins – electronic couplings in modified cytochrome *c* and myoglobin derivatives. *J. Phys. Chem.* **97**: 13078–13082.

Skourtis SS, Regan JJ, Onuchic JN. (1994) Electron transfer in proteins – a novel approach for the description of donor acceptor coupling. *J. Phys. Chem.* **98**: 3379–3388.

Skourtis SS, Beratan DN, Onuchic JN. (1995) A method to analyze multi-pathway effects on protein mediated donor acceptor coupling interactions. *Inorg. Chim. Acta* (in press).

Stemp EDA, Hoffman BM. (1993) Cytochrome *c* peroxidase binds 2 molecules of cytochrome *c* – evidence for a low affinity, electron transfer active site on cytochrome *c* peroxidase. *Biochemistry* **32**: 10848–10865.

Therien MJ, Bowler BE, Selman MA, Gray HB, Chang IJ, Winkler JR. (1991) Long range electron transfer in heme proteins: porphyrin–ruthenium electronic couplings in three Ru(His)cytochromes *c*. In: *Electron Transfer in Inorganic, Organic and Biological Systems, Advances in Chemistry Series 228* (eds JR Bolton, N Mataga, G McLendon). ACS Press, Washington, DC, pp. 191–199.

Tiede D, Chang CH. (1988) The cytochrome *c* binding surface of reaction centres from *Rhodobacter sphaeroides*. *Isr. J. Chem.* **28**: 183–191.

Turró C, Chang CK, Leroi GE, Cukier RI, Nocera DG. (1992) Photoinduced electron transfer mediated by a hydrogen-bonded interface. *J. Am. Chem. Soc.* **114**: 4013–4015.

Winkler JR, Gray HB. (1992) Electron transfer in ruthenium-modified proteins. *Chem. Rev.* **92**: 369–379.

Wuttke DS, Bjerrum MJ, Winkler JR, Gray, HB. (1992) Electron tunneling paths in cytochrome *c*. *Science* **256**: 1007–1009.

Zusman LD. (1994) Dynamical solvent effects in electron transfer reactions. *Z. Phys. Chem.* **186**: 1–29.

Interprotein electron transfer

D.S. Bendall

3.1 Introduction

Many electron transfer reactions occurring *in vivo* involve pairs of proteins in which at least one is free to diffuse. Small, electron-carrying proteins (10–15 kDa) like cytochrome *c*, plastocyanin or ferredoxin allow the distribution of reducing power over large distances within the cell, and to a variety of metabolic acceptor systems. The general principles which have been expounded in the first two chapters apply to these diffusional reactions, but there are additional considerations which are the subject of the present chapter.

Specific recognition and binding between protein molecules occur widely in biology. In many cases (antigen–antibody complexes, for example), the complexes formed are both highly specific and very stable. In others, the complex has a more or less transient existence and there has to be a balance between specificity on the one hand and the need to limit the binding strength on the other. This is especially true of enzymes, which need rapid turnover of the Michaelis complex to achieve high overall rates of reaction, and redox proteins. Too high an affinity between an enzyme and its substrate is deleterious because the rate of reaction depends on the concentration of *free* enzyme. Similarly, the affinity between a redox protein and its reaction partner must be high enough to achieve rapid electron transfer, but not so high as to prevent rapid dissociation of the products and turnover of the chain of carriers as a whole.

The aim of this chapter is to show how theory and experiment can be used to develop a physical picture of the interaction between two redox proteins. A simple kinetic model for a bimolecular reaction is described first to provide a framework for the subsequent discussion. We proceed to discuss (i) the formation of the complex by diffusion and the role of long-range electrostatic forces, (ii) the specific binding forces and energies that may be involved in bringing the proteins in close apposition for rapid electron transfer, and (iii) the structure of the reaction complex and the degree to which it may be static or dynamic in character. In conclusion, some possible alternative binding strategies are suggested. Some of the ideas developed are examined more deeply in subsequent chapters, especially Chapters 4 and 5.

Protein Electron Transfer, Edited by D.S. Bendall
© 1996 BIOS Scientific Publishers Ltd, Oxford

3.2 Kinetics of electron transfer reactions in solution

The physical theory described in Chapters 1 and 2 leads to the conclusion that a key factor in determining the electronic coupling between two redox centres is the distance between them. Two diffusible proteins must therefore form a specific complex in such a way that the centres are brought sufficiently close together for the electron to jump. The simplest kinetic model to represent this is shown in Scheme 3.1:

$$A + B \underset{k_{-a}}{\overset{k_a}{\rightleftharpoons}} (AB) \overset{k_{et}}{\rightarrow} (A^-B^+) \rightarrow A^- + B^+$$

Scheme 3.1.

Two major assumptions are made in this scheme. One is that the stable form of the complex is the reactive form. Cases in which this is not true are discussed in Section 3.5.2. The other is that the back reaction is negligible, which is usually justified when ΔG° has a sufficiently large negative value ($\Delta E_m > 100$ mV).

The rate of an electron transfer reaction is usually measured in terms of a first order rate constant, k_1, even when the reaction is bimolecular. If the concentration of one reactant (say the oxidant A, in Scheme 1.1) is in large excess, the concentration of the other (B) will follow an exponentially decaying time course to give a pseudo-first order rate constant. The reaction can be followed either with a stopped flow spectrophotometer, which has a specially designed flow cell to mix the two proteins in about 1 ms, or by a flash photolysis technique which brings about very rapid and preferential reduction of one of the two proteins, both of which are initially in the oxidized form (Hahm *et al.*, 1992; McLendon and Miller, 1985; Pan *et al.*, 1990; Tollin and Hazzard, 1991; Tollin *et al.*, 1993). Nuclear magnetic resonance (NMR) can also give kinetic information in some circumstances (King *et al.*, 1985; Satterlee *et al.*, 1987; Whitford *et al.*, 1991).

An expression for k_1 can be derived by assuming that the rate of formation of the complex (AB) equals its rate of decay (see Appendix B for discussion of the steady-state approximation). The result obtained is

$$k_1 = \frac{k_a k_{et}[A]}{k_{-a} + k_{et} + k_a[A]} \tag{3.1}$$

Equation 3.1 has the form of a rectangular hyperbola when k_1 is plotted against [A] and is closely related to the Michaelis equation of enzyme kinetics. When [A] is so large that the term $k_a[A]$ dominates the denominator of Equation 3.1, rate saturation sets in and k_1 tends towards a maximum value equal to k_{et}. On the other hand, when [A] is small, $k_a[A] \ll (k_{-a} + k_{et})$ and k_1 becomes proportional to [A] (in practice [A] must be large enough for the reaction to remain pseudo-first order). The proportionality constant, k_2, is the second order rate constant for the reaction of *free* B (which is essentially total B under this condition) with A, and

$$k_2 = \frac{k_a k_{et}}{k_{-a} + k_{et}} = \frac{k_{et}}{K_M} \qquad (3.2)$$

If an independent measurement of K_A (k_a/k_{-a}) can be made, then a complete kinetic description according to Scheme 3.1 has been achieved, with separate evaluation of the 'on' (k_a) and 'off' (k_{-a}) rate constants.

Frequently there are practical reasons why measurements cannot be made at concentrations of A beyond the initial linear region. All that one can then obtain from k_1 is an estimate of k_2; independent measurements of one rate constant, k_{et}, k_a or k_{-a}, must be sought in addition to K_A. However, when k_{-a} and k_{et} are of different orders of magnitude, Equation 3.1 simplifies to one of two forms.

(i) When $k_{et} \gg k_{-a} + k_a[A]$, $k_1 = k_a[A]$ and $k_2 = k_a$. This is the diffusion-limited regime. The rate of reaction cannot be faster than the rate at which the two proteins form a successful encounter complex.

(ii) When $k_{-a} \gg k_{et} + k_a[A]$, $k_1 = k_a k_{et}[A]/k_{-a} = K_A k_{et}[A]$ and $k_2 = K_A k_{et}$. This is an activation-controlled reaction which occurs when the relatively small value of k_{et} allows a genuine equilibrium to exist between the free reactants and the complex. These are often spoken of as rapid pre-equilibrium conditions.

3.3 Diffusional kinetics of complex formation

3.3.1 *The nature of diffusional encounter*

Diffusion in solution bears little resemblance to diffusion in the gas phase. While the latter involves long mean free paths between encounters, a protein molecule in solution diffuses by a jittery Brownian motion as it is buffeted continually by a myriad of much smaller water molecules. The result is a random walk, involving steps that have an average size smaller than the diameter of a water molecule. Eventually the protein molecule will explore whatever volume of solution it is confined to. The rate at which two molecules collide is proportional to the product of r (the sum of their radii) and the sum of the individual diffusion constants (Berg and von Hippel, 1985), and is modified by long-range electrostatic forces (see Equation 4.5, Chapter 4).

To get an impression of the order of magnitude of diffusional rate constants, consider two proteins of about the same size, cytochrome c and plastocyanin, treating them at first as uncharged spherical molecules. The diffusion equation:

$$k_a = 4\pi N_A r (D_A + D_B) \times 10^3 \qquad (3.3)$$

(N_A is Avogadro's constant and D_A, D_B, are the diffusion constants in $m^2 s^{-1}$) yields $k_a = 6.9 \times 10^9$ M^{-1} s^{-1}. Real values of k_a are likely to be considerably smaller, however, because only a small proportion of initial collisions give productive conformations of the complex. Two effects tend to offset this negative steric factor leading to diffusion-limited rates in the range 10^8–10^9 M^{-1} s^{-1}. The more obvious is the effect of long-range electrostatic forces. The net charges on the proteins we are considering (cytochrome c positive and plastocyanin negative) will not only enhance the rate of collision by diffusion, but also cause a 'steering effect' because the charges are not distributed randomly, giving each protein a dipole moment (Koppenol and Margoliash, 1982; Margoliash and Bosshard, 1983; Roberts et al., 1991). The proteins tend to associate with their dipole moments lined up in an orientation which, in this case at least, is favourable for electron transfer.

The other factor is the solvent cage effect and the principle of the reduction of dimensionality (Adam and Delbrück, 1968), which applies to all macromolecular associations. When two protein molecules initially collide by diffusion they tend to be trapped by the surrounding solvent and undergo numerous short-range collisions within one 'encounter complex'. This considerably enhances the lifetime of the encounter complex and allows the two proteins to explore a large number of relative conformations by surface diffusion before effectively separating again (Northrup and Erickson, 1992).

In the remaining parts of this section, we will consider the theoretical and experimental analysis of long-range electrostatic forces, which often dominate the diffusional interactions between redox proteins in solution. For electron transfer actually to occur, however, the two proteins must come together in a more intimate and specific conformation than is defined by simple coulombic forces. At close range, hydrogen bonding, van der Waals interactions and entropic effects involving exclusion of bound water play an important role. These are discussed in Section 3.4. Exclusion of water from the interface is also likely to have the important effect of reducing the solvent contribution to the overall reorganization energy (Durham et al., 1995; Gray and Malmström, 1989).

3.3.2 Ionic strength effects

The importance of electrostatics in the interaction between two proteins is most simply demonstrated by the effects of ionic strength on k_2. Figure 3.1a shows an experiment carried out with proteins purified from Brassica komatsuna (Niwa et al., 1980). The strong decrease in k_2 as the ionic strength, I, increases, shows that complementary charges on the two proteins are involved. Since all higher plant plastocyanins carry a large net negative charge at neutral pH, the interaction must involve basic groups on cytochrome f. The soluble, haem-containing domain of cytochrome f has a large number of both acidic and basic residues, with only a slight excess of the latter, so that it is probably the local charge at the interaction site of cytochrome f which is important.

Figure 3.1b shows the results of a related experiment carried out with a preparation of the cytochrome bf complex isolated from a cyanobacterium, Phormidium laminosum (Wagner et al., 1995). Here, contrasting electron acceptors for cytochrome f have been employed, namely mammalian cytochrome c (net positive charge), and a higher plant plastocyanin (net negative charge). Increasing salt concentration has correspondingly complementary effects, the reaction with plastocyanin being stimulated and that with cytochrome c inhibited. The significant local charges on the cyanobacterial cytochrome f are therefore negative rather than positive. In agreement with this conclusion, plastocyanin from the same cyanobacterium is much less strongly acidic than the higher plant proteins (Varley et al., 1995).

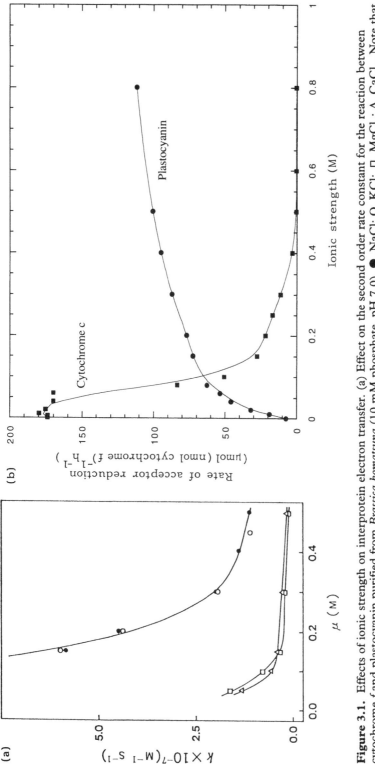

Figure 3.1. Effects of ionic strength on interprotein electron transfer. (a) Effect on the second order rate constant for the reaction between cytochrome f and plastocyanin purified from *Brassica komatsuna* (10 mM phosphate, pH 7.0). ●, NaCl; ○, KCl; □, MgCl$_2$; △, CaCl$_2$. Note that the divalent cations supress the rate considerably more effectively than the monovalent cations, even though the rate constants have been expressed in terms of ionic strength. Reproduced from Niwa *et al.* (1980) Electron transfer reactions between cytochrome f and plastocyanin from *Brassica komatsuna*. *J. Biochem.* **88:** 1177–1183 with permission from the authors. (b) Effect on the rate of reduction of spinach plastocyanin or horse heart cytochrome c by the cytochrome bf complex purified from *Phormidium laminosum*. Reductant: 10 μM plastoquinol-1.

The interpretation of ionic strength effects is based on the Debye–Hückel theory which imagines an ion, or charge on the surface of a protein, as being surrounded by a cloud of mobile ions with those of opposite sign in slight excess. This screens charges from each other and reduces the strength of their interaction. The effects are often expressed in terms of the Debye–Hückel factor, κ, which in SI units is defined by

$$\kappa^2 = \frac{2000 N_A e^2}{\varepsilon_0 \varepsilon k_B T} \times I \tag{3.4}$$

where I is ionic strength, k_B is Boltzmann's constant and e the charge on an electron (see Section 3.3.4 for the significance of ε_0 and ε, the permittivity of vacuum and the dielectric constant, respectively). The reciprocal of κ is known as the Debye length or screening distance, which gives a measure of the way in which the field around an ion is compressed by other ions in the solution (the distance over which the field is reduced to a value $1/e$ of what it would have been in the absence of screening). The value of $1/\kappa$ is about 10 Å at $I = 0.1$ M and 5 Å at $I = 0.37$ M.

Further insight into the effects of ionic strength depends on a quantitative analysis in terms of the structure of the two proteins, and the system may be modelled at various levels of complexity. In essence, the problem is how to define the electrostatic field of a protein. Broadly speaking, there are two kinds of approach, depending on how the solvent molecules are treated. The most satisfactory, in principle, is the microscopic approach once the three-dimensional structure of the protein is known, but the enormous computational power required for a rigorous treatment has so far limited its use. The principle of microscopic modelling will be described first, and then some of the more widely used macroscopic models. What the latter have in common is the use of continuum approximations for both solvent and protein, that is to say the effects of polarization and ionic strength are dealt with by introducing a dielectric constant, which is likely to be position dependent. They may or may not treat the protein at an atomic level of detail. Good general reviews are those of Harvey (1989) and Honig and Nicholls (1995).

3.3.3 *Microscopic electrostatic methods*

In microscopic models, the system is treated as a system of charges, each one of which interacts with every other one according to Coulomb's law. The dielectric constant has a value of 1, because the screening effect of orientable dipoles is automatically taken into account by considering them as contributing individually to the overall system of charges. The calculation is seriously complicated by the existence of induced dipoles in polarizable groups. The strength of any one induced dipole is dependent on the local field established by all the other charges of the system, including other inducible dipoles that have not yet been calculated. The problem can be solved numerically by a computer program acting in iterative fashion, but the procedure is very demanding of computer power and time. The net polarizations resulting from reorientation of the solvent water molecules pose a similar, but even larger, computational problem, which cannot yet be dealt with except by some simplifying approximations.

The most successful microscopic approach is the protein dipoles/Langevin dipoles (PDLD) method for which a computer program, POLARIS, is available (Warshel and Aqvist, 1991). The protein dipoles and charges are calculated from the X-ray structure. The Langevin dipoles represent the surrounding water molecules as a grid of point dipoles at the same density. Other microscopic methods employ molecular dynamics simulations to determine the average orientations of the permanent dipoles of the protein and of water molecules. A rigorous use of this method is impossible because the long times required to reach true equilibrium cannot be achieved with current computational limitations.

3.3.4 *Macroscopic electrostatic methods*

Macroscopic models are based on the classical method of treating the solvent as a continuum, with a dielectric constant (ε) to represent the overall effect of screening of charges by alignment of solvent dipoles. The protein is also given a dielectric constant. At the microscopic level, the system is heterogeneous and it may be necessary to introduce several different values of ε (such as 78.5 for the bulk solvent and 4 for the interior of the protein), or even a continuously varying value (at the protein surface).

The most familiar expression for electrostatic interactions is Coulomb's law, according to which the force between two point charges falls off with the inverse square of the distance, r, between them. This is equivalent to saying that the electrostatic potential energy, U (the work required to bring the charges from infinity to separation r), varies as $1/r$:

$$U = \frac{Q_1 Q_2}{4\pi\varepsilon_0 \varepsilon r} \qquad (3.5)$$

Equation 3.5 is given in its SI form which employs the constant $1/4\pi\varepsilon_0$ to get the units right, ε_0 being the permittivity of vacuum (8.854×10^{-12} C^2 J^{-1} m^{-1}). For dealing with complex structures such as those of proteins, a more general expression than Coulomb's law is required. In electrostatic theory the concept of charge density, ρ, is preferred to that of point charges, and the fundamental law can be expressed as Poisson's equation:

$$\nabla^2 \phi = -\frac{\rho}{\varepsilon_0 \varepsilon} \qquad (3.6)$$

where ϕ is the electric potential at a particular point in space and ∇^2 (del squared, the Laplacian) is the second differential of the spatial coordinates. The solution of Poisson's equation for the spherical distribution of potential around a point charge is a form of Coulomb's law:

$$\phi = \frac{Q}{4\pi\varepsilon_0 \varepsilon r} \qquad (3.7)$$

because electrostatic potential is potential energy per unit charge ($\phi = U/Q$). A further term, given by the Boltzmann distribution, must be introduced to include the screening effect of mobile ions. The resulting Poisson–Boltzmann equation:

$$\nabla^2 \phi = \kappa^2 \sinh\phi - \frac{\rho}{\varepsilon_0 \varepsilon} \qquad (3.8)$$

is the basis for various macroscopic methods that describe the dependence of bimolecular rate constants on ionic strength. It can only be solved analytically for simple geometries, but numerical integration allows the real shape of proteins to be taken into account.

Simple electrostatic models. The first electrostatic treatment of electron transfer reactions involved the application of the Brønsted–Bjerrum theory of ionic reactions. The basic idea, which can be derived from transition state theory, is that for charged reactants the rate constant has to be corrected by the activity coefficients of the two reactants and the activated complex:

$$k = k_0 \frac{\gamma_A \gamma_B}{\gamma_{AB}^{\ddagger}} \qquad (3.9)$$

The Debye–Hückel theory (which involves the solution of a linearized form of the Poisson–Boltzmann equation) is used to provide an ionic strength-dependent expression for the activity coefficients, which yields an expression relating the rate constant at ionic strength I to that at zero ionic strength.

Another approach uses the Debye–Hückel theory to derive an expression for the potential energy, U, of interaction, that is to say the electrostatic work done in bringing the reactants together. This is used as a component of the total activation energy in the Marcus equation. The rate at ionic strength I may then be expressed in terms of U and the rate at infinite ionic strength, $k(I_\infty)$, when electrostatic effects are completely shielded:

$$k(I) = k(I_\infty)e^{-U/RT} \qquad (3.10)$$

These methods give numbers for the total charge and radius of each reactant, but they must be used with caution because they are based on a simple model in which each protein is treated as a uniformly charged sphere. The rapid development in knowledge of protein structures has made it possible to take into account the real distribution of charges within a protein.

Structure-based models giving analytical solutions. For analytical solutions to the Poisson–Boltzmann equation to be possible, the protein must still be treated as a sphere, but the distribution of charges can be related to that of the real protein rather than being assumed to be uniform. The distribution at a restricted reaction site is the most important consideration, and the net charge at this site may differ from that of the protein as a whole. The full structure of a protein can only be taken into account by using numerical techniques for solving the Poisson–Boltzmann equation.

A simple way of allowing for interactions between local charges is to restrict the calculation of the interaction energy to that of a set of ion pair bonds (salt bridges) at the reaction site between the two proteins. Each ion pair must be assumed to be uninfluenced by the others, which requires a separation of about

10 Å (at $I = 0.1$ M). A plot of $\ln k$ against $I^{1/2}$ then allows a three-parameter fit to the curve to give values for $k(I_\infty)$, r, assumed to be the same for each ion pair, and n, the number of charge pairs involved. Again the values obtained for r and n must be interpreted cautiously. A useful application of the model, which can be extended readily to others, is the assessment of the value of U_i, the interaction energy for any particular ion pair. If a single charge pair is disrupted, either by site-directed mutagenesis or chemically, Equation 3.10 leads to the following:

$$U_i = -RT \ln k(I)/k'(I) \qquad (3.11)$$

$k'(I)$ being the rate constant for the modified or mutant protein. Replacement of $k'(I)$ by $k(I_\infty)$ allows calculation of the total interaction energy at any particular ionic strength. The classic example of this method is the mapping of the interaction site of cytochrome c by modification of individual lysine residues by trifluoroacetylation (see *Figure 3.2*) (Smith *et al.*, 1981).

A non-uniform distribution of charge on a protein gives it a dipole moment. Two proteins that interact by diffusion are likely to encounter each other with the dipole moments lined up, and this can cause a considerable enhancement of the overall rate of electron transfer. The significance of dipole moments was first clearly recognized by Margoliash and colleagues (Koppenol and Margoliash, 1982; Margoliash and Bosshard, 1983) who studied forms of cytochrome c chemically modified at specific lysine residues. Modification of residues on the surface opposite to the exposed haem edge had a small but significant effect on activity.

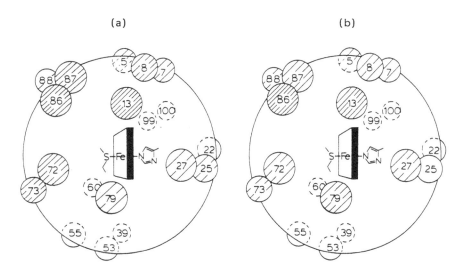

Figure 3.2. Contributions of individual lysine residues of cytochrome c to the electrostatic interaction energy with cytochrome oxidase (a) and cytochrome c_1 (b). Each hatch mark corresponds to -0.1 kcal mol^{-1}. Reproduced from Smith *et al.* (1981) Electrostatic interaction of cytochrome c with cytochrome c_1 and cytochrome oxidase. *J. Biol. Chem.* **256**: 4984–4990 with permission from The American Society for Biochemistry and Molecular Biology and the authors.

Van Leeuwen (1983) developed analytical equations that allow for the effects of protein dipoles on the second order rate constant. The orientations of protein dipoles are calculated from the spatial distribution of charges known from the structure, but the protein is still treated as spherical. The equations include terms for the influence of monopole–dipole and dipole–dipole interactions in addition to the monopole–monopole interactions. Van Leeuwen pointed out that the reaction site of a protein need not necessarily coincide with the region of the protein surface defined by the direction of the dipole moment. Rush *et al.* (1988), using this kind of analysis, reached the conclusion that plastocyanin reacts with cytochrome *c* through its northern hydrophobic site, rather than the remote eastern site, but other evidence suggests the opposite conclusion (Modi *et al.*, 1992).

An alternative model, which may be more realistic but still provides a geometry amenable to analytical treatment, is one which represents the active site of the protein as a uniformly charged disc of radius ρ (Watkins *et al.*, 1994). Strictly speaking, this 'parallel plate' model applies only to diffusion-limited reactions, but has been used with some success more widely. Electron transfer takes place when the plates of two protein molecules come sufficiently close together. An interesting feature of the model is that the uniform distribution of charge over the plates might be viewed as a way of representing a dynamic interaction between the two proteins in which the probability of electron transfer is averaged over a number of different relative orientations (see below). When the monopole–monopole interaction energy only is taken into account, the model yields an equation of similar form to that of the ion pair model described above (Equation 3.11), but the physical interpretation of the parameters is different, which again indicates that the numbers obtained should be interpreted cautiously. The parallel plate and ion pair models, despite their obvious limitations, have proved useful for describing the effects of modifying specific residues of proteins.

Further discussion of the equations generated by the models described above can be found in Appendix B.

Finite difference methods for calculating protein electrostatic fields. To progress beyond the relatively simple continuum models discussed above, realistic shapes as well as charge distributions must be introduced. Poisson's equation or the Poisson–Boltzmann equation must be solved numerically by computer, but the computational load is kept manageable by retaining dielectric constants. The most commonly used algorithm is that of Warwicker and Watson (1982). The system of protein and solvent is divided up into a regular grid defining small cubes, and charge densities are assigned to each grid point according to the structure of the protein. Each cube is assigned a dielectric constant, one value for a cube lying in the solvent and another for one in the protein. The electrostatic potential at each point on the grid is calculated in an iterative fashion by using a finite difference form of Poisson's equation to relate it to the potentials at neighbouring grid points. A computer program called DelPhi has been written by Honig and colleagues (Honig and Nicholls, 1995) to calculate electrostatic potentials of proteins at different values of the ionic strength.

Calculations of protein electrostatic potentials can be used to aid imaginary docking experiments. Two kinds of more objective, analytical procedure have been used. Warwicker has developed a static method for calculating the electrostatic interaction energies over the full range of six-dimensional configuration space of a pair of proteins, and has applied it to the interaction between cytochrome c and cytochrome c peroxidase (Warwicker, 1989). This study revealed a rather large band of negative potential on the surface of the peroxidase which would allow many different but energetically favourable configurations of the complex with cytochrome c. This approach is complementary to that of Northrup and colleagues who have used the Brownian dynamics algorithm to study the kinetics of interaction of several pairs of redox proteins at the submolecular level (see Chapter 4).

3.4   Binding forces and binding energies

Once the initial encounter has occurred, the two proteins must form a more intimate and specific complex to produce a configuration that favours rapid electron transfer. This can be regarded as a second stage of complex formation. The initial encounter is dominated by relatively long-range electrostatic forces that vary as $1/r^2$ (and the potential energy as $1/r$), but the second stage involves additional short-range forces which make important contributions both to the overall binding energy and to the specificity of the interaction. These forces are the same as for any interaction between polypeptide chains, but the balance between them can vary markedly from one system to another, and full understanding depends on a quantitative assessment. Unfortunately, rigorous calculations are not yet possible, mainly because of the difficulty of coping satisfactorily with the large number of solvent molecules.

3.4.1 *Non-covalent sources of binding energy*

The non-covalent forces between polypeptide chains or proteins are conventionally described as electrostatic, hydrogen bonds, van der Waals forces and the hydrophobic effect, although all are electrostatic in origin.

Electrostatic interactions. While interactions between the fixed charges of acidic and basic residues dominate the initial encounter, as discussed in Section 3.3, at short range, dipolar effects arising from the partial charges associated with side chain hydroxyl and amide groups or the peptide bond also become significant. Dipolar interactions are complex because both attractive and repulsive forces are involved, and orientation is important unless the molecules are free to rotate. In general, whereas the interaction energy between a pair of monopoles varies as $1/r$, that between a monopole and a dipole, or between a pair of dipoles, involves higher powers of r ($1/r^6$ in the extreme case). Dipole–dipole interactions are usually considered separately under the heading of either van der Waals forces or hydrogen bonds because of their importance in proteins.

Hydrogen bonds. A hydrogen bond is an attractive interaction between two electronegative atoms as a result of the presence of a hydrogen atom covalently linked to one of them. The main source of binding energy is the electrostatic interaction between a pair of dipoles in a favourable orientation, and strong hydrogen bonds also have some covalent character. A salt bridge between a carboxylate group and a side chain amine group of lysine or a guanidino group of arginine often includes a hydrogen bond. The energies of hydrogen bonds in the vapour phase vary between about 13 and 95 kJ mol^{-1} (3–23 kcal mol^{-1}), with the larger values applying to charged groups (Rose and Wolfenden, 1993). In aqueous solution, the effective values are much smaller, however, because the formation of a hydrogen bond between two protein groups is an exchange reaction in which protein–water bonds are broken and a water–water bond is formed (Fersht, 1987).

Van der Waals forces. All atoms and molecules attract one another weakly, even when no permanent dipoles are involved. In the case of non-polar atoms, the attractive force (the London dispersion force) results from a dipole–dipole effect involving the fluctuating distribution of electrons in atoms which have no net charge. The energy of these induced dipole effects varies as $1/r^6$. Although a large number of individual interactions is often involved in the association between two proteins, the net contribution to binding energy can be small because, as with hydrogen bonds, interactions with the solvent are lost at the same time.

Hydrophobic effect. The hydrophobic interaction is the least well understood and most controversial component of binding energy. It is not a discrete force or bond, but the net result of several different interactions, some involving hydrogen bonds and van der Waals forces, and some being entropic effects derived from changes in the freedom of movement of molecules. These depend upon the clustering of water molecules around the surface of a hydrophobic group to form a more ordered, or ice-like, layer compared with the bulk water (Creighton, 1993). In the case of proteins, a useful feature is that the hydrophobic free energy is additive and can be defined in terms of contributions from individual side chains and the area of van der Waals surface exposed to solvent (Lee and Richards, 1971; Richards, 1977). Estimates of these atomic solvation parameters ($\Delta\sigma$) are shown in *Table 3.1*. Carbon and sulphur atoms show a significant negative free energy change favouring transfer from water to a non-polar environment, while charged O and N atoms show an unfavourable positive change. O and N atoms that are formally uncharged show a weak 'anti-hydrophobic' effect (Eisenberg and McLachlan, 1986).

3.4.2 *Specificity and binding energy*

The standard free energy change for protein association is obtained directly from the measured binding constant

$$\Delta G^\circ = -RT \ln K_A \qquad (3.12)$$

ΔG° has two components, the enthalpy change (ΔH°) derived from changes in bond energy, and an entropic term ($-T\Delta S$) derived from changes in the freedom of motion of the molecules involved. Proteins can exist free in solution largely because the loss of translational and rotational entropy when two molecules

Table 3.1. Atomic solvation parameters[a]

Atom type	Atomic solvation parameter ($\Delta\sigma$)	
	cal Å^{-2} mol^{-1}	J Å^{-2} mol^{-1}
C	16 ± 2	67 ± 8
N/O	-6 ± 4	-25 ± 17
O^{-}	-24 ± 10	-100 ± 42
N^{+}	-50 ± 9	-209 ± 38
S	21 ± 10	88 ± 42

[a] Adapted from Eisenberg and McLachlan (1986).

associate presents an energy barrier to association which has been estimated as approximately 60 kJ mol^{-1} (15 kcal mol^{-1}) at 300 K (Finkelstein and Janin, 1989). The barrier is overcome by a favourable balance between numerous non-covalent interactions, which will have both enthalpic and entropic components, and which individually may aid or hinder association.

A common feature of protein association reactions is the exclusion of water from the interface (Chothia and Janin, 1975). Free protein molecules are covered with a layer of bound water molecules in solution. Liberation of these into the bulk phase when the proteins associate causes an increase in entropy which is usually sufficient to offset the decreased translational and rotational entropy. The amount of water liberated is roughly proportional to the area of the interface. Analysis of the known structures of protease–inhibitor complexes and antibody–antigen complexes shows that the solvent-accessible surface area that has been buried in these tightly associating complexes amounts to 1200–2000 Å^2 (Chothia and Janin, 1975; Janin and Chothia, 1990). Much larger areas may be buried in oligomeric proteins (Jones and Thornton, 1995). The little structural information available for the complexes formed between soluble redox proteins suggests that smaller areas are involved (800–1500 Å^2, see Chapter 5).

Direct evidence for loss of water also comes from the effects of high pressure, which has a tendency to dissociate protein complexes because the complex has a larger volume than the separate proteins. One cause is an imperfect fit between the two proteins (Paladini and Weber, 1981). Another is electrostriction, that is to say a water molecule is bound more tightly to a charged group on a protein than it is to another water molecule in the bulk, and so occupies a smaller volume (Rodgers and Sligar, 1991). In the case of the complex between cytochrome b_5 and cytochrome c, ΔV was found to be 122 ml mol^{-1} (Rodgers et al., 1988). Similar measurements with a series of site-directed mutants of cytochrome b_5 identified the acidic residues involved in binding to cytochrome c (Rodgers and Sligar, 1991). The loss of water on association of the two cytochromes has been confirmed by the effects of glycerol, which causes osmotic stress and has a tendency to dehydrate protein surfaces and to encourage complex formation (Kornblatt et al., 1993). A similar study on the binding of ferredoxin to ferredoxin-NADP$^+$ reductase found a volume loss of 70–126 ml mol^{-1}, corresponding to the release of 4–7 water molecules (Jelesarov and Bosshard, 1994).

As indicated above, the entropic gain from water exclusion is not confined to the hydrophobic effect. The energy of ion pairs in solution is largely entropic (Ross and Subramanian, 1981) and a hydrogen bond exchange equation tends to favour association because of the higher entropy of water molecules in the bulk (Fersht et al., 1985). From this, one would expect association of proteins in general to be driven by a large entropy increase, with only a small ΔH because the number of non-covalent contacts is more or less conserved. In fact, the balance of enthalpic and entropic factors varies widely in different protein–protein complexes. Summaries of available thermodynamic data (excluding redox proteins) are to be found in Hibbits et al. (1994), Tello et al. (1993) and Ross and Subramanian, (1981). While a number of associations fall into the simple pattern just described, a significant number show a negative entropy change offset by a large negative enthalpy. A few cases show a positive ΔH. Two explanations for negative entropy changes have been put forward. One possibility is that vibrational motion of part of the protein structure is lost in the complex (Hibbits et al., 1994; Ross and Subramanian, 1981). Another explanation is based on the fact that X-ray crystallography of some antigen–antibody complexes reveals that the interface retains bound water molecules, which would be more highly ordered than in the free proteins (Bhat et al., 1994). This is an example of water molecules improving the fit between two protein surfaces. Another is the water molecule that occupies a depression at the centre of the hydrophobic patch in the crystal structure of azurin, a blue copper protein, and may mediate electron transfer to other proteins (Mikkelsen et al., 1993).

Large negative enthalpies are less easy to explain, but are most likely to arise from particularly favourable hydrogen bonds and a close fit between the two proteins (van der Waals interaction). Hydrogen bonds formed in a hydrophobic environment have a larger enthalpy (Ross and Subramanian, 1981). On the other hand, a positive enthalpy can be an indicator of a lack of strict complementarity between the two protein surfaces. Misfit disturbs the balance of van der Waals contributions (protein–protein versus protein–water). Furthermore, an unmatched hydrogen bond causes a significant loss of binding energy, which is of the order of 4 kJ mol^{-1} (1 kcal mol^{-1}) when neutral groups are involved and up to 20 kJ mol^{-1} (5 kcal mol^{-1}) if either or both groups are charged (Fersht, 1987). Thus, hydrogen bonds and van der Waals interactions play important roles in the specificity of binding.

Thermodynamic analysis of the binding between redox proteins has been made in only a small number of cases, and the relevant data are collected in Table 3.2. In three out of the four cases, the association is entropically driven, conforming to the simple model described above. In the case of cytochrome c_6 reacting with photosystem I, however, the positive entropy change is very small and association depends on a negative ΔH.

A special feature of many (but by no means all) interactions between redox proteins is the large number of charged groups thought to be involved in the contact region, and hence the presence of putative salt bridges. The energetic contribution of a single salt bridge depends on the distance between the charges and the ionic strength. An estimate of a maximum value of about 5 kJ mol^{-1} (1.2 kcal mol^{-1}) is obtained for charges in van der Waals contact if the dielectric

Table 3.2. Thermodynamic quantities for association between electron transfer proteins

Reaction partners	K_A (M)	$\Delta G°$ (kJ mol^{-1})	$\Delta H°$ (kJ mol^{-1})	$T\Delta S$ (kJ mol^{-1})	Conditions	Ref.
Cytochrome c– cytochrome b_5	3×10^6	–37	4	41	$I = 1$ mM, pH = 7.0, T = 298 K	b
Cytochrome c_3– ferredoxin	2×10^7	–39.5	24	63.4	$I = 10$ mM, pH = 6.6, T = 283 K	c
Cytochrome c_6– photosystem I	3.2×10^4	–25.6	–24.7	0.9	$I = 10$ mM, pH = 7.0, T = 298 K	d
Ferredoxin/ FNR[a]	6.5×10^6	–39.1	–1.3	37.8	$I = 41$ mM, pH = 7.5, T = 300 K	e

[a] Ferredoxin-NADP$^+$ oxidoreductase; [b] Mauk *et al.* (1982); [c] Guerlesquin *et al.* (1987); [d] Díaz *et al.* (1994); [e] Jelesarov and Bosshard (1994).

constant is taken as that of water. Values can be measured for protein complexes by applying Equation 3.11 to binding constants measured for mutant (or chemically modified) proteins in which a single acidic or basic residue has lost its charge. In a study of the interaction between cytochrome b_5 and cytochrome c, the maximum contribution was found to be 2 kJ mol^{-1} (0.5 kcal mol^{-1}) (Rodgers and Sligar, 1991). This agrees with the observation that engineering a single solvent-exposed salt bridge into a protein makes only a small contribution to stability (0–2 kJ mol^{-1}) (Fersht and Serrano, 1993).

A representative association constant of 10^5 M^{-1} for a pair of redox proteins corresponds to a total binding energy of about 30 kJ mol^{-1} (7 kcal mol^{-1}). In the case of the cytochrome b_5–c couple, Rodgers and Sligar (1991) concluded that electrostatic interactions made a modest contribution to the total binding energy of 36 kJ mol^{-1}, despite their obvious importance in guiding the formation of productive complexes. The corollary is that hydrogen bonds and hydrophobic interactions are likely to make important contributions. Nevertheless, the presence of several salt bridges in a relatively small area will ensure that the interface has a polar nature and possibly that the exclusion of water is limited to an even smaller area at any one moment. The presence of several salt bridges in an interface may represent a device by which the interaction between the two proteins is kept dynamic and transient.

3.5 Statics and dynamics of reaction complexes

One of the most striking features of electron transfer proteins is that their redox centres are invariably buried beneath the surface of the protein. Such an arrangement not only helps to minimize unwanted reactions in the cell, but also minimizes the reorganization energy, because solvent is excluded and a rigid structure can be provided around the redox group. When two redox proteins encounter each other by diffusion, they must seek out a relative configuration that provides a sufficiently small distance between the two centres for efficient electron transfer to occur. What is the structure of the complex in this reactive configuration?

3.5.1 *The static view*

The development of ideas about the structures of reaction complexes between pairs of redox proteins was stimulated by the computer models generated by Salemme (1976) for the cytochrome c–cytochrome b_5 complex and by Poulos and Kraut (1980) for cytochrome c and cytochrome c peroxidase. These were based on the crystal structures of the individual proteins and the assumption that binding between the pairs of proteins would depend mainly on electrostatic forces. The first X-ray crystal structures for complexes only became available recently. The structure of the complex between cytochrome c and yeast cytochrome c peroxidase determined by Pelletier and Kraut (1992) shows an interaction domain on the peroxidase which is a little different from, but overlapping, that proposed earlier by Poulos and Kraut (1980). The use of X-ray crystallography to determine the structures of complexes is discussed in Chapter 5.

NMR is also capable of providing structural information about protein complexes. No complete structure of a redox complex has been determined in this way because, until recently, the upper size limit has been about 12 kDa. However, the introduction of multidimensional, heteronuclear methods for the analysis of proteins doubly labelled with ^{13}C and ^{15}N brings a small complex (~20 kDa) into the realms of possibility. What may be more important, however, is that presently available techniques can provide information about which groups or residues are involved directly in the interaction without the need for knowledge of the complete structure of the complex. Furthermore, NMR has the advantage over X-ray crystallography that proteins are studied in solution under conditions that are flexible and relevant to studies of electron transfer.

A number of studies have attempted to exploit the possibilities of NMR. The most direct effect of binding one protein to another is a change in chemical shift as a result of a change in the immediate environment of atoms at the interface. This has been observed with certain hyperfine shifted resonances of haem protons (Burch *et al.*, 1990; Whitford *et al.*, 1990, 1991) and of protons of side chains lying in the haem pocket (Moench *et al.*, 1992; Yi *et al.*, 1993). Any shifts of proton resonances in the diamagnetic region of a protein, which would help to define a binding site, seem to be small, but might be detectable in suitable two-dimensional experiments (Ubbink *et al.*, 1995). Larger shifts can be observed with ^{13}C-labelled protein and have provided some experimental support for the Salemme model (Salemme, 1976) for binding of cytochrome b_5 to cytochrome c (Burch *et al.*, 1990).

Paramagnetic relaxation effects and H/D exchange measurements have been used successfully to give information about binding sites. The paramagnetic competition experiment (Bagby *et al.*, 1990; Eley and Moore, 1983; Hartshorn *et al.*, 1987; Whitford, 1992) exploits the fact that paramagnetic complexes such as $[Cr(oxalate)_3]^{3-}$ and $[Cr(en)_3]^{3+}$ cause specific line broadening of groups to which they are bound. Protection within an electrostatic complex provides evidence that a group occurs at the binding interface between the two proteins. Care must be taken, however, that the paramagnetic probe does not disturb the protein complex significantly. This problem does not arise when the solvent itself is the probe. The rate of exchange of backbone amide protons depends on their degree of exposure, so protection from solvent by a reaction partner can cause significant decreases in the measured rate of exchange with D_2O. This method has been used to explore the regions of the surface of cytochrome c that are involved in binding to cytochrome c peroxidase (Jeng et al., 1994; Yi et al., 1994). The results of these experiments tend to confirm that, in solution, the peroxidase binds to the same region on the surface of cytochrome c as in the crystalline complex of Pelletier and Kraut (1992). However, the protection of some residues that lie outside this region suggests that the interaction is not well described by a single, fixed mode of binding.

3.5.2 *The dynamic view*

Crystal structures of complexes are most readily interpreted in terms of the classical lock-and-key model of Emil Fischer and give detailed information about one configuration, which is one of minimum energy under the conditions used for crystallization. By contrast, Brownian dynamics simulations, discussed extensively in Chapter 4, have shown that relatively long-range electrostatic forces allow a large number of different configurations to be explored within the lifetime of a single encounter complex. The suggestion has been made that recognition and binding between redox proteins involves complementary sticky patches (the 'Velcro model') rather than association into a single, precise configuration (McLendon, 1991). While this may be true as far as the coulombic interactions alone are concerned, short-range forces might define a more limited set of configurations. Several types of experiment, inevitably less precise than X-ray crystallography, have been used in attempts to obtain direct evidence about the dynamics of interaction.

Mobility of complexes. A simple NMR experiment which can give direct evidence for relative mobility involves the measurement of line widths or transverse relaxation times (T_2). T_2 depends on the rotational correlation time, which in turn depends on the size of the molecule. In general, the larger the molecule the slower it tumbles and the smaller is T_2. The value of T_2 for a tight complex cannot be predicted theoretically (although it may be calculated approximately if the individual proteins and the complex are treated as spherical), but an experimental value can be obtained for reference purposes by cross-linking the two proteins together by treatment with 1-ethyl-3-(3-dimethylaminopropyl)carbodiimide. One-dimensional experiments are adequate if a single line can be picked out. Although an experiment of precisely this form has not been reported, a study of the effect of ionic strength on the line width of the C12 methyl resonance of cytochrome b_5 in the presence of an equimolar concentration of cytochrome c has been described (Whitford et al., 1990). The interesting feature of this experiment was that the line width decreased in a stepwise manner, suggesting that the diprotein complex existed in more than one form with different mobilities.

Evidence of a different kind has come from study of the kinetics of energy transfer between cytochrome c and cytochrome c peroxidase. For this experiment, it was necessary to substitute a fluorescent Mg-porphyrin for the haem group of the peroxidase and to use a pulsed dye laser to excite the fluorescence, the decay of which was measured over a period of a few nanoseconds (Zhang *et al.*, 1990). Energy transfer from the Mg-porphyrin to the haem of cytochrome *c* stimulated the decay. A solution of the complex frozen to 77 K showed multiphasic quenching kinetics, which gave way to quenching by a single exponential term when the sample was warmed to room temperature. The experiment suggested that the protein complex exists in a set of different conformational states which are trapped at low temperature but which are equilibrated with each other within about 5 ns at room temperature.

Mapping of binding sites by NMR methods can also be taken to imply mobility within the complex if the area occupied is larger than expected for a single static complex. An example is the finding that at least six lysine residues of cytochrome *c* are involved in the interaction with cytochrome b_5 rather than the four predicted by the Salemme model (Burch *et al.*, 1990).

Two binding sites. For some of the best studied systems, evidence now exists that the larger redox partner has at least two binding sites for the smaller. A surprising feature is that the most stable configuration may not be the most active. This possibility has been invoked to explain the fact that an optimum in the curve relating the second order rate constant to ionic strength has been found for the reduction of cytochrome *c* peroxidase by cytochrome *c* (Hazzard *et al.*, 1988). A similar effect for the reduction of plastocyanin by cytochrome *f* is shown in *Figure 3.3*. A likely explanation is that the proteins tend to get locked into an unproductive configuration under the influence of strong electrostatic

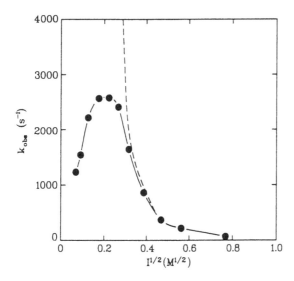

Figure 3.3. Effect of ionic strength on the reoxidation of 14.7 µM spinach cytochrome *f* by equimolar spinach plastocyanin. The buffer contained 0.5 mM EDTA, 0.8 mM phosphate and NaCl to give the specified ionic strength. The broken line represents a fit of the parallel plate model (Watkins *et al.*, 1994) to experimental points at $I^{1/2} = 0.387$ M$^{1/2}$ and above. Reprinted with permission from Meyer *et al.* (1993) Transient kinetics of electron transfer from a variety of *c*-type cytochromes to plastocyanin. *Biochemistry* **32**: 4552–4559. Copyright 1993 American Chemical Society.

forces at very low ionic strength, and that formation of the more reactive configuration requires a rearrangement by surface diffusion. In support is the finding that the complex formed by covalent cross-linking of the two proteins shows only a slow rate of intracomplex electron transfer (Qin and Kostic, 1993). An alternative explanation is that the optimum results from the increase in Debye length at low ionic strength. This might lead to more charged groups, possibly of opposite sign, being incorporated into the effective binding site.

The latter explanation cannot apply to the following experiments on the electron transfer kinetics between zinc-substituted cytochrome c and either plastocyanin or cytochrome b_5 (Qin and Kostic, 1994; Zhou and Kostic, 1992, 1993) studied by flash photolysis. The flash generates a long-lived triplet state of the Zn-porphyrin, which can be quenched by electron transfer to the other member of a pre-formed complex. This reaction differs from that of the native cytochrome c by having a very high driving force and so is much faster and can reveal other rate-limiting steps such as surface diffusion. Varying the viscosity showed the presence of two interconverting conformers, a dominant one with a relatively slow rate of electron transfer, and a minor one with a rapid rate. If this reaction is a good model for the physiological one, rearrangement may also occur with the native proteins.

The two binding sites may or may not overlap. However, recent experiments by Hoffman and colleagues have provided clear evidence that cytochrome c peroxidase has two separate binding sites for cytochrome c (Stemp and Hoffman, 1993; Zhou and Hoffman, 1993, 1994). The principle of electron transfer quenching of the triplet state of Zn-substituted cytochrome c was again exploited in a series of titration experiments illustrated in *Figure 3.4*. Titration of Zn-cytochrome c (Znc) with increasing amounts of the peroxidase (Cp) at low ionic strength gave a quenching curve with an optimum at a ratio of Cp/Znc of 0.5. The result was explained by a model in which the peroxidase has two independent binding sites for cytochrome c, a high affinity site with a low rate of electron transfer and a low affinity site with fast electron transfer. The two-site model has been confirmed in other types of experiment (Mauk *et al.*, 1994; Wang and Margoliash, 1995).

At first glance, it seems paradoxical that the site to which a redox partner is bound most strongly should be relatively inactive. However, low affinity for the active site could allow the rapid dissociation of the products if the off rate constant were enhanced, especially under steady-state rather than single turnover conditions. An adjacent high affinity site could have the function of providing a relatively high concentration of protein in the reactive configuration at the low affinity site. The disadvantage of this arrangement is that the principle of microscopic reversibility demands that the equilibrium constant between the two configurations favours the high affinity site. Intuition alone is unlikely to be capable of understanding such a complex kinetic situation. Kinetic modelling with the help of a numerical integration program such as Kinsim (Barshop *et al.*, 1983; Frieden, 1993) can be helpful, and shows that, under laboratory conditions with low concentrations of protein, the introduction of a second site can, indeed, have a kinetic advantage. Whether or not this would remain true under *in vivo* conditions is more difficult to decide.

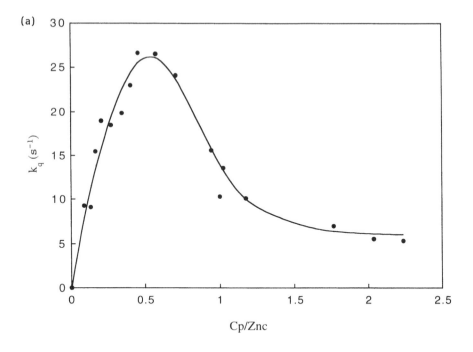

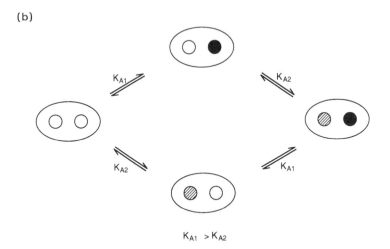

Figure 3.4. Two binding sites of yeast cytochrome c peroxidase for cytochrome c. (a) Experimental evidence. The rate constant for quenching of the triplet state of Zn-substituted cytochrome c (^{3}Znc) by cytochrome c peroxidase (Cp), in 2.5 mM phosphate, pH 7.0. Znc (8.5 μM) was titrated with Cp. The solid line is the theoretically calculated total quenching according to the two-binding site model with appropriate parameters. (b) The model. The ellipse represents Cp and the small circles Znc. Filled circles, high affinity site ($K_{A1} = 4.9 \times 10^5$ M^{-1}); hatched circles, low affinity site ($K_{A2} = 1.2 \times 10^4$ M^{-1}). Redrawn from Zhou and Hoffman (1993).

3.6 Conclusions

In conclusion, we return to the question raised at the beginning of this chapter. How do diffusing proteins achieve a balance of binding forces that allows both a rapid rate of electron transfer within the complex, and a rapid turnover? Knowledge of the specific details of the structural and dynamic features of interactions between redox proteins is not yet well developed, but a picture is beginning to emerge in which one can detect different strategies for solving the problem. The most direct would be one in which there is a good fit between the surfaces of the two proteins that allows the two redox centres to be brought close together. The relative orientation can also, in principle, play a part in optimizing the electronic coupling. The discussion in Section 3.4 leads one to suggest that a strong, specific interaction is likely to be predominantly hydrophobic in nature. A good example may be provided by the reaction between plastocyanin and photosystem I in which close apposition between the northern hydrophobic patch of plastocyanin and the reaction centre polypeptide is thought to be essential. Modification of residues within the hydrophobic patch by site-directed mutagenesis has shown the importance of surface complementarity for binding (Haehnel *et al.*, 1994; Nordling *et al.*, 1991). The intracomplex k_{et} has been estimated to be approximately 5.8×10^4 s^{-1} (Haehnel *et al.*, 1994; Nordling *et al.*, 1991) and, at the same time, the turnover under steady-state conditions is known to be very fast, with k_2 being of the order of 10^8 M^{-1} s^{-1} (Wood and Bendall, 1975). A careful study of the reaction has shown that the midpoint redox potential of plastocyanin in the complex is 50 mV more positive than in the free state (Drepper *et al.*, 1995). The implication of this observation is that the photosystem has a lower affinity for the oxidized form of plastocyanin than the reduced, a conclusion confirmed by the kinetic analysis. In this system, then, it seems that a redox-linked conformational change of plastocyanin, which aids dissociation of the oxidized protein, offsets the danger of having a high affinity between the reactant proteins.

There is no evidence for a redox-linked change in binding constant for the reaction between plastocyanin and cytochrome f in solution, at least at pH 6 (He *et al.*, 1991). In this case, there are some strong hints of a different kind of strategy in which there is more than one configuration of the complex. The pathway taken by the electron in plastocyanin, on its way from the haem of cytochrome f to the Cu atom, is thought to involve the side chain of Tyr83, which lies between two acidic patches on the eastern face of the molecule (He *et al.*, 1991). However, the basic patch on cytochrome f that is likely to be involved in the association lies at least 25 Å from the haem edge (Martinez *et al.*, 1994). This suggests that there may be two exchanging modes of binding, only one of which gives a high rate of electron transfer. Such behaviour has been explored most thoroughly for the interaction between cytochrome c and cytochrome c peroxidase and is discussed in Section 3.5, but the kinetic significance of the presence of strong and weak binding is not entirely clear.

Recent refinement of the crystal structure of cytochrome f has shown the presence of a chain of bound water molecules that seem to link the haem with the positive patch to which plastocyanin binds. It has been suggested that these water molecules define the terminal part of the pathway for protons liberated into the thylakoid lumen by quinol oxidation *in vivo* (Martinez *et al.,* 1996). If this is true, it could provide a subtle mechanism by which reduced plastocyanin is encouraged to dissociate from oxidized cytochrome *f in vivo*, but this effect would not be apparent when soluble forms of cytochrome *f* and plastocyanin are studied *in vitro*.

Photosynthetic systems are capable of high rates of turnover, and it may well be that many electron transfer reactions between proteins have less stringent kinetic requirements. The above discussion is by no means exhaustive, however, and the study of the recognition processes between pairs of redox proteins is still in its infancy. The wide variety of physiological systems in which electron transfer is important and the structural complexity of proteins allow many possible ways of achieving similar functional objectives.

Acknowledgements

I am indebted to Dr M. Ubbink for many fruitful discussions, and to Dr W.A. Cramer for sending a manuscript in advance of publication.

References

Adam G, Delbrück M. (1968) Reduction of dimensionality in biological diffusion processes. In: *Structural Chemistry and Molecular Biology* (eds A Rich, N Davidson). Freeman, San Francisco, pp. 198–215.

Bagby S, Driscoll PC, Goodall KG, Redfield C, Hill HAO. (1990) The complex formed between plastocyanin and cytochrome *c*. Investigation by NMR spectroscopy. *Eur. J. Biochem.* **188**: 413–420.

Barshop BB, Wrenn RF, Frieden C. (1983) Analysis of numerical methods for computer simulation of kinetic processes: development of KINSIM – a flexible, portable system. *Anal. Biochem.* **130**: 327–344.

Berg OG, von Hippel PH. (1985) Diffusion-controlled macromolecular interactions. *Annu. Rev. Biophys. Biophys. Chem.* **14**: 131–160.

Bhat TN, Bentley GA, Boulot G, Greene MI, Tello D, Dall'Acqua W, Souchon H, Schwarz FP, Mariuzza RA, Poljak RJ. (1994) Bound water molecules and conformational stabilization help mediate an antigen–antibody association. *Proc. Natl Acad. Sci. USA* **91**: 1089–1093.

Burch AM, Rigby SEJ, Funk WD, MacGillivray RTA, Mauk MR, Mauk AG, Moore GR. (1990) NMR characterization of surface interactions in the cytochrome b_5–cytochrome *c* complex. *Science* **247**: 831–833.

Chothia C, Janin J. (1975) Principles of protein–protein recognition. *Nature* **256**: 705–708.

Creighton TE. (1993) *Proteins. Structures and Molecular Properties*, 2nd Edn. Freeman, New York.

Díaz A, Hervás M, Navarro JA, De la Rosa MA, Tollin G. (1994) A thermodynamic study by laser-flash photolysis of plastocyanin and cytochrome c_6 oxidation by photosystem I from the green alga *Monoraphidium braunii. Eur. J. Biochem.* **222**: 1001–1007.

Drepper F, Hippler M, Haehnel W. (1995) Dynamic interactions between plastocyanin and photosystem I. In: *Photosynthesis: from Light to Biosphere* (ed. P Mathis). Kluwer Academic Publishers, Dordrecht, Vol. II, pp. 697–700.

Durham B, Fairris JL, McLean M, Millett F, Scott JR, Sligar SG, Willie A. (1995) Electron transfer from cytochrome b_5 to cytochrome *c. J. Bioenerg. Biomembr.* **27**: 331–340.

Eisenberg D, McLachlan AD. (1986) Solvation energy in protein folding and binding. *Nature* **319**: 199–203.

Eley CGS, Moore GR. (1983) [1]H-n.m.r. investigation of the interaction between cytochrome *c* and cytochrome b_5. *Biochem. J.* **215**: 11–21.

Fersht AR. (1987) The hydrogen bond in molecular recognition. *Trends Biochem. Sci.* **12:** 301–304.

Fersht AR, Serrano L. (1993) Principles of protein stability derived from protein engineering experiments. *Curr. Opin. Struct. Biol.* **3:** 75–83.

Fersht AR, Shi J-P, Knill-Jones J, Lowe DM, Wilkinson AJ, Blow DM, Brick P, Carter P, Waye WMY, Winter G. (1985) Hydrogen bonding and biological specificity analysed by protein engineering. *Nature* **314:** 235–238.

Finkelstein AV, Janin J. (1989) The price of lost freedom. Entropy of bimolecular complex formation. *Prot. Eng.* **3:** 1–3.

Frieden C. (1993) Numerical integration of rate equations by computer. *Trends Biochem. Sci.* **18:** 58–60.

Gray HB, Malmström BG. (1989) Long-range electron transfer in multisite metalloproteins. *Biochemistry* **28:** 7499–7505.

Guerlesquin F, Sari J-C, Bruschi M. (1987) Thermodynamic parameters of cytochrome c_3–ferredoxin complex formation. *Biochemistry* **26:** 7438–7443.

Haehnel W, Jansen T, Gause K, Klösgen RB, Stahl B, Michl D, Huvermann B, Karas M, Herrmann RG. (1994) Electron transfer from plastocyanin to photosystem I. *EMBO J.* **13:** 1028–1038.

Hahm S, Durham B, Millett F. (1992) Photoinduced electron transfer between cytochrome *c* peroxidase and horse cytochrome *c* labeled at specific lysines with (dicarboxybipyridine)(bisbipyridine)ruthenium(II). *Biochemistry* **31:** 3472–3477.

Hartshorn RT, Mauk AG, Mauk MR, Moore GR. (1987) NMR study of the interaction between cytochrome b_5 and cytochrome *c*. Observation of a ternary complex formed by the two proteins and $[Cr(en)_3]^{3+}$. *FEBS Lett.* **213:** 391–395.

Harvey SC. (1989) Treatment of electrostatic effects in macromolecular modeling. *Proteins* **5:** 78–92.

Hazzard JT, McLendon G, Cusanovich MA, Tollin G. (1988) Formation of electrostatically-stabilized complex at low ionic strength inhibits interprotein electron transfer between yeast cytochrome *c* and cytochrome *c* peroxidase. *Biochem. Biophys. Res. Commun.* **151:** 429–434.

He S, Modi S, Bendall DS, Gray JC. (1991) The surface-exposed tyrosine residue Tyr83 of pea plastocyanin is involved in both binding and electron transfer reactions with cytochrome *f*. *EMBO J.* **10:** 4011–4016.

Hibbits KA, Gill DS, Willson RC. (1994) Isothermal titration calorimetric study of the association of hen egg lysozyme and the anti-lysozyme antibody HyHEL-5. *Biochemistry* **33:** 3584–3590.

Honig B, Nicholls A. (1995) Classical electrostatics in biology and chemistry. *Science* **268:** 1144–1149.

Janin J, Chothia C. (1990) The structure of protein–protein recognition sites. *J. Biol. Chem.* **265:** 16027–16030.

Jelesarov I, Bosshard HR. (1994) Thermodynamics of ferredoxin binding to ferredoxin: NADP+ reductase and the role of water at the complex interface. *Biochemistry* **33:** 13321–13328.

Jeng MF, Englander SW, Pardue K, Rogalskyj JS, McLendon G. (1994) Structural dynamics in an electron-transfer complex. *Struct. Biol.* **1:** 234–238.

Jones S, Thornton JM. (1995) Protein–protein interactions: a review of protein dimer structures. *Prog. Biophys. Mol. Biol.* **63:** 31–59.

King GC, Binstead RA, Wright PE. (1985) NMR and kinetic characterization of the interaction between French bean plastocyanin and horse cytochrome *c*. *Biochim. Biophys. Acta* **806:** 262–271.

Koppenol WH, Margoliash E. (1982) The asymmetric distribution of charges on the surface of horse cytochrome *c*. *J. Biol. Chem.* **257:** 4426–4437.

Kornblatt JA, Kornblatt MJ, Hui Bon Hoa G, Mauk AG. (1993) Responses of two protein–protein complexes to solvent stress: does water play a role at the interface? *Biophys. J.* **65:** 1059–1065.

Lee B, Richards FM. (1971) The interpretation of protein structures: estimation of static accessibility. *J. Mol. Biol.* **55:** 379–400.

Margoliash E, Bosshard HR. (1983) Guided by electrostatics, a textbook protein comes of age. *Trends Biochem. Sci.* **8:** 316–320.

Martinez SE, Huang D, Szczepaniak A, Cramer WA, Smith JL. (1994) Crystal structure of chloroplast cytochrome f reveals a novel cytochrome fold and unexpected heme ligation. *Structure* **2**: 95–105.

Martinez SE, Huang D, Ponomarev M, Cramer WA, Smith JL. (1996) The heme redox center of chloroplast cytochrome f is linked to a buried five-water chain. *Prot. Sci.* (in press).

Mauk MR, Reid LS, Mauk AG. (1982) Spectrophotometric analysis of the interaction between cytochrome b_5 and cytochrome c. *Biochemistry* **21**: 1843–1846.

Mauk MR, Ferrer JC, Mauk AG. (1994) Proton linkage in formation of the cytochrome c–cytochrome c peroxidase complex: electrostatic properties of the high- and low-affinity cytochrome binding sites on the peroxidase. *Biochemistry* **33**: 12609–12614.

McLendon G. (1991) Control of biological electron transport via molecular recognition and binding: the "Velcro" model. *Struct. Bonding* **75**: 159–174.

McLendon G, Miller JR. (1985) The dependence of biological electron transfer rates on exothermicity: the cytochrome c/cytochrome b_5 couple. *J. Am. Chem. Soc.* **107**: 7811–7816.

Meyer TE, Zhao ZG, Cusanovich MA, Tollin G. (1993) Transient kinetics of electron transfer from a variety of c-type cytochromes to plastocyanin. *Biochemistry* **32**: 4552–4559.

Mikkelsen KV, Skov LK, Nar H, Farver O. (1993) Electron self-exchange in azurin: calculation of the superexchange electron tunneling rate. *Proc. Natl Acad. Sci. USA* **90**: 5443–5445.

Modi S, He S, Gray JC, Bendall DS. (1992) The role of surface-exposed Tyr-83 of plastocyanin in electron transfer from cytochrome c. *Biochim. Biophys. Acta* **1101**: 64–68.

Moench SJ, Chroni S, Lou B-S, Erman JE, Satterlee JD. (1992) Proton NMR comparison of noncovalent and covalently cross-linked complexes of cytochrome c peroxidase with horse, tuna, and yeast ferricytochromes c. *Biochemistry* **31**: 3661–3670.

Niwa S, Ishikawa H, Nikai S, Takabe T. (1980) Electron transfer reactions between cytochrome f and plastocyanin from *Brassica komatsuna*. *J. Biochem.* **88**: 1177–1183.

Nordling M, Sigfridsson K, Young S, Lundberg L, Hansson Ö. (1991) Flash-photolysis studies of the electron transfer from genetically modified spinach plastocyanin to photosystem I. *FEBS Lett.* **291**: 327–330.

Northrup SH, Erickson HP. (1992) Kinetics of protein–protein association explained by Brownian dynamics computer simulation. *Proc. Natl Acad. Sci. USA* **89**: 3338–3342.

Paladini AA Jr, Weber G. (1981) Pressure-induced reversible dissociation of enolase. *Biochemistry* **20**: 2587–2593.

Pan LP, Frame M, Durham B, Davis D, Millett F. (1990) Photoinduced electron transfer within complexes between plastocyanin and ruthenium bispyridine dicarboxybipyridine cytochrome c derivatives. *Biochemistry* **29**: 3231–3236.

Pelletier H, Kraut J. (1992) Crystal structure of a complex between electron transfer partners, cytochrome c peroxidase and cytochrome c. *Science* **258**: 1748–1755.

Poulos TL, Kraut J. (1980) A hypothetical model of the cytochrome c peroxidase–cytochrome c electron transfer complex. *J. Biol. Chem.* **255**: 10322–10330.

Qin L, Kostic NM. (1993) Importance of protein rearrangement in the electron-transfer reaction between the physiological partners cytochrome f and plastocyanin. *Biochemistry* **32**: 6073–6080.

Qin L, Kostic NM. (1994) Photoinduced electron transfer from the triplet state of zinc cytochrome c to ferricytochrome b_5 is gated by configurational fluctuations of the diprotein complex. *Biochemistry* **33**: 12592–12599.

Richards FM. (1977) Areas, volumes, packing, and protein structure. *Annu. Rev. Biopyhs. Bioeng.* **6**: 151–176.

Roberts VA, Freeman HC, Olson AJ, Tainer JA, Getzoff ED. (1991) Electrostatic orientation of the electron-transfer complex between plastocyanin and cytochrome c. *J. Biol. Chem.* **266**: 13431–13441.

Rodgers KK, Pochapsky TC, Sligar SG. (1988) Probing the mechanisms of macromolecular recognition: the cytochrome b_5–cytochrome c complex. *Science* **240**: 1657–1659.

Rodgers KK, Sligar SG. (1991) Mapping electrostatic interactions in macromolecular associations. *J. Mol. Biol.* **221**: 1453–1460.

Rose GD, Wolfenden R. (1993) Hydrogen bonding, hydrophobicity, packing, and protein folding. *Annu. Rev. Biophys. Biomol. Struct.* **22**: 381–415.

Ross PD, Subramanian S. (1981) Thermodynamics of protein association reactions: forces contributing to stability. *Biochemistry* **20**: 3096–3102.

Rush JD, Levine F, Koppenol WH. (1988) The electron-transfer site of spinach plastocyanin. *Biochemistry* **27**: 5876–5884.

Salemme FR. (1976) An hypothetical structure for an intermolecular electron transfer complex of cytochromes c and b_5. *J. Mol. Biol.* **102**: 563–568.

Satterlee JD, Moench SJ, Erman JE. (1987) A proton NMR study of the non-covalent complex of horse cytochrome c–yeast cytochrome c peroxidase and its comparison with other interacting protein complexes. *Biochim. Biophys. Acta* **912**: 87–97.

Smith HT, Ahmed AJ, Millett F. (1981) Electrostatic interaction of cytochrome c with cytochrome c_1 and cytochrome oxidase. *J. Biol. Chem.* **256**: 4984–4990.

Stemp EDA, Hoffman BM. (1993) Cytochrome c peroxidase binds two molecules of cytochrome c: evidence for a low-affinity, electron-transfer-active site on cytochrome c peroxidase. *Biochemistry* **32**: 10848–10865.

Tello D, Goldbaum FA, Mariuzza RA, Ysern X, Schwarz FP, Poljak RJ. (1993) Three-dimensional structure and thermodynamics of antigen binding by anti-lysozyme antibodies. *Biochem. Soc. Trans.* **21**: 943–946.

Tollin G, Hazzard JT. (1991) Intra- and intermolecular electron transfer processes in redox proteins. *Arch. Biochem. Biophys.* **287**: 1–7.

Tollin G, Hurley JK, Hazzard JT, Meyer TE. (1993) Use of laser flash photolysis time-resolved spectrophotometry to investigate interprotein and intraprotein electron transfer mechanisms. *Biophys. Chem.* **48**: 259–279.

Ubbink M, Gong X-S, Gray JC, Bendall DS. (1995) Protein–protein interactions studied by NMR: does cytochrome c bind to plastocyanin on its acidic patch? *J. Inorg. Biochem.* **59**: 282.

Van Leeuwen JW. (1983) The ionic strength dependence of the rate of a reaction between two large proteins with a dipole moment. *Biochim. Biophys. Acta* **743**: 408–421.

Varley JPA, Moehrle JJ, Manasse RS, Bendall DS, Howe CJ. (1995) Characterization of plastocyanin from the cyanobacterium *Phormidium laminosum*: copper-inducible expression and SecA-dependent targeting in *Escherichia coli*. *Plant Mol. Biol.* **27**: 179–190.

Wagner MJ, Packer JCL, Howe CJ, Bendall DS. (1995) Some properties of cytochrome f in the cyanobacterium *Phormidium laminosum*. In: *Photosynthesis: from Light to Biosphere* (ed. P Mathis). Kluwer Academic Publishers, Dordrecht, Vol. II, pp. 745–748.

Wang Y, Margoliash E. (1995) Enzymic activities of covalent 1:1 complexes of cytochrome c and cytochrome c peroxidase. *Biochemistry* **34**: 1948–1958.

Warshel A, Åqvist J. (1991) Electrostatic energy and macromolecular function. *Annu. Rev. Biophys. Biophys. Chem.* **20**: 267–298.

Warwicker J. (1989) Investigating protein–protein interaction surfaces using a reduced stereochemical and electrostatic model. *J. Mol. Biol.* **206**: 381–395.

Warwicker J, Watson HC. (1982) Calculation of the electric potential in the active site cleft due to α-helix dipoles. *J. Mol. Biol.* **157**: 671–679.

Watkins JA, Cusanovich MA, Meyer TE, Tollin G. (1994) A "parallel plate" electrostatic model for bimolecular rate constants applied to electron transfer proteins. *Protein Sci.* **3**: 2104–2114.

Whitford D. (1992) The identification of cation-binding domains on the surface of microsomal cytochrome b_5 using ^{1}H-NMR paramagnetic difference spectroscopy. *Eur. J. Biochem.* **203**: 211–223.

Whitford D, Concar DW, Veitch NC, Williams RJP. (1990) The formation of protein complexes between ferricytochrome b_5 and ferricytochrome c studied using high-resolution ^{1}H-NMR spectroscopy. *Eur. J. Biochem.* **192**: 715–721.

Whitford D, Gao Y, Pielak GJ, Williams RJP, McLendon G, Sherman F. (1991) The role of the internal hydrogen bond network in first-order protein electron transfer between *Saccharomyces cerevisiae* iso-1-cytochrome c and bovine microsomal cytochrome b_5. *Eur. J. Biochem.* **200**: 359–367.

Wood PM, Bendall DS. (1975) The kinetics and specificity of electron transfer from cytochromes and copper proteins to P700. *Biochim. Biophys. Acta* **387**: 115–128.

Yi Q, Erman JE, Satterlee JD. (1993) Proton NMR studies of noncovalent complexes of cytochrome c peroxidase-cyanide with horse and yeast ferricytochromes c. *Biochemistry* **32**: 10988–10994.

Yi Q, Erman JE, Satterlee JD. (1994) Studies of protein–protein association between yeast cytochrome c peroxidase and yeast iso-1 ferricytochrome c by hydrogen–deuterium exchange labeling and proton NMR spectroscopy. *Biochemistry* **33:** 12032–12041.

Zhang Q, Marohn J, McLendon G. (1990) Macromolecular recognition in the cytochrome c–cytochrome c peroxidase complex involves fast two-dimensional diffusion. *J. Phys. Chem.* **94:** 8628–8630.

Zhou JS, Hoffman BM. (1993) Cytochrome c peroxidase simultaneously binds cytochrome c at two different sites with strikingly different reactivities: titrating a 'substrate' with an enzyme. *J. Am. Chem. Soc.* **115:** 11008–11009.

Zhou JS, Hoffman BM. (1994) Stern–Volmer in reverse: 2:1 stoichiometry of the cytochrome c–cytochrome c peroxidase electron-transfer complex. *Science* **265:** 1693–1696.

Zhou JS, Kostic NM. (1992) Photoinduced electron transfer from zinc cytochrome c to plastocyanin is gated by surface diffusion within the metalloprotein complex. *J. Am. Chem. Soc.* **114:** 3562–3563.

Zhou JS, Kostic NM. (1993) Gating of photoinduced electron transfer from zinc cytochrome c and tin cytochrome c to plastocyanin. Effects of solution viscosity on rearrangement of the metalloprotein complex. *J. Am. Chem. Soc.* **115:** 10796–10804.

Computer modelling of protein–protein interactions

Scott H. Northrup

4.1 Introduction

Protein–protein electron transfer is influenced by a host of important structural and dynamic factors. Theoretical studies have been used to complement experimental methods to address essentially four pertinent features of protein–protein interaction and electron transfer. Here we are less concerned with the intimate details of the electron transfer event itself, which is discussed in other chapters. Rather the primary focus is the events leading up to the formation of productive electron transfer complexes through rotational and translational diffusion, and influenced by predominantly electrostatic and steric forces. Four major topics we address here include:

(i) the role of electrostatic charge distribution in stabilizing individual electron transfer complexes;
(ii) dynamics of protein–protein association by a diffusional mechanism;
(iii) the possible existence of alternative docking geometries rather than a single complex through which electron transfer can occur;
(iv) the influence of protein flexibility and internal dynamics on an individual association complex.

4.1.1 *Model building of electron transfer complexes*

Protein–protein association plays a central role in biological electron transfer, with electrostatics providing the stability of complexes. Numerous theoretical studies fall into the category of static docking, which have focused on the construction and analysis of single representative electron transfer complexes, with a particular emphasis on the detailed electrostatic charge distribution and its role in facilitating the electron transfer between these proteins. Protein–protein complex structural models have been formulated in several

Protein Electron Transfer, Edited by D.S. Bendall
© 1996 BIOS Scientific Publishers Ltd, Oxford

studies through the use of computer graphics and optimization of ion pair interactions. The pioneering study of Salemme (1976) suggested a hypothetical structural complex of cytochrome c and cytochrome b_5 based on a least-squares fitting of crystallographic coordinates which enforced charge complementarity between highly conserved charged groups surrounding the haem crevices. Hypothetical structures of complexes cytochrome c–cytochrome c peroxidase (CCP) (Poulos and Kraut, 1980), cytochrome c–flavodoxin (Simondsen et al., 1982), cytochrome b_5–methaemoglobin (Poulos and Mauk, 1983), cytochrome P450–cytochrome b_5 (Stayton et al., 1989), putidaredoxin–putidaredoxin reductase (Geren et al., 1986) and flavocytochrome b_2–cytochrome c (Tegoni et al., 1993) subsequently have been constructed along similar lines facilitated by computer graphics display systems. The cytochrome c–flavodoxin complex was used subsequently (Matthew et al., 1983) to assess the role of complementary electrostatics in the pre-orientation of molecules enhancing the reaction kinetics. In a different type of static analysis, residue substitutions of 95 cytochrome c sequences have been categorized as either conservative or radical, colour-coded for instant visual recognition, and mapped on to the three-dimensional structure of yeast iso-1-cytochrome c (CY) (Meyer et al., 1994). All of these studies have been useful in establishing the role of the conserved charged amino acids surrounding the prosthetic group in the interactions between electron transfer species, and in some cases establishing a plausible orientation, prosthetic group separation distance and other criteria in metalloprotein pairs. However, considerations of dynamic aspects of protein electron transfer have required more robust simulation methodology, which we now discuss.

4.1.2 Diffusional dynamics of protein–protein association

A second feature of protein electron transfer which has been addressed by theoretical simulations is the dynamics of formation of productive electron transfer complexes. A multitude of references and detailed discussion of these is provided in subsequent sections of this chapter. Protein–protein association reactions require the initial diffusional encounter of species, which in many cases may limit or at least partially influence the bimolecular rate constant (Berg and von Hippel, 1985). Thus a treatment of the dynamics of associative encounters is of fundamental importance in the full modelling of protein electron transfer, and is amenable to study by the methodology called Brownian dynamics (BD). Since the surface through which electron transfer takes place may be typically small relative to the overall size of the protein, strict orientational criteria for reaction may exist. It has been hypothesized (Northrup et al., 1988) that protein–protein reactions with severe orientational constraints may employ a 'reduction in dimensionality' principle (Adam and Delbrück, 1968). In this mechanism, the proteins diffuse in three dimensions and initially associate in unproductive orientations by fairly non-specific electrostatic forces of attraction. Loosely associated pairs may then rotationally diffuse until eventual production of a properly oriented pair or until escape. The forces promoting the initial non-specific association must be of such magnitude as to allow particles to remain in an encountered state for time scales sufficient for the rotational reorientation event, but not so strong that particles are stuck in unproductive orientations. In addition to this effect, for highly dipolar species, such as the cytochromes, a 'steering' effect may be in operation which enhances the possibility of productive first encounter of species.

4.1.3 *Alternative docking geometries for electron transfer*

Despite the recent observation of a well-defined complex between cytochrome *c* and CCP in crystallographic studies of Pelletier and Kraut (1993), there is still strong evidence in the mechanism of interaction of the cytochromes that an ensemble of near-optimal docking geometries exist and contribute to electron transfer rather than a single complex. Theoretical studies of the electrostatic interactions of three different *c*-type cytochromes with flavodoxin (Mauk *et al.,* 1986; Weber and Tollin, 1985) reveal that variations in the computed electrostatic stability reflect differences in the distribution of all charged surface groups and not simply those localized at the site of intermolecular contact. On the basis of electrostatic stability alone, their calculations do not predict a dominant protein–protein interaction but rather a mixture of complexes whose relative populations change with pH and ionic strength. This is borne out by numerous experimental studies as described in Section 4.4.

4.1.4 *Influence of protein flexibility and internal dynamics*

A final feature amenable to theoretical studies is the influence of protein flexibility and internal dynamics on electron transfer within an association complex. Wendoloski and co-workers (1987) have performed molecular dynamics simulations of a cytochrome $c-b_5$ complex which model a flexible association complex able to sample a range of interhaem geometries. The Fe–Fe distance decreased by as much as 2 Å after less than 5 ps of molecular dynamics simulation due to dynamic relaxation of the complex. Substantial movement of Phe82 of cytochrome *c* from its crystallographic location occurred. The phenyl side chain, initially packed near the haem, moved to a position where it could conceivably bridge the two haem group π orbital systems. The time evolution of these geometrical details during molecular dynamics simulation is shown in *Figure 4.1*. Flexibility within complexes could also contribute to the relative lack of recognition specificity between electron transfer proteins, and explain why some non-physiological partners such as cytochrome $c-b_5$ react at physiological rates.

Internal flexibility and dynamics of this same complex have also been studied by Guillemette *et al.* (1994). They have performed energy minimizations and molecular dynamics of a couple of representative complexes previously generated by BD simulation (Northrup *et al.,* 1993) and have observed the tightening up of the protein–protein interfaces and production of more stable and intimate complexes. This has been an important supplement to BD, in that it overcomes the limitation of these simulations to rigid body docking. This study is discussed in greater detail in Section 4.3.8.

In the remainder of this chapter, we shall discuss in detail studies pertaining to the overview in Sections 4.1.2–4.1.4, since the greatest theoretical attention has been focused there.

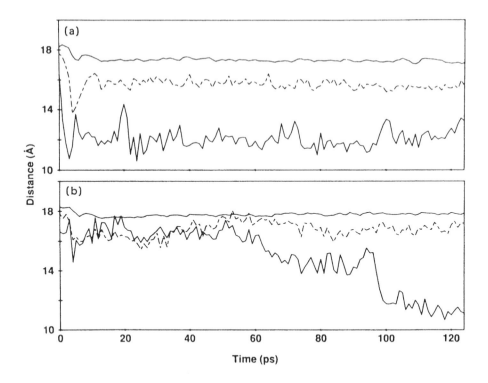

Figure 4.1. Time evolution of representative geometrical interactions during (a) molecular dynamics simulation with a distance-dependent dielectric constant approximation, and (b) the simulation with the complex solvated with a 6 Å thick water shell. In each panel, the upper solid curve shows the complex radius of gyration. The dashed curve is the Fe–Fe distance, and the lower solid curve is the distance between the centre of the phenyl ring of the Phe82 of cytochrome c and the haem iron of cytochrome b_5. Reprinted with permission from Wendoloski *et al.* (1987) Molecular dynamics of a cytochrome c–cytochrome b_5 electron transfer complex. *Science* **238:** 794–797. Copyright 1987 American Association for the Advancement of Science.

4.2   Theoretical methodologies

4.2.1 *Modelling of protein structure*

The current wealth of available high-resolution X-ray crystallographic information on proteins provides the structural basis for simulational modelling of protein–protein interactions. Structural models have been constructed at varying levels of complexity of the protein shapes and placement of charges on the reactants. Early Brownian studies treated proteins as simple spherical models where each sphere has an imbedded monopole and dipolar charges, while more robust studies have been performed in which every atom is modelled as a partial charge in its crystallographic position and the irregular surface topography is treated.

4.2.2 Calculation of protein electrostatic fields

The first step in the treatment of electrostatic interactions in proteins is the assignment of charges on atoms. The pK_as and, ultimately, the protonation state of each titratible amino acid residue in its protein environment, pH and solution ionic strength needs to be estimated. At or near neutral pH it is fair to assume that lysines and arginines are fully protonated and aspartic and glutamic residues are fully dissociated. The more difficult assignments of the protonation state of amino termini and histidines depend on these environmental parameters and are estimated by calculations. A simple and quick estimate can be obtained by performing a Tanford–Kirkwood calculation (Matthew, 1985; Shire et al., 1974; Tanford and Kirkwood, 1957; Tanford and Roxby, 1972). Tanford and Kirkwood treat the electrostatic free energy for a set of discrete point charges on a spherical surface, inside of which is a continuum low dielectric region. A thin layer 1–2 Å outside this sphere is included as the electrolyte ion exclusion radius. LaPlace's equation is applied within the ion exclusion region where no diffusible ions exist, while the Poisson–Boltzmann equation is applied outside with the same approximations as the Debye–Hückel treatment. A static-accessibility modification is also included, an empirical modification which accounts for the degree of surface exposure of titratable residues and is included to reproduce experimental data. Tanford and Kirkwood obtain an analytical solution for the electrostatic interaction energies for pairwise interactions of charge sites. Such calculations can be used easily to compute titration curves for proteins and fractional charge on each titratable site.

More sophisticated methods of calculation of pK_as in proteins are now available (Bashford and Karplus, 1990; Yang et al., 1993) based on the numerical solution of the finite difference Poisson–Boltzmann equation on a cubic grid (see next paragraph). These methods make use of detailed structural information and treat solvation self-energies of titratable groups at their point of burial within the low dielectric region of the protein and also account for interactions arising from permanent partial charges and titratable charges.

Now let us consider the treatment of the interprotein electrostatic force, which has also been accomplished by the numerical solution of the discretized Poisson–Boltzmann equation on a cubic lattice. This equation has the most general non-linear form as follows, in which the dielectric constant ε, electrolyte ion charge density and fixed protein charge density ρ are spatially varying:

$$\nabla.[\varepsilon(r)\nabla\phi(r)] - \bar{\kappa}^2(r)\sinh[\phi(r)] + 4\pi\rho(r) = 0 \qquad (4.1)$$

Here, $\bar{\kappa}$ is the modified Debye–Hückel screening parameter, $\bar{\kappa} = \varepsilon^{1/2}\kappa$, and ϕ is the electrostatic potential. The linearized form of this equation is often used for convenience, under conditions of low ionic strength and low charge density:

$$\nabla.[\varepsilon(r)\nabla\phi(r)] - \bar{\kappa}^2(r)\phi(r) + 4\pi\rho(r) = 0 \qquad (4.2)$$

These equations can be solved analytically only for very simple geometries, but must be approximated numerically in discretized form for structurally realistic models of proteins and complexes. The use of the numerical Poisson–Boltzmann equation allows one to account for (i) the atomic scale irregularity of the protein surfaces, (ii) the field-dampening influence of diffusible electrolytes in solvent medium, and (iii) the existence of a low dielectric region inside proteins.

For Brownian simulations, one of the proteins, typically the larger, is chosen as the central protein, designated Protein I. The electrostatic potential field surrounding Protein I is computed by iterating finite-difference solutions of the Poisson–Boltzmann equation on a cubic grid by the method originally proposed by Warwicker and Watson (1982) as adapted subsequently by numerous other workers (see, for example, Holst and Saied, 1995). Protein I is represented as an irregularly shaped cavity of low dielectric constant (typically $\varepsilon = 2$ or 4) and zero internal ionic strength and having fixed imbedded charges in the crystallographic configuration. Surrounding this cavity is a continuum dielectric with $\varepsilon \sim 80$ representative of bulk water and having appropriate ionic strengths. A 'focusing' method is used in which first a solution is iterated on an 'outer' coarse-grained lattice with the boundary potential set to the Coulomb/Debye result. Having obtained a solution on an outer coarse grid, an additional calculation is performed on subsequently refined grids of higher resolution.

To calculate the direct interprotein electrostatic force and torque in BD docking simulations, the second protein (Protein II) is treated as an array of test charges in the field of Protein I. A simplification is made by ignoring the low dielectric in the Protein II interior as well as allowing the electrolyte screening effect to operate throughout its interior. Thus, the direct force between the two proteins, and the torque operating on Protein II, are determined at each dynamic simulation time step by placing the Protein II array of charges into the field around the Protein I cavity, consulting the stored grid of forces, and performing a summation over all Protein II charges. A more self-consistent electrostatic treatment has been developed by Zhou (1993) as described in Section 4.3.5.

The above procedure treats translation and rotation of Protein II in a field of force and torque generated by Protein I. The rotation of Protein I in the field of Protein II may be included by generating an inner and outer force lattice around Protein II. These lattices rotate in rigid body rotation with Protein II. A limited set (e.g. dipolar pair) of charges are included on Protein I to serve as test charges which interact with the field around Protein II, and are used to compute the approximate torque on Protein I in the field of Protein II. This feature of two rotating proteins is essential in treating protein pairs of similar size.

4.2.3 *Brownian dynamics equation of motion*

The molecular dynamics method has received the greatest attention in computational modelling of proteins. The internal dynamics of a solvated protein can be simulated conveniently over tens of picoseconds by this method. However, to treat the diffusional dynamics and interactions between two whole proteins in solution over nanoseconds of time, a more coarse-grained dynamic approach has been devised called BD (Allison *et al.*, 1985, 1986; Ermak and McCammon 1978; Ganti *et al.*, 1985; Northrup *et al.*, 1984, 1986a,b,c). This theoretical technique is capable of generating a representative set of dynamic trajectories of bimolecular diffusion and association of two proteins over time scales which include full exploration of mutual rotational degrees of freedom of the molecules. Utilizing stored potential energy grids, the complicated electrostatic force and torque field arising from the actual charge distribution on proteins can be treated. Excluded volume interactions are also treated by a stored spatial occupation grid which accounts for the full atomic resolution irregular surface topography of the molecules. Electron transfer reactivity criteria at varying levels of complexity are easily incorporated into the model.

The BD method has been applied to the study of the association and electron transfer in a variety of systems: horse heart cytochrome c (Fe^{2+}) and yeast CCP (Northrup et al., 1987a,b, 1988), horse heart cytochrome c (Fe^{3+}) and bovine cytochrome b_5(Fe^{2+}) (Eltis et al., 1991), CY and bovine cytochrome b_5 (Northrup et al., 1993), self-exchange of Pseudomonas aeruginosa cytochrome c-551 (Herbert and Northrup, 1989) and self-exchange in cytochrome c and cytochrome b_5 (Andrew et al., 1993).

The BD method treats the diffusive motion of interacting proteins in solution by generating trajectories composed of a series of small displacements obeying the diffusion equation with forces. The algorithm for free displacements Δr in the relative separation vector r of protein centres of mass in time step Δt is (Ermak and McCammon, 1978)

$$\Delta r = DF(k_B T)^{-1} \Delta t + S \qquad (4.3)$$

Here, D is the relative translational diffusion coefficient, F is the systematic interparticle coulombic force, $k_B T$ is the Boltzmann constant times absolute temperature, and S is the stochastic component of the displacement arising from collisions with solvent molecules. This is the stochastic or random component, which is generated by taking normally distributed random numbers obeying the average relationship $<S^2> = 2D\Delta t$. A similar equation governs the independent rotational Brownian motion of each particle, where force is replaced by torque, and D is replaced by an isotropic rotational diffusion coefficient D_{ri} for each particle i. Proteins are treated as diffusing rigid bodies with no internal dynamics, one of the fundamental drawbacks of this method which must be remedied in future developments.

4.2.4 *Extraction of rate constants*

In order to forge the important link between Brownian simulations and experimental data, a scheme has been devised which enables the computation of bimolecular rate constants from trajectory fate statistics (Northrup et al., 1984). This is depicted schematically in *Figure 4.2*. The diffusion problem is broken into two pieces conveniently by defining a surface at protein–protein separation $r = b$ taken to be just outside the region of strong asymmetric electric fields. Trajectories of diffusing proteins are initiated with random orientation at this surface, and are truncated at an outer spherical surface $r = c$ (typically 200 Å). Values chosen for b are typically 60 Å. This separates the problem into the analytically tractable one of centrosymmetric diffusion to a starting surface b from outside, followed by diffusion in a complicated force field inside b, described by BD. Several thousand trajectories are typically monitored to obtain the probability p of association of pairs in favourable geometries for reaction

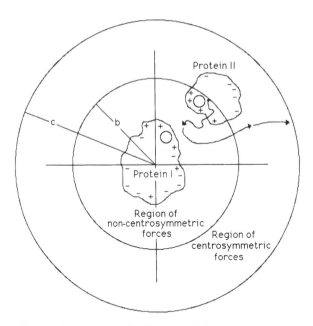

Figure 4.2. Schematic diagram of the BD separation of space into regions for trajectory simulation and extraction of rate constants. Spherical surfaces b and c are starting and truncating surfaces for trajectories, respectively.

prior to ultimate separation to distance $r = c$. The diffusion-controlled bimolecular reaction rate constant k may then be obtained from p through Equation 4.4:

$$k = k_D(b) \frac{p}{1 - (1 - p)k_D(b)/k_D(c)} \tag{4.4}$$

Here, $k_D(s)$ is the diffusive rate constant for first arrival at an arbitrary spherical surface s given by the Smoluchowski–Debye (Debye, 1942; Smoluchowski, 1917) expression when centrosymmetric forces apply in the region $r > s$

$$k_D(s) = \frac{4\pi N_A}{\int_s^\infty dr \left[D(r)r^2\right]^{-1} e^{u(r)/k_B T}} \tag{4.5}$$

where s is a starting or truncation surface radius, $u(r)$ is a centrosymmetric potential energy of interaction between the diffusing pair, N_A is Avogadro's number, and $D(r)$ is the relative translational diffusion coefficient, which could be, in principle, spatially varying at short range owing to hydrodynamic forces, but is generally taken to be a constant.

A large set of reaction boundary conditions can be considered simultaneously in a single simulation by monitoring the survival of the trajectory with respect to each of these criteria. (Northrup et al., 1986a).

Trajectories are not truncated when any given reactive criterion is met, but are continued until outer surface c is reached. With each trajectory, a parallel set of survival probabilities are monitored which correspond to a parallel set of reaction criteria. Multistaging techniques have been developed (Allison *et al.*, 1985) to increase the efficiency of obtaining rate constants. These have been applied to problems that involve complicated short-ranged interactions between reactant and target, where the probability for a reactant to diffuse to a surface relatively close to the target is calculated, and then the probability of diffusing from this intermediate surface to the reaction site is computed (Allison *et al.*, 1988).

4.2.5 *Treatment of excluded volume effects*

Realistic atomic resolution models of the electrostatic fields and the irregular shapes of the proteins have been developed for use in BD simulations (Allison *et al.*, 1988; Ganti *et al.*, 1985; Northrup *et al.*, 1987a,b, 1988; Reynolds *et al.*, 1990; Sharp *et al.*, 1987). The excluded volume effect is treated by developing and storing for later use in the BD simulation program a cubic spatial exclusion grid of 1.0 Å resolution centred on Protein I which is used to define its excluded volume. To generate this grid, a probe atom is placed at grid points, and the distance from the probe to each Protein I atom is determined. If the probe falls within 2.7 Å of any atom, that grid cell is defined to be an excluded region. The probe represents any atom of Protein II, the shape of which is treated by selecting a limited number of surface atoms which approximately represent its shape and which will diffuse by rigid body motion and be tested for penetration with the Protein I exclusion grid.

4.2.6 *Reaction criteria*

Reaction criterion in the cytochrome studies have fallen into two categories. The first are based on the haem–edge distance d_{edge}, the minimum distance between porphyrin atoms on the two proteins. This value is compared with a series of possible cut-off distances which define whether reaction has occurred. This cut-off choice varies from system to system, and remains a somewhat arbitrary parameter of the model. An additional parameter of the reaction criterion is the angle ψ, which is the angle between haem plane normal vectors. This is included to determine whether haems must be co-parallel for efficient electron transfer.

A more advanced reaction criterion has also been explored which embodies the intrinsic electron transfer event itself, and allows a direct dynamic coupling between diffusion and reaction. This further allows treatment of reactions that are far from the diffusion-controlled limit. This is the exponential reactivity model, which has been employed in the study of self-exchange in cytochrome c and b_5 (Andrew *et al.*, 1993) and in the cytochrome c–b_5 reaction (Eltis *et al.*, 1991; Northrup *et al.*, 1993). This ideally provides a more physically realistic model of the electron transfer event in which the intrinsic spatially dependent electron transfer rate constant $k_{et}(r_{Fe})$ is given as a function of inter-iron distance by

$$k_{et}(r_{Fe}) = k_{et}^{o} \exp(-\beta(r_{Fe} - r_o)) \tag{4.6}$$

This formula is reminiscent of the theory of Siders *et al.* (1984) describing orientation and distance dependence of electron transfer between porphyrins. Pre-exponential factor k_{et}^o represents the unimolecular rate constant when porphyrins are in direct edge-on contact, and can be taken to be about 10^{11} s^{-1} (McLendon, 1988). In the CY–cytochrome b_5 mutant study and in the cytochrome c and cytochrome b_5 self-exchange studies described in Sections 4.3.6 and 4.3.7, this parameter was calibrated to give agreement with experiment. The minimum inter-iron distance of two porphyrins in an edge-on arrangement is $r_o = 11.7$ Å. The electron transfer distance-dependent factor β is a quantity of intense interest (see, for example, Farid *et al.*, 1993) and is an adjustable parameter which can be varied to fit the experimental data. Thus, this important parameter can be estimated from BD. The electron transfer unimolecular rate constant k_{et} can be folded into BD by the following procedure. The spatially dependent probability P that the reactant pair survive an individual Brownian step Δt without reaction is $P = \exp(-k_{et}\Delta t)$. The product of this factor over all the steps of the trajectory gives the probability of non-reaction for that trajectory.

4.3   Results

4.3.1   *Cytochrome reaction rate constants with distance cut-off model*

Figure 4.3 shows the reproduction of the experimental ionic strength dependence of rate constants obtained in BD simulations with the simple distance cut-off model of reactivity of the cytochrome $c(Fe^{2+})$–CCP reaction (Northrup and Herbert, 1990). The experimental ionic strength dependence is steeper than any of the theoretical models. In *Figure 4.4* we see that for the cytochrome $c(Fe^{3+})$–cytochrome $b_5(Fe^{2+})$ pair, the BD theory gives a better representation of the ionic strength dependence. A very weak and positive ionic strength dependence is observed experimentally for the *P. aeruginosa* cytochrome c-551 self-exchange reaction, which was also predicted by BD simulations (see *Table 4.1*).

The effects of various choices of reaction parameters on the simulated rate constants have also been explored. Variations in the required haem plane orientation ψ, which is a parameter of the distance cut-off reaction model, have been studied. For similar choices of ψ [30° in the cytochrome $c(Fe^{3+})$–cytochrome $b_5(Fe^{2+})$ reaction, and 40° in the cytochrome $c(Fe^{2+})$–CCP and c-551 self-exchange reaction], the appropriate magnitudes of the experimental values are reproduced as well as the slope of the ionic strength plot. Ionic strength dependence agreement is especially excellent in the cytochrome $c(Fe^{3+})$–cytochrome $b_5(Fe^{2+})$ reaction.

The choice for the required haem–haem edge distance has always been a problematic issue in these simulations, and remains as a somewhat arbitrary parameter. However, it has been observed generally that the best comparison with experiment is always obtained by a choice which is about 2 Å larger than the minimum possible distance allowed by excluded volume interactions. Inside this distance, the available volume of orientation space rapidly vanishes. By allowing a 2 Å tolerance, one guarantees having proteins that have been brought to a point inside which strong hydrophobic forces, hydrogen bond contacts,

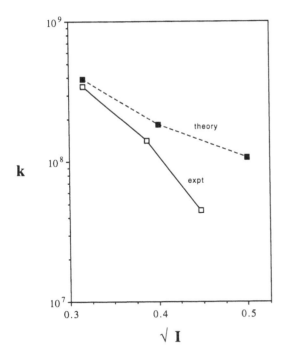

k

√ I

Figure 4.3. Bimolecular electron transfer rate constants (in M^{-1} s^{-1}) versus $\sqrt{I}$ for the reaction cytochrome $c(Fe^{2+})$–CCP at pH 6.0 and 298 K. BD results (Northrup *et al.*, 1987a) with cut-off distance d_{edge} = 20 Å and ψ = 40°(■) are compared with experiment (Kang *et al.*, 1978) (□).

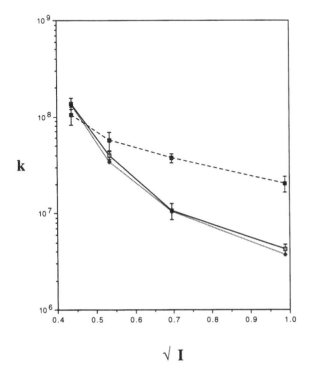

k

√ I

Figure 4.4. Bimolecular electron transfer rate constants (in M^{-1} s^{-1}) versus $\sqrt{I}$ for the reactions cytochrome $c(Fe^{3+})$–cytochrome b_5 at pH 7.0 and 298 K. BD results (Northrup *et al.*, 1993) with exponential reactivity criterion using β = 1.2 $Å^{-1}$(······); and also with a simple uniform dielectric model (- - -) are compared with experiment of Eltis *et al.*, (1991) (———). Reprinted with permission from Northrup *et al.* (1993). Effects of charged amino acid mutations on the bimolecular kinetics of reduction of yeast iso-1-ferricytochrome *c* by bovine ferrocytochrome b_5. *Biochemistry* **32**: 6613–6623. Copyright 1993 American Chemical Society.

Table 4.1. Bimolecular rate constants (M^{-1} s^{-1}) for the self-exchange reaction of *Pseudomonas aeruginosa* cytochrome c-551 at two ionic strengths

	$I = 0.1$ M	$I = 0.6$ M
Experiment[a]	1.2×10^7	2.0×10^7
Smoluchowski–Debye isotropically reacting monopolar charged spheres[b]	3.2×10^9	4.7×10^9
BD[c] distance cut-off model with $d_{edge} = 8$ Å and $\psi = 40°$	1.3×10^7	1.6×10^7

pH = 7.0; $T = 313$ K.
[a] Timkovich *et al.* (1988); [b] Debye (1942), Smoluchowski (1917); [c] Northrup and Herbert (1990).

flexibility of the surfaces and expulsion of water are likely to begin to operate to create a more intimate contact capable of electron transfer. An underlying assumption required to validate BD is that proteins brought to this point will be drawn together inexorably by the assistance of neglected hydrophobic forces and a relaxation of the interfacial contacts.

4.3.2 *Protein–protein docking in the cytochrome c–CCP reaction*

The cytochrome $c(Fe^{2+})$–CCP pair successfully satisfy realistic geometric criteria for electron transfer from a large ensemble of electrostatically stable encounter complexes rather than from a single optimal protein–protein complex (Northrup *et al.*, 1987b, 1988). This multitude of potential complexes occurs in three distinct domains on CCP in approximately the haem plane as shown in *Figure 4.5*. These coincide with the electrostatically attractive regions observed in that plane, as seen in *Figure 4.6*, which is an interprotein potential energy contour map around CCP in its haem plane as a function of the centre of mass of the incoming cytochrome c. At each relative position of the centres of mass of the two proteins, a Boltzmann average by Monte Carlo sampling was performed of the potential energy over all accessible rotational orientations of the incoming cytochrome c protein. An extensive electrostatic channel of a depth of 1–2 $k_B T$ spans the three docking regions. The deepest part of the channel matches the region around Asp34 of CCP hypothesized in molecular graphics model building (Poulos and Kraut, 1980). The second region is substantially removed from the first region and is centred around Asp148. A third less populated region lies intermediate between these two dominant areas near Asp217. This channel provides a lower-dimensional search conduit by which the incoming translating and rotating cytochrome c may engage in an extended exploration of the surface of CCP in search of a productive electron transfer configuration. The typical encounter duration of these proteins is quite prolonged (10^2 ns), such that cytochrome c can undergo numerous rotational reorientations (rotational relaxation time 6 ns) as it explores the CCP surface (Northrup *et al.*, 1988). It is important to note that the encounter time of 100 ns estimated by BD only takes into account the loosely docked complexes which are able to form, and does not provide for formation of more long-lived

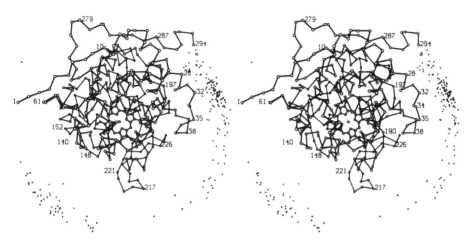

Figure 4.5. The α-carbon skeleton of CCP surrounded by centres of mass of incoming cytochrome $c(Fe^{2+})$ at instants during Brownian simulation when haem–edge distances are sufficiently small for reaction (d_{edge} = 20 Å). Accordingly, electron transfer may occur from a wide range of docking geometries rather than from a single dominant encounter complex. Reprinted with permission from Northrup *et al.* (1988) Brownian dynamics of cytochrome c and cytochrome c peroxidase association. *Science* **241:** 67–70. Copyright 1988 American Association for the Advancement of Science.

tightly bound ones that would result from relaxation of the surface groups and inclusion of short-ranged hydrophobic forces. For this reason, the BD encounter time does not provide a good estimate of the k_{off} rate constant, which could be reasonably estimated as $1/k_{off}$ ~100 µs.

The frequencies of close ionic contacts were analysed in 243 docked cytochrome c–CCP complexes with favourable geometric criteria for electron transfer (Northrup *et al.*, 1987b). Lysines 13, 25, 27, 72, 79, 86 and 87 were found to be most frequently involved in favourable electrostatic interactions with CCP, with lysines 7, 8 and 73 also making favourable contacts but to a much lesser extent. Negatively charged Asp50, also located on the front of cytochrome c, makes frequent contact but of a destabilizing nature, most often with the negatively charged CCP residues Glu35 and Glu290. Similarly, negatively charged cytochrome c residue Glu90 makes frequent destabilizing contacts, most regularly with Glu35 on CCP.

The triads of ionic contacts that occur most frequently in these same 243 complexes were analysed also. The frequency distribution of triads is given in *Table 4.2*, and implies a non-specific coulombic binding process in which a predominantly positive surface of cytochrome c interacts with several predominantly negative surfaces of CCP. *Table 4.2* clearly shows that there is no dominant complex which forms, but a surprisingly even distribution over a variety of complexes. The most frequent complex which forms is the triad in which CCP residues Asp34/Glu35, Glu209 and Glu290 contact cytochrome c Lys79, Lys13 and Lys25/27, respectively.

These contact analysis studies of complexed states revealed that electrostatic stabilities typically arise from the interaction of four ion pairs with

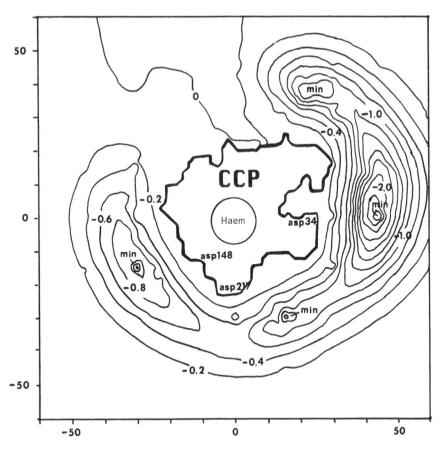

Figure 4.6. A haem-plane cross-section of the Boltzmann-averaged total electrostatic potential energy of interaction between CCP and cytochrome c in units of $k_B T$ as a function of the centre of mass position of cytochrome c. The view is in the same orientational perspective as *Figure 4.5*. Monte Carlo sampling has been used to average the potential over all sterically accessible orientations of cytochrome c at each centre of mass position. Reprinted with permission from Northrup *et al.* (1988) Brownian dynamics of cytochrome c and cytochrome c peroxidase association. *Science* **241:** 67–70. Copyright 1988 American Association for the Advancement of Science.

varying degrees of complementarity. This supports a possible explanation of the ionic strength dependence of the rate constant as shown in *Figure 4.3*. The ionic strength dependence is much weaker than what one would expect based on the large net monopole on these proteins (−12/+8) and reflects the interaction of approximately four elementary charges of opposite sign. This is also borne out in the contact study, and lends support to the hypothesis that it is the net electrostatic charge in the neighbourhood of the interaction domain and not the global electrostatic properties of the proteins that primarily dictates the ionic strength dependence of the association rate. This was also the finding of the Brownian simulation study of Zhou (1993) on the same system.

Table 4.2. The most frequent occurrences of triads of contacts in docked complexes of cytochrome c and CCP

CCP residue	Cytochrome c residue	Frequency
Asp34/Glu35	Lys79	
Glu209	Lys13	11
Glu290	Lys25/Lys27	
Asp34/Glu35	Lys13	
Glu188	Lys25/Lys27	8
Asp217	Lys25/Lys27	
Glu201	Lys79	
Glu209	Lys79	8
Glu290	Lys72	
Asp34/Glu35	Lys13	
Glu201	Lys79	7
Glu209	Lys79	
Asp34/Glu35	Lys79	
Glu201	Lys25/Lys27	7
Glu290	Lys25/Lys27	
Asp34/Glu35	Lys13	
Glu201	Lys79	6
Glu290	Lys72	
Asp34/Glu35	Lys13	
Asp34/Glu35	Lys86	6
Glu201	Asp79	

Reprinted with permission from Northrup, S.H., Luton, J.A., Boles, J.O. and Reynolds, J.C.L., *J. Comput.-Aided Mol. Design*, **1**: (1987) 291–311. Copyright 1988 ESCOM Science Publishers B.V.

4.3.3 *CDNP-derivatized cytochrome* c *reactions with CCP*

To map out the binding domain of cytochrome c, kinetic experiments have been performed (Kang *et al.*, 1978) in which various carboxydinitrophenyl (CDNP) derivatives of horse cytochrome c were reacted with yeast CCP. Lys8, 13, 25, 27, 60, 72, 73, 86 and 87 were derivatized by modifying the lysyl side chain. These modified proteins have structural and redox properties which are unchanged from the native horse heart cytochrome c. Brownian simulations of the association rate (Northrup *et al.*, 1987b) of CCP reacting with this series of CDNP-modified cytochrome c proteins have been performed, including additional simulations of modified residues Lys7, 22, 39 and 79. Additionally the $-1e$ charge of Asp50 has been switched to $+1e$ to confirm its negative influence on the association process, as deduced from the contact analysis. CDNP-derivatized lysines were approximated simply by changing the charge on the lysine nitrogen from $+1e$ to $-1e$ without changing its location, thus ignoring the steric effect of the chemical modification. Results at ionic strength 0.1 M are tabulated in *Table 4.3* in terms of both association rate constants and relative activity, which is defined as the ratio of the rate constant of the modified protein to that of the native protein.

The lowest relative activities (meaning largest negative influences in the docking rate) were obtained experimentally when derivatizing lysines surrounding the haem crevice of cytochrome c, including Lys72, 86, 87, 27 and 13. Only moderate effects were observed for modifications of Lys73, 25 and 8, while no perturbation was observed for Lys60, which is on the side of cytochrome c opposite the haem edge. These effects were reproduced qualitatively in the simulations, as shown in *Table 4.3*. Since Asp50 is a prominent negative residue in the region of the haem edge and is anticipated to have a negative influence on complexation as seen in the contact study, an additional simulation was performed in which the charge of Asp50 on cytochrome c was changed from $-1e$ to $+1e$, while the charge location was not changed. The simulated rate doubled from the native simulations. This suggests a strong incentive to explore new site modifications to alter a protein's effectiveness to form complexes with other proteins, and reveals the large role that specific electrostatic contacts play in rates of electron transfer reactions.

4.3.4 *Extension to surface side chain flexibility*

Limited side chain flexibility has been incorporated into Brownian simulations of the diffusional association of the cytochrome c–CCP reaction pair (Northrup and Herbert, 1990). To include side chain mobility, each molecule is partitioned into a rigid and dynamic portion. The dynamic portion is a set of surface-charged side chains on each molecule expected to have a high degree of mobility most likely to influence docking, typically including a number of the lysine, arginine, glutamate and aspartate side chains. Motional freedom of the side chain atoms is treated in the following fashion. The time scale separation between torsional motion within surface side chains and relative diffusional motion is exploited. At each time step ($\sim$10 ps) a Monte Carlo algorithm is used

Table 4.3. Chemical modification studies of cytochrome c reaction with CCP at 0.1 M ionic strength

Modification	BD relative activity	Experimental relative activity
Lys27 → −1e	0.38	0.24
Lys79 → −1e	0.41	–
Lys72 → −1e	0.52	0.13
Lys86 → −1e	0.54	0.16
Lys13 → −1e	0.59	0.24
Lys25 → −1e	0.64	0.57
Lys87 → −1e	0.70	0.19
Lys 8 → −1e	0.78	0.63
Lys73 → −1e	0.81	0.33
Lys60 → −1e	0.82	1.00
Lys 7 → −1e	0.91	–
Native horse cytochrome c	1.00	1.00
Lys22 → −1e	1.01	–
Lys39 → −1e	1.11	–
Asp50 → +1e	2.00	–

Comparison is made with experimental results (Kang *et al.*, 1978). BD results (Northrup *et al.*, 1987b) are shown for a haem plane criterion of $\psi = 60°$. Relative activity is defined as the ratio of the rate constant of the modified species to the native.

to choose a new torsional conformation of mobile side chains from an equilibrium distribution of conformers subject to the instantaneous electrostatic potential field at the side chain charge site and a sinusoidal torsion potential energy term. Every torsion angle of each side chain is given the opportunity to take a random step, subject to excluded volume interactions and, when the new trial conformation is thus generated, the standard Monte Carlo energy criterion is used to accept or reject the step.

In preliminary tests, rate constants with flexibility were essentially unchanged compared with the rigid study on this system (Northrup *et al.*, 1987a), indicating that the more computationally convenient rigid body model gives reasonable results, at least for association rate constants in this system. Even when only a small effect is registered in the rate constant, the inclusion of flexibility of charged surface groups is expected to modify the interaction potentials and distribution of the important ionic contacts in a meaningful way. More intimate encounters are likely to occur between proteins because of the increased opportunity for interdigitation of surface groups.

4.3.5 *Treatment of the cytochrome c–CCP reaction with self-consistent electrostatics*

One of the major drawbacks of the topologically sophisticated models of protein–protein BD employed by Northrup and co-workers has been that the two proteins are not treated on an equal footing from an electrostatic standpoint. For instance, Protein II is treated as a collection of test charges diffusing in the high dielectric surroundings of Protein I rather itself being a low dielectric region also. This imbalance affects the interaction potential significantly, as was demonstrated in BD simulations by Zhou (1993). By treating the proteins as simple diffusing spheres bearing charges and extending the Tanford–Kirkwood model such that both spheres are low dielectric cavities, Zhou was able to model the ionic strength dependence accurately. Additionally he was able to explain mathematically why mutations in the protein–protein interface have strong influences on the association rate while those far removed from the interface have minimal effects.

4.3.6 *Self-exchange reactions using the exponential reactivity model*

The first studies with rigorous BD employing the exponential reactivity model were the self-exchange reactions of cytochrome c and b_5 (Andrew *et al.*, 1993). Since the parameters β and k_{et}^o of the exponential reactivity model are unknown, the combination giving the best fit to experiment over a range of ionic strength was determined. *Figure 4.7* depicts the comparison of theory and experiment over a range of ionic strengths using this choice. As expected for self-exchange reactions, where like-charged surfaces must dock, the rate increases with ionic strength as the dielectric screening increasingly shields the electrostatic repulsion. The best fit is obtained using the reactivity parameters $k_{et}^o = 3.43 \times 10^8$ s^{-1} and $\beta = 0.9$ Å^{-1} (cytochrome b_5) and $k_{et}^o = 1.18 \times 10^{10}$ s^{-1} and $\beta = 0.9$ Å^{-1} (cytochrome c). Based upon the Marcus equation and optimum values of k_{et}^o, an estimate of the reorganization energy $\lambda = 1.06$ eV for cytochrome b_5 and 0.69 eV for cytochrome c was extracted. These values compare well with

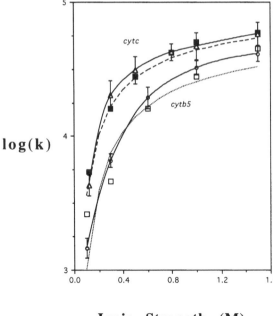

Ionic Strength (M)

Figure 4.7. Ionic strength dependence of the bimolecular rate constants k for self-exchange reactions of cytochrome b_5 and c. Experimental values (Dixon *et al.*, 1989, 1990) for cytochrome b_5 ($\square$) and c ($\blacksquare$) are compared with rigorous BD-simulated values (Andrew *et al.*, 1993) for cytochrome b_5 (O) and c (Δ) and van Leeuwen theory for cytochrome b_5 (*cytb5*)(.....) and c (*cytc*) (- - -). Best-fit parameters for the BD and van Leeuwen theories are given in the text and in Dixon *et al.* (1989, 1990), respectively. Reprinted with permission from Andrew *et al.* (1993) Simulation of electron transfer self-exchange in cytochrome c and b_5. *J. Am. Chem. Soc.* **115:** 5516–5521. Copyright 1993 American Chemical Society.

estimates of 1.2 and 0.7 eV made by Dixon *et al.* (1989, 1990) but are obtained with fewer estimated parameters, since the diffusion–collision part of the problem is treated explicitly by the BD theory and steric factors such as association constant and haem exposure are included explicitly. *Figure 4.7* also gives the best-fit theoretical curves of the van Leeuwen (1983) theory, with parameters specified in Dixon *et al.* (1989, 1990). The van Leeuwen theory predicts a comparable ionic strength dependence but with a greater number of independent parameters to be specified, including the net monopole charges of the proteins, the components of the dipole moments throughout the exposed haem edge, the sum of spherical radii of the two partners and the rate constant at infinite ionic strength.

4.3.7 *Site-directed mutants of CY: reactions with cytochrome* b_5

The reduction of wild-type CY and several of its mutants by trypsin-solubilized bovine liver ferrocytochrome b_5 has been studied under conditions in which the electron transfer reaction is bimolecular (Northrup *et al.*, 1993). The effect of electrostatic charge modifications and steric changes on the kinetics has been determined by experimental and theoretical observations of the electron transfer rates of CY mutants K79A (Lys79→Ala79), K'72A (Tml72→Ala72) K79A/K'72A (Lys79,Tml72→Ala79,Ala72) and R38A (Arg38→Ala38). Here Tml stands for the amino acid trimethyllysine. This study represents the first investigation of the effects of cytochrome c site-directed mutations on the bimolecular kinetics of electron transfer with another electron transfer protein. As the mutational effects are subtle in nature, detailed mechanistic interpretation of these results is virtually impossible without the assistance of relatively sophisticated modelling procedures. For this reason, this study has employed the most advanced and rigorous BD simulation procedure available for the calculation of diffusional docking trajectories between two whole translating and rotating proteins, incorporating the intrinsic electron transfer unimolecular rate constant which varies exponentially with haem–haem distance. By comparison with bimolecular rate constants obtained experimentally for reaction of the wild-type and mutant cytochromes, it was demonstrated that the BD simulation method is able to predict bimolecular rate constants quantitatively for the electron transfer reaction of these cytochromes over a considerable range of ionic strengths. Quantitative agreement was made possible only by using the exponential electron transfer model and by suitable adjustment of reactivity parameters. These parameters include the distance decay parameter β, which was determined theoretically for this reaction pair to have a value of 1.0 Å^{-1}, a value in reasonable agreement with estimates made for these types of systems.

In order to obtain truly quantitative predictions of mutant influences on the rates, it was found that one must account for the perturbing influence of the mutations on the redox potentials as well as on the electrostatics and dynamics of docking. Using appropriately varying potentials E^{o} and the Marcus equation assuming a fixed reorganization energy of 0.7 eV, the variation in the intrinsic electron transfer rate constant factor k_{et}^{o} was estimated. BD simulations were performed with these varying values of k_{et}^{o} (see *Table 4.4*) for the different mutants. Resulting bimolecular rate constants descend in the following fashion: native CY>K79A>K'72A>K79A/K'72A in both theory and experiment (see *Table 4.5*). This exemplifies the substantial predictive power of these theoretical simulations.

The proteins dock through essentially a single domain, in contrast to the cytochrome c–CCP study, with a distance of closest approach of the two haem groups in rigid body docking around 12 Å. *Figure 4.8* shows the docking profile for this reaction. Two predominant classes of complexes were observed to dock through this domain, the most frequent involving the quartet of ferrocytochrome b_5–CY interactions Glu48–Lys13, Glu56–Lys87, Asp60–Lys86 and haem–Tml72, having an average electrostatic energy of −13.0 kcal mol^{-1} (−54.4 kJ mol^{-1}). The second most important type of complexes were of the type previously postulated (Rodgers *et al.*, 1988; Salemme, 1976) with interactions Glu44–Lys27, Glu48–Lys13, Asp60–Tml72 and haem–Lys79, and having an energy of −6.4 kcal mol^{-1} (−26.8 kJ mol^{-1}). These complexes are shown in *Figure 4.9*. These complexes have been subjected to further energy refinement in subsequent studies, which we discuss in the next section.

Table 4.4. Net charge, dipole moments, and redox potential of cytochrome b_5 and various CY mutants at pH = 7.0 and I = 0.19 M

Species	Net charge (e)	Dipole magnitude (Debye)	γ	$\Delta(\Delta G^o_{el})$ (kcal mol^{-1})	E^o (mV)
Cytochrome b_5	−8.15	597	144°		
Native CY	7.31	485	78°	0.000	290
CY K79A	6.35	506	86°	−0.046	292
CY K'72A	6.32	467	85°	+0.138	284
CY K79A/K'72A	5.36	497	93°	+0.046	288
CY R38A	6.39	524	80°	+1.176	239

Reprinted with permission from Northrup *et al.* (1993) Effects of charged amino acid mutations on the bimolecular kinetics of reduction of yeast iso-1-ferricytochrome *c* by bovine ferrocytochrome b_5. *Biochemistry* **32**: 6613–6623. Copyright 1993 American Chemical Society.
γ = angle of dipole moment relative to Fe position vector in centre of mass frame.
$\Delta(\Delta G^o_{el})$ = change in free energy of reduction in kcal mol^{-1} relative to native CY derived from observed reduction potentials shown in the last column.
E^o = potential of the reduction half-reaction of wild-type or mutant CY determined from potentiometric titrations (A.G. Mauk, personal communication).

4.3.8 *Energy refinement and analysis of Brownian dynamics-generated complexes*

Docked complexes of cytochrome $c(Fe^{3+})$–cytochrome $b_5(Fe^{2+})$ generated by Brownian simulation (Northrup *et al.*, 1993) have been analysed extensively in a subsequent energy refinement study (Guillemette *et al.*, 1994). These proteins

Table 4.5. Theoretical and experimental bimolecular rate constants for the oxidation of cytochrome b_5 by CY and four mutants

Species	Experimental	BD, distance cut-off model[a]	BD[b], expo. model, fixed k^o_{et}	BD[c], expo. model driving force corrected k^o_{et}	k^o_{et} (ps^{-1})
Native CY	136 ± 3	172 ± 63	136 ± 17	136 ± 17	0.130
CY K79A	67.82 ± 3.02	62 ± 40	110 ± 10	112 ± 9	0.133
CY K'72A	37.21 ± 3.11	55 ± 15	63 ± 3	59 ± 6	0.121
CY K79A/K'72A	30.14 ± 1.71	15 ± 23	47 ± 6	46 ± 6	0.127

Reprinted with permission from Northrup *et al.* (1993) Effects of charged amino acid mutations on the bimolecular kinetics of reduction of yeast iso-1-ferricytochrome *c* by bovine ferrocytochrome b_5. *Biochemistry* **32**: 6613–6623. Copyright 1993 American Chemical Society. All rates given in 10^6 M^{-1} s^{-1} units.
[a] Reaction criterion = distance cut-off model with d_{edge} = 12 Å, ψ = 30°.
[b] Reaction criterion = exponential reactivity model with β = 1.0 Å^{-1}, fixed k^o_{et} = 1.3 x 10^{11} s^{-1} (no mutant perturbation of redox potential).
[c] Reaction criterion = exponential reactivity model with β = 1.0 Å^{-1}, k^o_{et} varying according to Marcus equation (mutant perturbation of redox potential).
pH = 7.0; T = 25°C; ionic strength = 0.19 M.

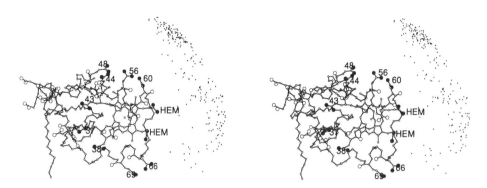

Figure 4.8. Docking profile showing the target protein cytochrome b_5 surrounded by dots representing centres of mass of incoming CY molecules in reactive docked complexes (T = 298 K; pH = 7.0). Reprinted with permission from Northrup *et al.* (1993) Effects of charged amino acid mutations on the bimolecular kinetics of reduction of yeast iso-1-ferricytochrome *c* by bovine ferrocytochrome b_5. *Biochemistry* **32:** 6613–6623. Copyright 1993 American Chemical Society.

associate for electron transfer through a single domain, but not a single conformation, with a distance of closest approach of the two haem groups in rigid body docking typically around 12 Å. Within this domain are found basically two types of mutual orientations of the proteins which could contribute to electron transfer (*Figure 4.9*). One of these is very similar to the Salemme complex with interactions Glu44–Lys27, Glu48–Lys13, Asp60–Tml72 and haem–Lys79, and having an electrostatic binding energy of –6.4 kcal mol^{-1} (27 kJ mol^{-1}). Another complex, called the 'dominant' complex has also been predicted by BD, with interactions Glu48–Lys13, Glu56–Lys87, Asp60–Lys86 and haem–Tml72, having an average electrostatic energy of –13.0 kcal mol^{-1} (–54.4 kJ mol^{-1}). Subsequent refinement (Guillemette *et al.*, 1994) of the complexes generated by the BD study have been performed, in which these two complexes have been subject to energy minimization, and more detailed comparison has been made between Salemme's complex and the dominant one. The energy minimization calculations reported here for the two encounter complexes generated by the BD simulations are an attempt to produce models that more accurately reflect the structures of these complexes that are expected to occur as the initial encounter process evolves to produce a putative transition state complex. Examination of the two minimized complexes reveals several similarities with the crystallographic structure reported for the cytochrome *c*–CCP complex (Pelletier and Kraut, 1992). For example, all three complexes exhibit formation of salt bridges between the interacting proteins as well as the presence of hydrophobic residues at the protein–protein interface. Nevertheless, the number of salt bridges predicted for the cytochrome *c*–b_5 complex is greater than that found in the structure of the cytochrome *c*–CCP complex. In addition, Phe82 of cytochrome *c* occupies a central position between the donor and acceptor centres in all three complexes that suggests at the least a structural or functional role for this residue in electron transfer between the two proteins. Finally, it is notable that the angles formed by the

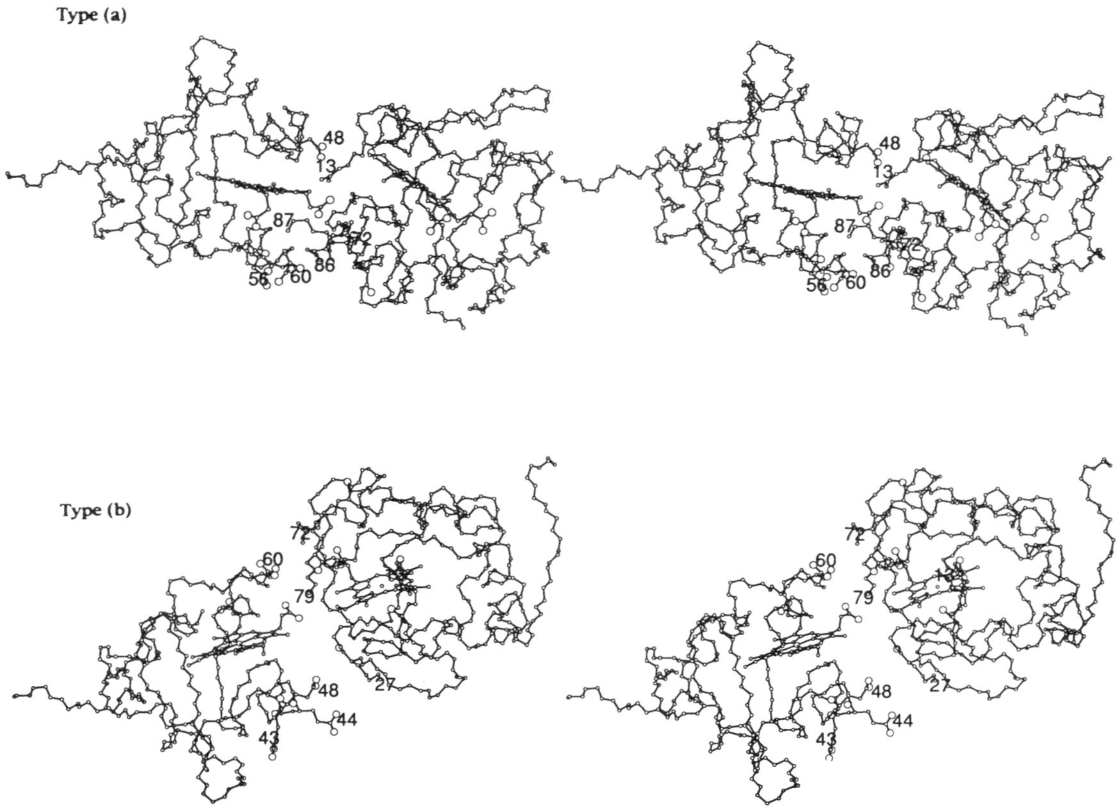

Figure 4.9. Stereo projections of representatives from two predominant types of reactive docked complexes between cytochrome b_5 and CY generated (Northrup et al., 1993) in BD simulations ($I = 0.19$ M, pH = 7.0, $T = 298$K). Type (a) is the dominant one in frequency and electrostatic stability; type (b) is next most frequent and is of the Salemme (1976) variety. The cytochrome b_5 species is on the left. Reprinted with permission from Northrup et al. (1993) Effects of charged amino acid mutations on the bimolecular kinetics of reduction of yeast iso-1-ferricytochrome c by bovine ferrocytochrome b_5. *Biochemistry* **32:** 6613–6623. Copyright 1993 American Chemical Society.

average plane of the haem groups in the crystallographically determined structure of the cytochrome c–CCP complex and the minimized models for the cytochrome c–b_5 complex are all greater than 20°, in contrast to the original models proposed for these complexes (Poulos and Kraut, 1980; Salemme, 1976) that placed the haem groups in parallel planes.

It is interesting to note that Guillemette *et al.* (1994) also performed molecular dynamics simulations on the Salemme-type complex similar to that performed by Wendoloski *et al.* (1987). No comparable movement of the Phe82 side chain was observed. One possibility is that the 1987 study used the tuna cytochrome c rather than the yeast. Another more disturbing possibility is that difference in force fields used can account for this discrepancy.

4.4 Summary

Protein–protein electron transfer reaction rates are controlled by a host of physical factors. The initial stage of reaction, which has been long left out of considerations, is the translational and rotational diffusion of protein pairs to form associated complexes, a process influenced by electrostatic and steric forces communicated through the solvent electrolyte medium. The second stage of the reaction is the chemical event proper, which is the electron transfer from pre-formed complexes. This stage depends on a variety of factors, including the geometric disposition of donor and acceptor sites in the complex, electromotive force, reorganization energy and the existence of electron transfer pathways, to mention a few. The disentangling of the roles of these various factors is an important goal in the understanding of biological electron transfer.

The initial theoretical studies of metalloprotein complexation concerned the model building of plausible electrostatically stable complexes through optimization of ionic contacts by fitting procedures or computer graphics. These studies are still profoundly useful for rationalizing the role of electrostatics and geometric effects in the existence of stable complexes from which electron transfer is likely to occur. However, studies of this type ignore other important dynamic features which must be recognized for a complete understanding of electron transfer rates. These have been the primary focus of this chapter and include (i) dynamics of protein–protein association by a diffusional mechanism, (ii) the possible existence of alternative docking geometries rather than a single complex through which electron transfer can occur, and (iii) the influence of protein flexibility and internal dynamics on an individual association complex.

BD simulation has demonstrated itself to be an important theoretical tool for including these ignored dynamical features. Through use of BD simulations, a number of essential principles have been identified which account for the behaviour of proteins in electron transfer. The strong ionic strength dependence in the bimolecular rates confirms that the electrostatic forces play a vital role in protein–protein interaction. Brownian simulations coupled with a rigorous Poisson–Boltzmann electrostatic treatment with full charge distribution is consistently capable of making quantitative predictions of the ionic strength dependence. The BD of charge-modified proteins offers an immediate look at the effects of individual amino acids on association rates. The charge mutation study provides additional evidence that global dipole–dipole and monopole–dipole 'steering' interactions between these proteins are of much less consequence relative to the local ionic contacts between exposed haem regions.

The most recent BD simulations have made a true dynamic coupling between the diffusion event and the chemical event through the employment of an intrinsic electron transfer unimolecular rate constant in the diffusion dynamic scheme. BD theory was used in conjunction with an exponential distance–dependent model of electron transfer to first study self-exchange in cytochromes c-551, c and b_5. By fitting the BD-generated rate constants to experimental curves, reorganization energy estimates and electron transfer distance dependence factors have been extracted. Since BD provides a detailed description of the collision stage of the process, determined by the actual atomic scale irregularity of the proteins (steric factors) and the mutual electrostatic interactions, no estimates of the association constant and prosthetic group exposure are required, and so fewer parameters are necessary to make estimates of the reorganization energy.

The BD method is capable of predicting the effects of site-directed mutations on the rate of docking and electron transfer, thus determining the relative roles of individual charged amino acids. The effects of selective charge site perturbations have been determined by performing 'computer site-directed mutagenesis' on two residues on the haem-exposed face of CY into neutral residues. The experimental trend in rates of the various mutants of CY reacting with ferrocytochrome b_5 were well reproduced by the simulations. Using the finite difference-linearized Poisson–Boltzmann treatment of the electrostatic potentials to calculate the influence of charged amino acid mutations on the reduction potential of CY, and using Marcus theory and an estimated reorganization energy, the perturbations in the intrinsic electron transfer rate constant caused by the charge mutations were included. By including this additional effect, more quantitative agreement with experiment was obtained.

Besides predicting rate constants, BD provides a wealth of information on the docked complexes and ionic contacts which form during metalloprotein reactions. The complexes are not generated by hand-docking, which may be human-biased, but are a large representative sampling of configurations which meet geometric criteria for electron transfer and are statistically weighted by electrostatic forces. The cytochrome $c(Fe^{3+})$–cytochrome b_5 reaction and self-exchange reactions of cytochrome c, b_5 and c-551 meet geometric criteria through a single surface region of each protein. However, for the cytochrome $c(Fe^{2+})$–CCP reaction there exist three distinct domains on CCP approximately in its haem plane. These also match the electrostatically attractive regions, showing in the structure of these proteins a strong correlation between electrostatic details and electron transfer geometry restrictions. Furthermore, since CCP requires two electrons, one from each of two cytochrome c molecules, there may be mechanistic implications of this observation, with CCP able to bind simultaneously to two donor molecules.

The existence of multiple sites on CCP for cytochrome c binding has been the subject of hot debate in recent years. The first observation of 2:1 cytochrome c:CCP stoichiometry in solution was in fluorescence studies of Kornblatt and English (1986), who saw a strong site and a weak site.

Fluorescence quenching studies of Vitello and Erman (1987) were consistent with 1:1 binding. Proton nuclear magnetic resonance (NMR) studies of Moench *et al.* (1992) found the predominant complex in solution to be 1:1 as well. The noted crystal structure of these partners (Pelletier and Kraut, 1992) is also distinctly 1:1. These researchers believe in a highly specific electron transfer complex which takes advantage of a single salient pathway for electron transfer, and further speculate that the crystalline structure is also the one found in solution. On the other hand, studies of Hoffman's group (Stemp and Hoffman, 1993; Zhou and Hoffman, 1994) over a wide range of ionic strengths in solution reveal a persistent 2:1 stoichiometry in which the second site is a weak site which nevertheless transfers electrons at a high rate. A proton linkage study (Mauk *et al.,* 1994) found results consistent with 2:1 binding where the weak site is three orders of magnitude weaker. A very recent as yet unpublished study of Mauk's group at the University of British Columbia confirmed the location of the second site. As suggested by Northrup's BD simulations, the weak site is centred around Asp148. Mauk *et al.* prepared a site-directed mutant in which this residue was neutralized, and a fourfold weakening of the secondary site was manifested.

A whole ensemble of plausible electron transfer complexes with slightly different orientations, rather than a unique dominant stable reactive complex, are found in all BD simulations. This is further support for the hypothesis (Mauk *et al.,* 1986; Weber and Tollin, 1985) that in the mechanism of interaction of the cytochromes an ensemble of near-optimal docking geometries, rather than a single complex, exists and contributes to electron transfer. The simulated distribution of ionic contacts (with three or four interprotein salt bridges stabilizing encounter complexes) shows that there is not a strict charge complementarity in operation that locks the proteins into a single electron transfer arrangement, but that association is more non-specific. The electrostatic forces provide the appropriate range of attractive potential energy to allow proteins to explore a wide variety of mutual orientations in a single encounter rather than being steered deterministically by Coulomb forces over long distances into one selective docking arrangement (Northrup *et al.,* 1988). Recent experimental evidence in support of this hypothesis is now available. In studies of electron transfer reactions in electrostatic and covalently cross-linked complexes of plastocyanin and cytochrome *c* (Peerey and Kostic, 1989), the covalently linked complex is entirely unreactive, even though it is thought to be docked in the same fashion as the electrostatically stable complex. The tight cross-links seem to prevent cytochrome *c* and plastocyanin from the necessary exploration of a spectrum of nearby stable but less than energy-optimal docking geometries. A similar non-covalent and covalent cross-linking study of cytochrome *c*–CCP has shown the importance of conformational rearrangements of the two proteins within a complex to provide more favourable orientations for electron transfer (Hazzard *et al.,* 1988). Fast two-dimensional diffusion has been implicated in cytochrome *c*–CCP recognition by dynamic energy transfer measurements (Zhang *et al.,* 1990). An experimental study (Nocek *et al.,* 1990) of photo-initiated long-range electron transfer within the complex between zinc-substituted CCP and ferric cytochrome *c* has been used to probe the dynamics of intra-complex docking rearrangements. There it is found that the complex appears to undergo a docking conformational rearrangement to a low temperature electron transfer-inactive form over a very narrow temperature range which is independent of solvent composition. Thus the complex, when trapped in low lying energetic states, is unable to sample other less stable docking geometries which are electron transfer active. Evidence for a dynamic redistribution among docking conformations is also seen in energy transfer and fluorescence anisotropy measurements of McLendon *et al.* (1993).

The BD method for simulation of electron transfer rates between metalloproteins as it is currently implemented has a number of shortcomings, which warrant the following cautionary notes. BD as implemented here for protein–protein interactions must be regarded as a very coarse-grained theoretical approach. For example, the algorithm produces no more intimacy than that which the rigidly docking proteins attain when excluded volume overlap is encountered between only two atoms. Inside this point, strong hydrophobic forces, hydrogen bond contacts, flexibility of the surfaces and expulsion of water are likely to begin to operate to create a more intimate contact capable of electron transfer. By neglecting these, BD provides an incomplete picture of short-ranged events, and can be regarded as providing a reasonable theoretical description only for generating loosely docked complexes. While some form of hydrophobic interaction could be included in the force determination, this would be superfluous without also including provision for other short-ranged neglected factors, such as a spatially varying dielectric constant inside the interface, formation of orientation-specific hydrogen bonds, relaxation of surface charged groups and the like. These are beyond the scope of BD, since these forces vary rapidly on the spatial resolution scale of diffusive motion, requiring supplementary calculations based on molecular dynamics at short range. Still, there is room for future improvement within the BD context: (i) a more self-consistent treatment of low dielectric regions must be incorporated in the electrostatic treatment and, in some cases, the non-linear Poisson–Boltzmann equation must be used instead of the linearized version; (ii) more realistic internal protein flexibility within docked complexes must be incorporated directly into trajectory simulations; and (iii) more detailed modelling of the intrinsic electron transfer step must be incorporated to exploit the electron transfer pathways theory (Beratan *et al.*, 1992). With these improvements the simulation methods can be even more quantitative and comprehensive in treating protein–protein electron transfer and be of greater value in the guiding of and the interpretation of experiment.

References

Adam G, Delbrück M. (1968) Reduction of dimensionality in biological diffusion processes. In: *Structural Chemistry and Molecular Biology* (eds A Rich, N Davidson). Freeman Press, San Francisco, pp. 198–215.

Allison SA, Northrup SH, McCammon JA. (1985) Extended Brownian dynamics of diffusion controlled reactions. *J. Chem. Phys.* **83**: 2894–2899.

Allison SA, McCammon JA, Northrup SH. (1986) Dynamics of macromolecular interactions. *ACS Symp. Ser.* **302**: 216–231.

Allison SA, Bacquet RJ, McCammon JA. (1988) Simulation of the diffusion-controlled reaction between superoxide and superoxide dismutase. II. Detailed models. *Biopolymers* **27**: 251–269.

Andrew SM, Thomasson KA, Northrup SH. (1993) Simulation of electron transfer self-exchange in cytochrome c and b_5. *J. Am. Chem. Soc.* **115:** 5516–5521.

Bashford D, Karplus M. (1990) pK_a's of ionizable groups in proteins: atomic details from a continuum electrostatic model. *Biochemistry* **29:** 10219–10225.

Beratan DN, Betts JN, Onuchic JN. (1992) Tunneling pathway and redox-state-dependent electronic couplings at nearly fixed distance in electron-transfer proteins. *J. Phys. Chem.* **96:** 2852–2855.

Berg OG, von Hippel PH. (1985) Diffusion-controlled macromolecular interactions. *Annu. Rev. Biophys. Biophys. Chem.* **14:** 131–160.

Debye P. (1942) Reaction rates in ionic solutions. *Trans. Electrochem. Soc.* **82:** 265–272.

Dixon DW, Hong X, Woehler SE. (1989) Electrostatic and steric control of electron self-exchange in cytochromes c, c-551, and b_5. *Biophys. J.* **56:** 339–351.

Dixon DW, Hong X, Woehler SE, Mauk AG, Sishta BP. (1990) Electron-transfer self-exchange kinetics of cytochrome b_5. *J. Am. Chem. Soc.* **112:** 1082–1088.

Eltis LD, Herbert RG, Barker PD, Mauk AG, Northrup SH. (1991) Reduction of horse heart ferricytochrome c by bovine liver ferrocytochrome b_5. Experimental and theoretical analysis. *Biochemistry* **30:** 3663–3674.

Ermak DL, McCammon JA. (1978) Brownian dynamics with hydrodynamic interactions. *J. Chem. Phys.* **69:** 1352–1360.

Farid RS, Moser CC, Dutton PL. (1993) Electron transfer in proteins. *Curr. Opin. Struct. Biol.* **3:** 225–233.

Ganti G, McCammon JA, Allison SA. (1985) Brownian dynamics of diffusion-controlled reactions: the lattice method. *J. Phys. Chem.* **89:** 3899–3902.

Geren L, Tuls J, O'Brien P, Millett F, Peterson JA. (1986) The involvement of carboxylate groups of putidaredoxin in the reaction with putidaredoxin reductase. *J. Biol. Chem.* **261:** 15491–15495.

Guillemette JG, Barker PD, Eltis LD, Lo TP, Smith M, Brayer GD, Mauk AG. (1994) Analysis of the bimolecular reduction of ferricytochrome c by ferrocytochrome b_5 through mutagenesis and molecular modelling. *Biochimie* **76:** 592–604.

Hazzard JT, Moench SJ, Erman JE, Satterlee JD, Tollin G. (1988) Kinetics of intracomplex electron transfer and of reduction of the components of covalent and noncovalent complexes of cytochrome c and cytochrome c peroxidase by free flavin semiquinones. *Biochemistry* **27:** 2002–2008.

Herbert RG, Northrup SH. (1989) Brownian simulation of cytochrome c-551 association and electron self-exchange. *J. Mol. Liq.* **41:** 207–222.

Holst MJ, Saied F. (1995) Numerical solution of the nonlinear Poisson–Boltzmann equation: developing more robust and efficient methods. *J. Comp. Chem.* **16:** 337–364.

Kang CH, Brautigan DL, Osheroff N, Margoliash E. (1978) Definition of cytochrome c binding domains by chemical modification. Reaction of carboxydinitrophenyl- and trinitrophenyl-cytochromes c with Baker's yeast cytochrome c peroxidase. *J. Biol. Chem.* **253:** 6502–6510.

Kornblatt JA, English AM. (1986) The binding of porphyrin cytochrome c to yeast cytochrome c peroxidase. A fluorescence study of the number of sites and their sensitivity to salt. *Eur. J. Biochem.* **155:** 505–511.

Matthew JB. (1985) Electrostatic effects in proteins. *Annu. Rev. Biophys. Biophys. Chem.* **14:** 387–417.

Matthew JB, Weber PC, Salemme FR, Richards FM. (1983) Electrostatic orientation during electron transfer between flavodoxin and cytochrome c. *Nature* **301:** 169–171.

Mauk MR, Mauk AG, Weber PC, Matthew JB. (1986) Electrostatic analysis of the interaction of cytochrome c with native and dimethyl ester haem substituted cytochrome b_5. *Biochemistry* **25:** 7085–7091.

Mauk MR, Ferrer JC, Mauk AG. (1994) Proton linkage in formation of the cytochrome c–cytochrome c peroxidase complex: electrostatic properties of the high- and low-affinity cytochrome binding sites on the peroxidase. *Biochemistry* **33:** 12609–12614.

McLendon G. (1988) Long-distance electron transfer in proteins and model systems. *Acc. Chem. Res.* **21:** 160–167.

McLendon G, Zhang Q, Wallin SA, Miller RM, Billstone V, Spears KG, Hoffman BM. (1993) Thermodynamic and kinetic aspects of binding and recognition in the cytochrome c–cytochrome c peroxidase complex. *J. Am. Chem. Soc.* **115:** 3665–3669.

Meyer TE, Tollin G, Cusanovich MA. (1994) Protein interaction sites obtained via sequence homology. The site of complexation of electron transfer partners of cytochrome c revealed by mapping amino acid substitutions onto three-dimensional protein surfaces. *Biochimie* **76**: 480–488.

Moench SJ, Chroni S, Lou B-S, Erman JE, Satterlee JD. (1992) Proton NMR comparison of noncovalent and covalently cross-linked complexes of cytochrome c peroxidase with horse, tuna, and yeast ferricytochromes c. *Biochemistry* **31**: 3661–3670.

Nocek JM, Liang N, Wallin SA, Mauk AG, Hoffman BM. (1990) Low-temperature conformational transition within the [Zn-cytochrome c peroxidase, cytochrome c] electron-transfer complex. *J. Am. Chem. Soc.* **112**: 1623–1625.

Northrup SH, Herbert RG. (1990) Brownian simulation of protein association and reaction. *Int. J. Quant. Chem.* **17**: 55–71.

Northrup SH, Allison SA, McCammon JA. (1984) Brownian dynamics simulation of diffusion-influenced bimolecular reactions. *J. Chem. Phys.* **80**: 1517–1524.

Northrup SH, Curvin M, Allison SA, McCammon JA. (1986a) Optimization of Brownian dynamics methods for diffusion-influenced rate constant calculations. *J. Chem. Phys.* **84**: 2196–2203.

Northrup SH, Reynolds JCL, Miller CM, Forrest KJ, Boles JO. (1986b) Diffusion controlled association rate of cytochrome c and cytochrome c peroxidase in a simple electrostatic model. *J. Am. Chem. Soc.* **108**: 8162–8170.

Northrup SH, Smith JD, Boles JO, Reynolds JCL. (1986c) The effect of dipole moment on diffusion-controlled bimolecular reaction rates. *J. Chem. Phys.* **84**: 5536–5544.

Northrup SH, Boles JO, Reynolds JCL. (1987a) Electrostatic effects in the Brownian dynamics of association and orientation of haem proteins. *J. Phys. Chem.* **91**: 5991–5998.

Northrup SH, Luton JA, Boles JO, Reynolds JCL. (1987b) Brownian dynamics simulation of protein association. *J. Computer-Aided Mol. Des.* **1**: 291–311.

Northrup SH, Boles JO, Reynolds JCL. (1988) Brownian dynamics of cytochrome c and cytochrome c peroxidase association. *Science* **241**: 67–70.

Northrup SH, Thomasson KA, Miller CM, Barker PD, Eltis LD, Guillemette JG, Inglis SC, Mauk AG. (1993) Effects of charged amino acid mutations on the bimolecular kinetics of reduction of yeast iso-1-ferricytochrome c by bovine ferrocytochrome b_5. *Biochemistry* **32**: 6613–6623.

Peerey LM, Kostic NM. (1989) Oxidoreduction reactions involving the electrostatic and the covalent complex of cytochrome c and plastocyanin: importance of the protein rearrangement for the intracomplex electron-transfer reaction. *Biochemistry* **28**: 1861–1868.

Pelletier H, Kraut J. (1992) Crystal structure of a complex between electron transfer partners, cytochrome c peroxidase and cytochrome c. *Science* **258**: 1748–1755.

Poulos TL, Kraut J. (1980) A hypothetical model of the cytochrome c peroxidase–cytochrome c electron transfer complex. *J. Biol. Chem.* **255**: 10322–10330.

Poulos TL, Mauk AG. (1983) Models for the complexes formed between cytochrome b_5 and the subunits of methemoglobin. *J. Biol. Chem.* **258**: 7369–7373.

Reynolds JCL, Cooke KF, Northrup SH. (1990) Electrostatics and diffusional dynamics in the carbonic anhydrase active site channel. *J. Phys. Chem.* **94**: 985–991.

Rodgers KK, Pochapsky TC, Sligar SG. (1988) Probing the mechanisms of macromolecular recognition: the cytochrome b_5–cytochrome c complex. *Science* **240**: 1657–1659.

Salemme FR. (1976) An hypothetical structure for an intermolecular electron transfer complex of cytochromes c and b_5. *J. Mol. Biol.* **102**: 563–568.

Sharp K, Fine R, Honig B. (1987) Computer simulations of the diffusion of a substrate to an active site of an enzyme. *Science* **236**: 1460–1463.

Shire SJ, Hanania GIH, Gurd FRN. (1974) Electrostatic effects in myoglobin. Hydrogen ion equilibria in sperm whale ferrimyoglobin. *Biochemistry* **13**: 2967–2974.

Siders P, Cave RJ, Marcus RA. (1984) A model for orientation effects in electron-transfer reactions. *J. Chem. Phys.* **81**: 5613–5624.

Simondsen RP, Weber PC, Salemme FR, Tollin G. (1982) Transient kinetics of electron transfer reactions of flavodoxin: ionic strength dependence of semiquinone oxidation by cytochrome c, ferricyanide and ferric EDTA and computer modeling of reaction complexes. *Biochemistry* **21**: 6366–6375.

Smoluchowski M. (1917) Versuch einer mathematischer Theorie der Kougulationskinetik kolloider Lösungen. *Z. Phys. Chem.* **92:** 129–168.

Stayton PS, Poulos TL, Sligar SG. (1989) Putidaredoxin competitively inhibits cytochrome b_5–cytochrome P-450$_{cam}$ electron-transfer complex. *Biochemistry* **28:** 8201–8205.

Stemp EDA, Hoffman BM. (1993) Cytochrome c peroxidase binds two molecules of cytochrome c: evidence for a low-affinity, electron-transfer-active site on cytochrome c peroxidase. *Biochemistry* **32:** 10848–10865.

Tanford C, Kirkwood JG. (1957) Theory of protein titration curves. I. General equations for impenetrable spheres. *J. Am. Chem. Soc.* **79:** 5333–5339.

Tanford C, Roxby R. (1972) Interpretation of protein titration curves. Application to lysozyme. *Biochemistry* **11:** 2192–2198.

Tegoni M, White SA, Roussel A, Mathews FS, Cambillau C. (1993) A hypothetical complex between crystalline flavocytochrome b_2 and cytochrome c. *Proteins: Struct. Funct. Genet.* **16:** 408–422.

Timkovich R, Cai ML, Dixon DW. (1988) Electron self-exchange in *Pseudomonas* cytochromes. *Biochem. Biophys. Res. Commun.* **150:** 1044–1050.

van Leeuwen JW. (1983) The ionic strength dependence of the rate of a reaction between two large proteins with a dipole moment. *Biochim. Biophys. Acta* **743:** 408–421.

Vitello LB, Erman JE. (1987) Binding of horse heart cytochrome c to yeast porphyrin cytochrome c peroxidase: a fluorescence quenching study on the ionic strength dependence of the interaction. *Arch. Biochem. Biophys.* **258:** 621–629.

Warwicker J, Watson HC. (1982) Calculation of the electric potential in the active site cleft due to α-helix dipoles. *J. Mol. Biol.* **157:** 671–679.

Weber PC, Tollin G. (1985) Electrostatic interactions during electron transfer reactions between c-type cytochromes and flavodoxin. *J. Biol. Chem.* **260:** 5568–5573.

Wendoloski JJ, Matthew JB, Weber PC, Salemme FR. (1987) Molecular dynamics of a cytochrome c–cytochrome b_5 electron transfer complex. *Science* **238:** 794–797.

Yang A-S, Gunner MR, Sampogna R, Sharp K, Honig B. (1993) On the calculation of pKa's in proteins. *Proteins: Struct. Funct. Genet.* **15:** 252–265.

Zhang Q, Marohn J, McLendon G. (1990) Macromolecular recognition in the cytochrome c–cytochrome c peroxidase complex involves fast two-dimensional diffusion. *J. Phys. Chem.* **94:** 8628–8630.

Zhou H-X. (1993) Brownian dynamics study of the influences of electrostatic interaction and diffusion in protein–protein kinetics. *Biophys. J.* **64:** 1711–1726.

Zhou H-X, Hoffman BM. (1994) Stern–Volmer in reverse: 2:1 stoichiometry of the cytochrome c–cytochrome c peroxidase electron-transfer complex. *Science* **265:** 1693–1696.

Structure of electron transfer proteins and their complexes

F. Scott Mathews and Rosemary C.E. Durley

5.1 Introduction

Electron transfer between biological macromolecules usually occurs within a complex in which the participating redox centres maintain a certain geometrical relationship and the intervening protein material can promote and regulate electron flow between centres. In some cases, such as the photosynthetic reaction centre and several mitochondrial electron transport chains, these complexes are stable with time, allowing study of the factors controlling electron transfer rates within these fixed systems. In other cases, the complexes are only transiently stable and must first form before the electron transfer events can occur. Complex formation between redox partners will depend upon the nature of the surface topology, distribution of charged residues and other structural features. The rate of electron transfer between any pair of centres will be dependent on several factors, including the difference in free energy of the oxidized and reduced states, the activation energy of the transfer process, a 'reorganization' energy and the electronic coupling between the two redox centres. An understanding of the thermodynamics and kinetics of this process has been developed by Marcus and his co-workers as is discussed in Chapter 10.

Detailed structural analysis of electron transfer complexes can provide information about both complex formation and electron transfer. First, the interaction geometry of the components can be ascertained, indicating the most pertinent features of the protein for complex formation. Second, the relative geometry of the redox centres and a detailed description of the intervening medium can be established. This information will have a bearing on the electronic coupling between the two centres which may depend simply on their separation distance or on structural features such as the presence of one or more networks of covalent, non-covalent or van der Waals interactions linking the two centres.

Protein Electron Transfer, Edited by D.S. Bendall
© 1996 BIOS Scientific Publishers Ltd, Oxford

There are several categories of biological electron transfer proteins which vary in their structure and catalytic complexity. The simplest are the one-domain one-electron-carrier proteins in which the electron resides transiently on a special cofactor such as a metal atom or a haem. They are small, usually 8–15 kDa in molecular mass and carry out no other enzymatic functions. Often they can react with a number of different physiological partners. Examples of these electron carriers include cytochromes, cupredoxins and ferredoxins, each of which is discussed in other chapters. More intricate proteins may consist of one or several domains and carry out additional reactions such as oxidation or reduction of organic substrates. They can transduce the flow of charge from single electrons to pairs (or more) or vice versa by utilizing organic cofactors such as flavins and quinones, which can form relatively stable radical intermediates, or radicals of amino acid side chains such as tyrosine and tryptophan in conjunction with a metal centre.

In this chapter, we will first present a brief introduction to members of two of the most common types of electron carriers, cytochromes and cupredoxins. We will then consider two-site proteins containing pairs of redox cofactors bound either to a single domain, such as ascorbate oxidase (AO) or to separate domains such as the flavocytchromes. Next, two systems of weakly associating protein complexes, the methylamine dehydrogenase (MADH)–amicyanin–cytochrome system and the cytochrome c peroxidase (CCP)–cytochrome c system will be described. These sections will then be followed by discussions of the structures and dynamics of the complexes in solution and the enzymatic and electron transfer activities of the crystalline complexes.

5.2   Cytochromes and cupredoxins

5.2.1 *Cytochromes*

Although numerous and varied, cytochromes all contain a protohaem IX moiety and participate in electron transfer (for a review, see Mathews, 1985). The classes of haem-containing proteins are distinguished by the type of haem binding and secondary structure.

The c-type cytochromes are found in both eukaryotic and prokaryotic sources, and contain a covalently bound haem group. In mitochondria, they transfer electrons from the cytochrome reductase b–c_1 complex to the cytochrome oxidase a–a_3 complex. Both donor and acceptor complexes are integral membrane proteins, but cytochrome c is soluble. The haem iron is axially coordinated to a histidine residue near the N terminus and a methionine residue near the C terminus (*Figure 5.1a*); the covalent attachment is achieved by two thioether bonds between the vinyl methylene atoms, CAB and CAC, and two cysteine side chains. Most of the haem group is in a hydrophobic environment, often with one or both propionic acid groups shielded from solvent, but with the CBC methyl group exposed to solvent. The protein secondary structure is very simple, yet characteristic; there are usually three or four helices and no β-strands or sheets. The first helix comes before the His ligand and the last helix comes after the Met ligand. Mitochondrial cytochrome c contains two helices between the His and Met ligands, but some bacterial cytochromes contain only one. Mitochondrial cytochrome c is very basic, with

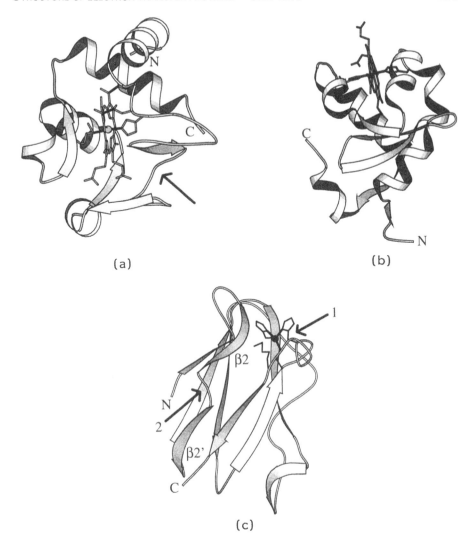

Figure 5.1. (a) Ribbon diagram for eukaryotic cytochrome c from tuna [Protein Database (PDB) entry 3CYT]. The arrow indicates the approximate homologous binding site for amicyanin in the MADH–amicyanin–cytochrome c-551$_i$ complex. (b) Ribbon diagram for cytochrome b_5 (PDB entry 1CYO). (c) Ribbon diagram for plastocyanin (PDB entry 1PLC). Arrow 1 indicates the region of the cupredoxin which is most accessible for electron transfer and corresponds to the MADH binding site in the MADH–amicyanin complexes. The binding site for the cytochrome in the MADH–amicyanin–cytochrome c-551$_i$ complex is indicated by arrow 2. The diagrams were prepared using MOLSCRIPT (Kraulis, 1991).

several positive residues clustered around the mouth of the haem crevice, possibly for recognition by the cytochrome oxidase and cytochrome reductase. Bacterial cytochromes c are often neutral or acidic, but in some cases also maintain a positive patch of side chains surrounding the haem edge.

Cytochrome b_5 contains a non-covalently bound haem group and is bound to membranes in liver cells of birds and mammals. It transfers electrons from the NADH-dependent cytochrome b_5 reductase to a fatty acid desaturase or to P450. Membrane binding is achieved by a hydrophobic 40-residue C-terminal polypeptide which can be cleaved proteolytically to give a soluble 90-residue cytochrome domain. The cytochrome domains of several multifunctional enzymes such as flavocytochrome b_2, sulphite oxidase and nitrate reductase are members of the b_5 family. The 'b_5 fold' is a haem-binding pocket formed by four α-helices and a mixed β-sheet (*Figure 5.1b*). Below the sheet, two or three additional short helices help stabilize the cytochrome structure. The haem group has two axial histidine ligands and is oriented with the propionic acid portions directed outward from the molecule while the remainder is buried in the protein interior. In cytochrome b_5, one of the haem propionic acid chains bends back onto the molecule, forming hydrogen bonds to the protein. Charged residues are distributed unevenly, with pronounced localization of negative charge near the haem crevice, possibly for partner recognition in electron transfer (Salemme, 1976).

5.2.2 *Cupredoxins*

The cupredoxins (blue copper proteins) are electron transfer proteins found in a variety of plants and bacteria (for a review, see Adman, 1991). They contain a single copper atom which cycles between the cupric and cuprous states. The most studied are plastocyanins and azurins. Plastocyanin transfers electrons from the cytochrome b_6f complex of photosystem II to pigment $P700^+$ of photosystem I in chloroplasts; azurin is thought to transfer electrons from cytochrome c-551 to cytochrome oxidase in bacteria. The common secondary structure of cupredoxins consists of eight β-strands which form two twisted β-sheets sandwiching a hydrophobic interior (*Figure 5.1c*). The two sheets are joined on one side by β-strand 2 which first forms part of sheet 1 then twists and forms part of sheet 2. The copper ion is held in a pocket formed by three loops at the 'northern' end of the sandwich, with ligands contributed from both β-sheets. In plastocyanin, the copper atom is bound by four side chains, two histidines, a cysteine and a methionine. The copper-bound atoms of the ligands form a distorted tetrahedron. In azurin, there is a fifth copper ligand, a peptide carbonyl, giving rise to a distorted, axially elongated trigonal bipyramid geometry. The intricate hydrogen bonding network which involves the copper coordination sphere forms a very rigid structure. The copper ion is shielded from solvent but one of the ligand histidine residues is surface accessible and is surrounded by a ring of hydrophobic residues. This surface patch has been implicated as a major binding site for electron transfer partners (Chen *et al.*, 1992; Farver *et al.*, 1982). In plastocyanin, a second binding site for electron transfer has been postulated, an acidic patch consisting of six negatively charged groups clustered around a tyrosine residue (He *et al.*, 1991). In azurin or other cupredoxins, no such prominent surface identifies a second putative electron transfer site.

5.3 Two-site proteins

5.3.1 *Ascorbate oxidase*

This multicopper oxidase, discussed in Chapter 7, is a dimeric protein of M_r 130 kDa containing two cofactors in each subunit. The enzyme catalyses the oxidation of ascorbate concomitant with the reduction of molecular oxygen to water. Each subunit, shown in *Figure 5.2a*, consists of three closely associated domains of approximately equal size and structure which are also linked by two disulphides, one connecting domains 1 and 2 and the other domains 1 and 3 (Messerschmidt *et al.*, 1992). Each domain forms a β-sandwich with a cupredoxin-type fold. The first cofactor, a single copper (Cu1), is located within domain 3 and has typical cupredoxin coordination (His, Cys, His, Met) and distorted tetrahedral geometry. The second cofactor is a novel cluster of three copper ions bound to eight histidine ligands contributed equally by the first and third domains. Two of the coppers (Cu2 and Cu3), each with three histidine ligands, are bridged by a hydroxyl or other oxygen species (Messerschmidt *et al.*, 1992). These copper atoms are thought to give rise to the type-3 spectra. The third copper of the cluster (Cu4) has two histidine ligands and a third ligand, either a water or hydroxyl ion. This copper is believed to give rise to the type-2 spectra.

All of the histidine ligands to the trinuclear copper centre are found in polypeptide triplets, His–X–His, with each histidine binding to different coppers. Two of these triplets are found in domain 3 and two in domain 1. One of these triplets, in domain 3, has Cys in the central X position which forms the cysteinyl sulphur ligand of Cu1, the type-1 copper (*Figure 5.2b*). The distance between the type-1 copper and the trinuclear cluster centre is around 12 Å.

Oxidation of the organic substrate, ascorbate, generates a single electron with release of a semidehydroascorbate radical. The primary electron acceptor is the type-1 copper. Ascorbate has been modelled into a solvent-filled depression adjacent to the cupredoxin site (Messerschmidt *et al.*, 1992). Three solvent sites could be identified which mimic the oxygens of the lactone ring of the hypothetical substrate complex. These solvent molecules are within hydrogen bonding range of the solvent-exposed histidine ligand of the cupredoxin site.

The second substrate, molecular oxygen, binds near the trinuclear copper site and requires four electrons for reduction to H_2O. The trinuclear centre is postulated to store up to three electrons (during three catylic turnovers) and then transfer them along with a fourth electron (after the fourth turnover) to the bound dioxygen followed by the release of water. Two solvent-filled channels lead to the trinuclear site, a broad one open to the type-3 copper pair and a narrow one extending to the type-2 site.

Intramolecular electron transfer between the mono and trinuclear centres is restricted to domain 3 of each subunit. Kinetic experiments (Farver and Pecht, 1992; Meyer *et al.*, 1991) suggest that this transfer may be the rate-limiting step of the catalytic cycle. The principal pathways from the cupredoxin site to the copper cluster follow the cysteinyl side chain from the sulphur ligand to the backbone and then along the backbone chain in either direction to one of the

(a)

(b)

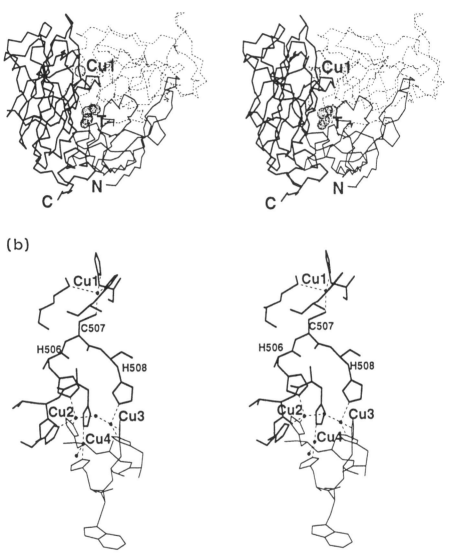

Figure 5.2. (a) Stereoview of the Cα backbone of one subunit of ascorbate oxidase (PDB entry 1AOZ). Heavy lines represent domain 3, medium lines represent domain 1 and dashed lines represent domain 2. The mononuclear copper site is labelled 'Cu1'; the trinuclear cluster, labelled 'T', is visible at the interface between domains 3 and 1. (b) Detailed view of the copper sites of ascorbate oxidase with residues from domain 3 in heavy lines and those from domain 1 in lighter lines. The copper atoms are represented by small spheres as are the bridging oxygen, between Cu2 and Cu3, and the oxygen bonded to Cu4. The diagrams were prepared using TURBO-FRODO (Roussel and Cambillau, 1991).

adjacent histidines which are bound to the type-3 copper atoms of the cluster (*Figure 5.2b*; see Chapter 7 for a more complete discussion of electron transfer in AO).

5.3.2 *Flavocytochromes*

Overview. As the name implies, these proteins contain two cofactors, a flavin (FMN or FAD) and a haem group. Each is located on a separate domain or subunit and thus closely resembles a transient complex between separate partner molecules. During substrate oxidation, two electrons are transferred from the substrate to the flavin ring to form a hydroquinone. Reoxidation of the flavin then occurs stepwise, by two single electron transfers to the endogenous haem group, with formation of a flavin semiquinone intermediate. The crystal structures of three flavocytochromes are known; they contain considerable structural information and provide a basis for evaluation of some of the theories discussed in other chapters of this book.

Flavocytochrome b_2. FCB2 (L-lactate:ferricytochrome c oxidoreductase) is a tetrameric enzyme, of M_r 230 kDa, located in the intermembrane space of yeast mitochondria (for a review, see Lederer, 1991). Both cofactors, FMN and haem, are bound non-covalently. The enzyme catalyses the oxidation of lactate to pyruvate with subsequent transfer of electrons to cytochrome c.

FCB2 from *Saccharomyces cerevisiae* forms a tetramer whose four subunits are related by a molecular fourfold symmetry axis (Xia and Mathews, 1990). One subunit of FCB2 is shown in *Figure 5.3a*. Each subunit is composed of two domains, an N-terminal cytochrome domain of M_r 11 kDa and a C-terminal flavin-binding domain of M_r 45 kDa. They are connected by a flexible hinge of about 20 residues. In the electron density map, only two of the four cytochrome domains are visible, the other two being positionally disordered and their positions inferred from the molecular fourfold symmetry. In contrast, the four flavin-binding domains are quite similar in structure except that those which lack the ordered cytochrome domain each contain one molecule of pyruvate, the reaction product, bound at the active site adjacent to the flavin ring.

The flavin-binding domain is composed of a parallel $\beta_8\alpha_8$ barrel motif. The FMN is located at the C-terminal end of the central β-barrel. The plane of the pyruvate molecule is approximately parallel to that of the flavin ring. It is oriented so that the two carboxyl oxygen atoms can form hydrogen bonds to two side chains, Arg376 and Tyr143, of the flavoprotein subunit. Pyruvate appears to stabilize the semiquinone form of the enzyme by modulation of the redox potential of the flavin group.

The cytochrome domain is very similar in conformation to microsomal cytochrome b_5 (see Section 5.2.1). The β-sheet and two of the four helices which form the haem crevice are oriented very similarly in the two structures, but the other two helices are oriented somewhat differently so that the haem group appears to be more exposed in FCB2 than in cytochrome b_5.

The haem and flavin groups are nearly co-planar. The distance from the iron atom to the centre of the flavin ring is about 16 Å; the pyrrole and isoalloxazine rings are separated by about 10 Å. The haem propionate groups are located at the interface between the cytochrome and flavin-binding domains and extend toward the latter. The contact surface between the cytochrome domain and the flavin-binding domain is largely hydrophobic, although there

(a)

(b)

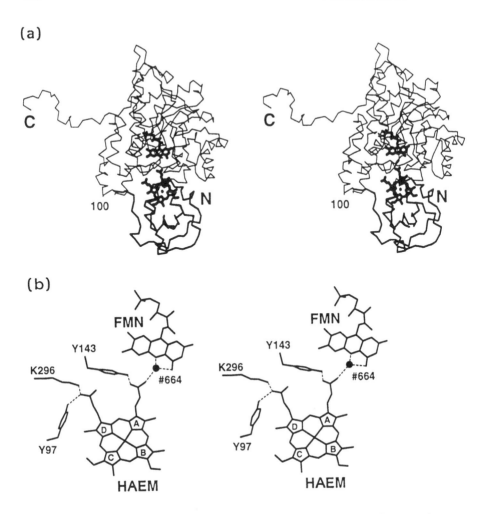

Figure 5.3. Stereoview of the Cα backbone of a single monomer of flavocytochrome b_2 (PDB entry 1FCB). The haem and flavin cofactors are shown by heavy lines as is the cytochrome domain (residues 1–99). The remaining flavin domain (residues 100–511) is shown by lighter lines. Residues 301–312 of the flavin-binding domain are disordered and are not included in the model. (b) Stereoview of the interaction between flavin and haem in flavocytochrome b_2. A sphere represents the water molecule (#664) which forms hydrogen bonds to both the flavin and haem moieties. The diagrams were prepared using TURBO-FRODO (Roussel and Cambillau, 1991).

are six direct hydrogen bonds, one salt bridge and several water molecules between them. The salt bridge links one of the haem propionic acids to a lysine side chain of the flavin-binding domain (*Figure 5.3b*). The other propionate is hydrogen bonded to Tyr143. Thus, in the crystal, Tyr143 binds alternately to pyruvate (where there is no ordered haem) or to a haem propionate (when there is no pyruvate).

The interaction between the cytochrome and the flavin-binding domains is weak, the interface between them occupying only about 400 Å². When the flavoprotein and cytochrome domains are separated by controlled proteolysis they no longer associate with one another (Gervais *et al.*, 1983). In solution, the cytochrome domain in FCB2 is at least partially mobile with respect to the flavin-binding domain according to proton nuclear magnetic resonance (NMR) studies (Labeyrie *et al.*, 1988).

Calculation of likely paths for electron flow using the program PATHWAYS II (Regan, 1993) indicates that the most favourable route from the flavin N5 to the porphyrin plane follows a haem propionate and passes through a water molecule (#664, *Figure 5.3b*) hydrogen bonded betweem them. The next most favourable route is through the hydroxyl of Y143. This latter route is about fivefold lower in coupling, but is about fivefold higher than the next lower coupling utilizing other nearby side chains. If the route involving the water is ignored, because of the possibilities of variable occupancy or a higher tunnelling barrier, the Y143 route becomes the most important. Mutagenesis of Tyr143 to Phe reduces the electron transfer rate to the cytochrome 20-fold, although electron transfer to ferricyanide, an artificial electron acceptor, is essentially unchanged (Miles *et al.*, 1992). Pathway calculations with the Y143F mutant indicate a fivefold reduction of coupling to the same level as the alternative routes.

Mutagenesis has indicated other key points in FCB2 important for electron transfer. Deletion of three residues from the flexible hinge between the cytochrome and flavin-binding domains reduces flavin to haem electron transfer rates fivefold (Sharp *et al.*, 1994). When the hinge is swapped with one from another, closely related flavocytochrome b_2 (from *Hansanula anomala*), the rate falls 300-fold (White *et al.*, 1993). These experiments show that orientation of the two domains is very important for electron transfer in the complex, since disruption of the hinge seems to harm flavin to haem electron transfer rates but not the rate of substrate oxidation nor the electron transfer rate to ferricyanide.

p-*Cresol methylhydroxylase.* PCMH is a flavocytochrome *c* of M_r 116 kDa found in the periplasmic space of *Pseudomonas putida* and related bacteria (Hopper and Taylor, 1977). It is an $\alpha_2\beta_2$ heterotetramer containing two flavoprotein subunits each of M_r 49 kDa and two cytochrome subunits each of M_r 9 kDa (Kim *et al.*, 1994). The flavoprotein subunit contains FAD bound covalently as an 8-α-*O*-tyrosyl-FAD. The enzyme catalyses the oxidation of *p*-cresol first to *p*-hydroxybenzyl alcohol and then to *p*-hydroxybenzaldehyde. In each step, two electrons are passed from the substrate to the flavin to yield the hydroquinone. The electrons are then transferred one at a time to the haem on the cytochrome subunit and then to an acceptor protein, possibly an azurin or another cytochrome (Causer *et al.*, 1984). The rate of electron transfer between the flavin and haem in PCMH is greater than 200 s^{-1} (Bhattacharyya *et al.*, 1985).

PCMH is organized with the flavoprotein subunits tightly packed about a molecular twofold axis; the cytochrome subunits lie on the periphery of the flavoprotein dimer, and are nestled into depressions on its surface on opposite sides of the molecule (Mathews *et al.*, 1990). A single $\alpha\beta$ dimer is shown in *Figure 5.4a*. The cytochrome subunit is similar in structure to mitochondrial cytochromes, with the haem covalently attached by thioether linkage to two cysteine side chains and the iron coordinated to a histidine and a methionine

(a)

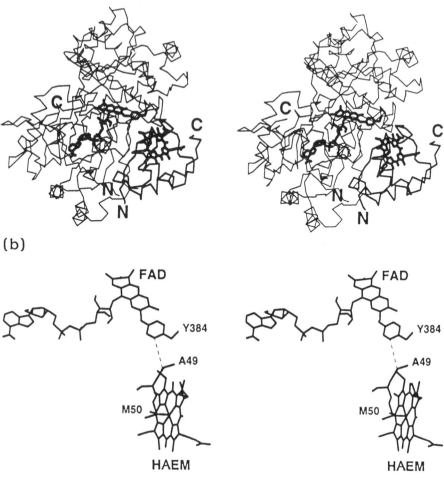

(b)

Figure 5.4. (a) The Cα trace of the flavoprotein (α) (light lines) and cytochrome (β) (heavy lines) subunits of p-cresol methylhydroxylase (PCMH). The flavin and haem cofactors as well as α:Y384 are shown by heavy lines. (b) Stereoview of the closest contacts between the flavin and haem subunits of PCMH. The Tyr α:Y384-bound flavin lies within 3.5 Å of β:A49 of the haem subunit; β:M50 is a ligand to the haem iron. The diagrams were prepared using TURBO-FRODO (Roussel and Cambillau, 1991).

side chain. The major difference is that the PCMH cytochrome has a large deletion in the centre of the sequence, between the two iron ligands. The main effect of this deletion is to leave the haem propionates relatively exposed to solvent. This region is also exposed in the intact flavocytochrome and may serve as the binding and/or electron transfer site for the natural electron acceptor protein for PCMH.

The flavoprotein subunit is folded into three domains. The N-terminal domain, of approximately 200 residues, envelopes the extended adenosine/ribityl phosphate portion of the FAD cofactor in a groove which runs parallel to several extended chains. This portion of FAD is also covered by the largely helical 40-residue C-terminal domain. The flavin ring protrudes into the central domain which is composed of about 210 residues and contains an 8-stranded antiparallel β-sheet forming a dome over one face of the flavin ring and encompassing the substrate binding site. The dome is covered on the outside by four antiparallel α-helices and contains a fifth α-helix on the inside. The site of covalent flavin attachment is located close to the central β-strand.

The interface between the flavoprotein and cytochrome subunits is quite extensive, covering an area of approximately 1200 $Å^2$. There are five hydrogen bonds and one salt bridge linking the cytochrome and flavoprotein subunits. Approximately half the residues in the interface are hydrophobic and another 30% are neutral hydrophilic. The relatively large area of the interface and strong interactions between subunits help explain the high stability of the complex which can only be resolved by isoelectric focusing (Koerber *et al.,* 1985) or by denaturation. The haem iron is about 14 Å from the flavin ring and the closest approach of the two groups is about 8 Å. The benzenoid portion of the isoalloxazine ring is oriented towards the thioether-containing edge of the haem group. The flavin and haem planes are distinctly non-co-planar, making an angle of about 65° to each other.

The covalent binding of FAD to the tyrosine side chain may be important for electron transfer from flavin to haem. The most direct path for electron transfer (*Figure 5.4b*) from the flavin N5 to the haem iron leads through the flavin 8α position and the phenolic ether bond of Tyr384 and then through three other bonds of the phenolic moiety. Following a 3 Å through-space jump to Ala49 of the cytochrome, the path follows Met50 to the haem iron to which it is liganded. Calculations using the program GREENPATH (Regan, 1994) indicate that this is by far the most efficient pathway. If the flavin were not bound covalently to the tyrosine side chain, a second through-space jump of roughly 3 Å would be introduced, markedly diminishing the electronic coupling of this path.

Flavocytochrome c *sulphide dehydrogenase.* FCSD from *Chromatium vinosum* is a periplasmic enzyme able to catalyse oxidation of sulphide ions to elemental sulphur *in vitro*, although its physiological role is not fully understood (Cusanovich *et al.,* 1991). The enzyme is a heterodimer of M_r 66 kDa consisting of a flavoprotein subunit of M_r 46 kDa and a dihaem cytochrome of M_r 21 kDa. The FAD is bound covalently to the flavoprotein subunit via an 8-α-methyl(*S*-cysteinyl)thioether linkage. The flavin of FCSD can bind sulphite and other ligands to form flavin N5 adducts which in turn form charge-transfer complexes with the protein. The redox potentials of the two haems are equal and unusually low (+15 mV at pH 7.0) while the two-electron redox potential of the flavin is unusually high (−26 mV at pH 7.0), compared with other members of the same class. Intramolecular electron transfer from flavin to haem is quite rapid ($>10^6$ s^{-1}). Sulphide dehydrogenase activity is supported by cytochrome *c* as an electron acceptor, and the two proteins form an electrostatically stabilized complex. FCSD also complexes with *Chromatium* cytochrome *c*-551 which is thought to be the electron acceptor *in vivo*.

The structure of FCSD is shown in *Figure 5.5a*. The cytochrome subunit of FCSD contains two domains and the flavoprotein subunit contains three domains (Chen, Z.W. *et al.,* 1994). The two domains of the cytochrome subunit are approximately equal in size and are similar in structure despite low sequence identity (only 7%)(Van Beeumen *et al.*, 1991). They are related by approximate twofold symmetry. The edges of the porphyrin rings are 11.4 Å apart and the two iron atoms are separated by 19.0 Å. The haem planes are inclined to each other by about 30° and point toward each other with their ring A propionic acid groups hydrogen bonded together in the protein interior. The other two

(a)

(b)

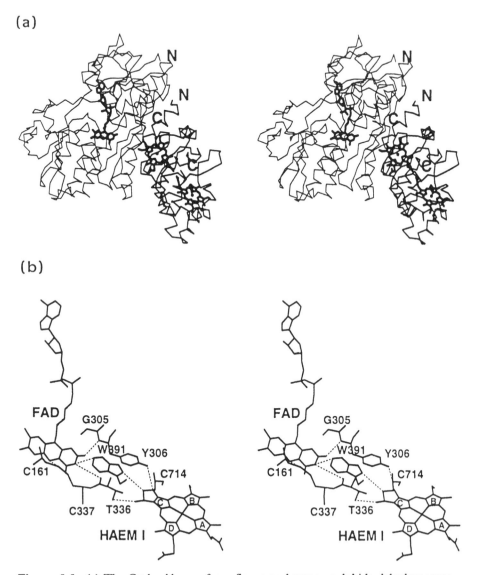

Figure 5.5. (a) The Cα backbone of one flavocytochrome *c* sulphide dehydrogenase monomer (PDB entry 1FCD). The flavin subunit is denoted by light lines, the cytochrome by heavy lines. The haem and flavin cofactors are also shown by heavy lines. (b) Stereoview of the interactions between the flavin and the closest haem from the cytochrome subunit in FCSD. Four potential electron transfer routes are indicated, involving G305/Y306, W391, T336 and C337/T336. The diagrams were prepared using TURBO-FRODO (Roussel and Cambillau, 1991).

propionates, on ring D, lie near the protein surface, but are only partially exposed to solvent.

The first two domains of the flavoprotein subunit closely resemble human glutathione reductase (GR) (Theime et al., 1981), each having a central parallel 5-stranded β-sheet flanked on one side by three α-helices and on the other side by a 3-stranded antiparallel β-sheet. The function of GR is to transfer reducing equivalents from NADPH, located in domain 2, to FAD, located in domain 1, and subsequently to glutathione via a redox-active disulphide, also located in domain 1. The FAD of FCSD is positioned almost identically to that in GR. Its site of covalent attachment is a cysteine, close to the amino terminus of domain 1, which corresponds approximately to the location of the redox-active disulphide in GR (Dolata et al., 1993). Above the pyrimidine portion of the flavin ring in FCSD there is a disulphide bridge which appears to be broken when the enzyme is reduced by a sulphite anion (Z.W. Chen and F.S. Mathews, unpublished results). This redox-active disulphide of FCSD links domains 2 and 3 and is located at the nicotinamide site of NADPH in GR. The third domain of FCSD consists of a 3-stranded antiparallel β-sheet followed by an α-helix. It is considerably smaller than the comparable domain of GR and serves to bind domain 1 of the cytochrome subunit in a concave depression on its surface.

In the intramolecular electron transfer complex of FCSD, the pyrimidine portion of the flavin ring lies closest to the haem of the N-terminal cytochrome domain with the flavin O2 9.9 Å away from a vinyl methylene atom of the haem. The planes of the haem and the flavin rings are inclined by about 20°. The interface between the flavoprotein and cytochrome subunits covers a surface area of about 1750 Å^2. The two surfaces are complementary, the cytochrome surface convex and the flavoprotein surface concave. There are 13 hydrogen bonds, including one salt bridge, connecting the two subunits. This sizeable interface and number of interactions may account for the inability of FCSD to be resolved into its subunits without denaturation.

Three side chains on the flavoprotein subunit may provide routes for electron flow from flavin to haem (*Figure 5.5b*). A tyrosine hydrogen bonded to the flavin O2 through its peptide nitrogen provides the most direct path to the haem, approximately 3.5 Å from its hydroxyl group. The indole ring of a tryptophan lies between the flavin and haem and provides an alternative path requiring two through-space jumps of about 3.5–4 Å. A threonine residue, adjacent to the disulphide bond, is situated at similar distances to the flavin and haem and provides a third route. A fourth path involves electron flow from the flavin N5 to the disulphide and then through the same threonine residue. Calculation of the electronic coupling of the four routes indicates that they are roughly the same.

Comparative aspects of the flavocytochromes. The interacting surfaces of the domains or subunits of the flavocytochromes are complementary to one another, with convex surfaces on the cytochrome nestled into concave surfaces on the flavoprotein. There is little ionic character to their interaction, there being only a single salt bridge linking the cytochrome- and flavin-binding moieties in all three cases. The extent of the interface area and/or the number of hydrogen bonds linking the domains or subunits varies with the strength of the interactions between them. FCB2 has the smallest interface (~400 Å²) but has six hydrogen bonds and a salt bridge. Its local interactions are weakest, as indicated by the lack of affinity of the isolated domains and the relative mobility of the cytochrome domain observed in the tetramer. Its global stability is provided by the polypeptide linking the two domains. PCMH has a larger interface than FCB2 (~1200 Å²), although the same number of intersubunit bonds, and can only be resolved into its components by isoelectric focusing or by denaturation. FCSD cannot be resolved without denaturation. Its interface is largest (~1750 Å²) and has twice the number of intersubunit links as FCB2 and PCMH, consistent with the very tight association of subunits in this complex.

There is considerable variability in the relationship of the flavin and haem in the three flavocytochromes. The angles between the haem and flavin planes are observed at values near 0° (FCB2), 20° (FCSD) and 60° (PCMH), demonstrating that the interplanar angle between redox cofactors can vary substantially in electron transfer complexes and is not constrained to co-planarity. The three flavocytochrome structures also show that there is variability in the pathway for electron flow through the flavin ring. The structures of all three flavocytochromes are consistent with reduction of the flavin by the substrate near the O4–N5 position. However, flavin reoxidation appears to involve electron flow through the O4–N5 position in FCB2, the C8α-methyl position in PCMH and the N1–O2 position in FCSD.

5.4   Complexes between redox proteins

5.4.1   *Complex of MADH with amicyanin and cytochrome c-551*ᵢ

The quinoprotein MADH is a periplasmic enzyme present in several methylotrophic and autotrophic bacteria (for review, see Davidson, 1993). It is a heterotetramer of M_r 125 kDa consisting of two heavy (H) and two light (L) chains of M_r 45 and 15 kDa, respectively. It contains, in each of the L subunits, the redox cofactor tryptophan tryptophylquinone (TTQ) consisting of two cross-linked tryptophan side chains, about 50 residues apart in sequence, one of which contains an orthoquinone. The enzyme converts methylamine to formaldehyde and ammonia, releasing two electrons which are passed along a soluble electron-transport chain to a membrane-bound terminal oxidase.

For MADH from *Paracoccus denitrificans* and related bacteria the physiological electron acceptor from MADH is amicyanin, a cupredoxin. The genes for MADH and amicyanin are located in the same operon, the expression of which is induced by growth on methylamine. Amicyanin, with $M_r = 12$ kDa, is most similar to plastocyanin (*Figure 5.1a*) and has virtually identical copper-binding geometry. It has one additional β-strand at the amino terminus which extends the size of the first sheet of the β-sandwich. Complex formation between MADH and amicyanin *in vitro* causes perturbation of the absorption spectrum of TTQ and a shift in the redox potential of the copper centre of amicyanin from 294 to 221 mV.

Electron transfer from amicyanin to the terminal oxidase is believed to involve one or more soluble *c*-type cytochromes. *In vitro* experiments indicate that three periplasmic cytochromes isolated from *P. denitrificans* may serve as electron acceptors for MADH through the mediation of amicyanin. Among them, cytochrome c-551$_i$ has been shown to be the most efficient electron acceptor. Cytochrome c-551$_i$ has a molecular weight of 17 kDa and a total of 155 amino acid residues (van Spanning *et al.*, 1991). It is very acidic (pI = 5.5) and contains approximately 40 additional residues at the amino terminus and 30 at the carboxyl terminus compared with most bacterial c-type cytochromes. The charged residues mostly cover the side and rear portions of the cytochrome surface (*Figure 5.1a*), leaving the front face, containing the exposed CBC methyl group, surrounded by hydrophobic residues. There is evidence that a ternary protein complex of MADH, amicyanin and cytochrome c-551$_i$ forms in solution before electron transfer from MADH to the cytochrome can occur, since the latter does not accept electrons directly from free MADH or from free amicyanin (see Davidson, 1993).

The crystal structures of the binary complex between MADH and amicyanin and of the ternary complex between MADH, amicyanin and cytochrome c-551$_i$ have been solved at 2.5 and 2.4 Å resolution, respectively (Chen, L. *et al.*, 1992, 1994). The binary complex contains one hetero-hexamer of the type $H_2L_2A_2$ in the crystallographic asymmetric unit while the ternary complex contains half the hetero-octamer of the type $H_2L_2A_2C_2$ (where A and C refer to the amicyanin and cytochrome c-551$_i$ respectively). In both cases, the crystals were prepared in approximately 2.4 M Na/K phosphate buffer, pH 5.5. The copper-free apo-binary and the apo-ternary complexes are virtually identical in structure to the holo-complexes.

The structure of MADH and of amicyanin are the same in both the binary and ternary complexes. The relationship of the cytochrome, amicyanin and the light subunit of MADH in the ternary complex is shown in *Figure 5.6a*. Amicyanin is in contact with both MADH and the cytochrome. The two molecules are oriented with the unmodified tryptophan of TTQ facing the exposed histidine ligand of copper. The distance from the O6 atom of TTQ, where substrate oxidation is believed to occur, to the copper is 16.8 Å. The closest distance from TTQ to the copper atom is 9.3 Å. Amicyanin and the cytochrome are oriented so that β-strand 2 of amicyanin (using the cupredoxin numbering scheme), which is shared between the two sheets of the β-sandwich (*Figure 5.1c*), makes contact with a peptide segment of the cytochrome. This segment is located between the histidine iron ligand and the start of a short helix located just before the methionine iron ligand (see *Figure 5.1a*), close to one of the haem propionates. The copper to iron distance is 24.7 Å and the distance from copper to the nearest atom of the haem is about 21 Å. Within the complex, the O6 of TTQ, the copper and the iron are nearly co-linear, with the O6 atom separated from the iron atom by 41.2 Å.

The MADH–amicyanin interface is largely hydrophobic and covers an area of approximately 750 Å². Of the MADH residues in the interface, 65% come from the L subunit and the remainder from the H subunit. The interface also contains the normally exposed histidine ligand to the copper of amicyanin as well as seven surrounding hydrophobic residues (three methionines, three prolines and a phenylalanine) which form a hydrophobic patch. A similar hydrophobic patch centred on an exposed histidine ligand is found in the other cupredoxins and has been implicated in electron transfer. About 40% of the residues in the interface are hydrophilic but only the aliphatic portions of their

(a)

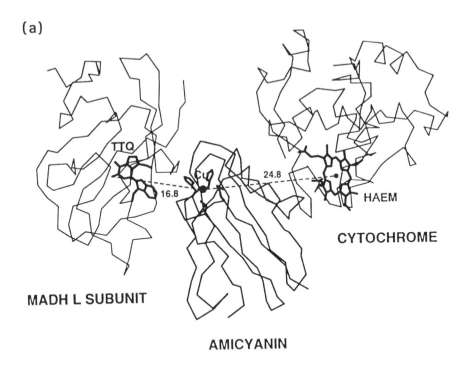

(b)

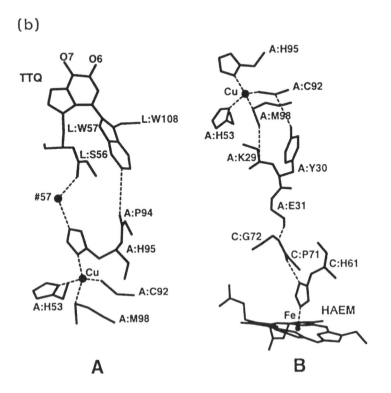

Figure 5.6. (a) The Cα backbone of three components of the MADH–amicyanin–cytochrome c-551$_i$ ternary complex (PDB entry 2MTA) showing the orientation of the three cofactors. The TTQ cofactor, the copper and its ligands and the haem group are drawn with heavy lines. Distances are shown in Å.
(b) Pathways calculated for electron transfer in the ternary complex. A shows two paths from the TTQ of MADH to the copper of amicyanin. A sphere represents the water (#57) which is hydrogen bonded to the L subunit of MADH and to a histidine ligand of amicyanin. B shows the branched path calculated from the copper ion to the cytochrome haem. The diagrams were prepared using TURBO-FRODO (Roussel and Cambillau, 1991).

side chains lie within the interface, leaving the polar portions of the residues pointing into solution. There are two salt bridges and three water molecules connecting the two proteins. The presence of the salt bridges connecting MADH with amicyanin, despite the largely hydrophobic nature of the interface, is consistent with the effects of ionic strength on cross-linking and on the kinetic properties of the complex (see Davidson, 1993). These effects indicate that the binary complex is stabilized by both hydrophobic and electrostatic forces. The electrostatic component of the interface may also contribute to the 73 mV decrease in the redox potential of amicyanin upon complex formation.

The amicyanin–cytochrome interface is smaller than the quino-protein–cupredoxin interface, covering approximately 425 Å^2. It is also more polar, with approximately 65% of its residues hydrophilic. The amicyanin and cytochrome are joined by one salt bridge, four hydrogen bonds and two solvent molecules. Based on its coordination, one of the solvent molecules may be a cation such as sodium or potassium.

Using the program PATHWAYS-II (Regan, 1993) as a guide, two main paths for electron flow were found leading from the quinone-containing indole ring of TTQ to the copper atom (*Figure 5.6b*). One of these goes through L:Trp108 then on to residue A:Pro94 and to the copper via A:His95. The other path follows the side and main chain of L:Trp57 to L:Ser56 and to the side chain of A:His95 via hydrogen bonds to an intra-complex water molecule. The electronic coupling of this path is about threefold more efficient than the former. However, the pathway depends critically on the presence of a water molecule, which corresponds to an intermediate level of electron density, and might be only partially occupied, thereby reducing the relative efficiency of this pathway. The efficiency of the former path is limited by a through-space jump of 3.5 Å from L:Trp108 of TTQ to A:Pro94. If electron delocalization extends throughout TTQ, then electron transfer directly from L:Trp108 to A:His95 might also be feasible. For electron transfer from copper to iron, the two most likely paths partially overlap (*Figure 5.6b*). The common portion involves main chains of A:Tyr30, A:Glu31, C:Gly72 and C:Pro71 and the side chain of C:His61, an iron ligand. This path includes passage through two hydrogen bonds, one from A:Glu31 O to C:Gly72 N and the other from C:Pro71 O to C:His61 ND1. The two paths differ in routing from the copper to the main chain of A:Tyr30. In one case, transfer is via the side chain of A:Cys92, a copper ligand, through the side chain of A:Tyr30 and on to the main chain. In the other, the path goes through A:Met98 CE, another copper ligand, and on to the main chain of A:Tyr30. Both of these paths include a through-space jump,

3.4 Å between A:Cys92 CB and A:Tyr30 OH in the first case, and 4.1 Å between A:Met 98 CB to A:Lys 29 C in the other. Both branches have approximately equal efficiency. However, the route through Cys92 may be favoured by the electronic structure of the copper (Solomon and Lowery, 1993).

5.4.2 *Cytochrome c peroxidase*

CCP is a 31 kDa haemoprotein which catalyses the reduction of alkyl hydroperoxides in yeast mitochondria (Bosshard *et al.*, 1991). The initial product of the two-electron oxidation of CCP by the substrate yields 'compound I', which contains a ferryl iron species [Fe(IV)] and a free radical tryptophan side chain (Trp*), both of which are reduced subsequently by ferrocytochrome *c* in two one-electron-transfer steps. Reduction of CCP is most efficiently achieved by yeast cytochrome *c*, but horse heart cytochrome *c* is also an effective reductant, although at a somewhat slower rate.

CCP is a highly helical structure with 10 α-helices, one 3-stranded antiparallel β-sheet, two β-hairpins and six extended chains (Finzel *et al.*, 1984). The structure forms two domains; domain 1 contains residues 1–144 and 254–294 and domain 2 consists of residues 145–253. The haem group is located in a cavity between the two domains, with a helix from the second part of domain 1 forming the back wall of the cavity. The haem moiety, including the two propionic acid groups, is almost totally buried. The haem iron is 5-coordinate with the proximal His175 forming the fifth ligand through its NE atom. The active site pocket for peroxide reduction is on the opposite side of the haem and contains the distal His52 and Arg48 as well as Trp51. His52 and Arg48 form hydrogen bonds to small ligands such as NO, CO, CN$^-$ and F$^-$ (Edwards and Poulos, 1990). The proximal pocket contains Trp191, the site of the free radical intermediate after substrate reduction (Sivaraja *et al.*, 1989). Trp191 is perpendicular to the haem and parallel to His175 and is in van der Waals contact with both. Asp235 is on the opposite edge of His175 and is hydrogen bonded to both His175 ND1 and Trp191 NE1.

A hypothetical model for the interaction of CCP with cytochrome *c* was proposed (Poulos and Finzel, 1984) based on the presence of a ring of positive charges surrounding the exposed haem edge of cytochrome *c* and a group of negative charges near the haem-binding crevice in CCP. Charge interactions of Asp33, Asp34, Asp37 and Asp217 in CCP with Lys8, Lys87, Lys13 and Lys72 in tuna cytochrome *c* were proposed. In the model, the two haem groups were approximately co-planar, with edge-to-edge and iron-to-iron separations of 18 and 25 Å, respectively.

There are now X-ray crystal structures of two complexes, one between CCP and yeast iso-1-cytochrome *c* (CY) (the CCP–CY complex) and the other between CCP and horse heart cytochrome *c* (CH) (the CCP–CH complex) (Pelletier and Kraut, 1992). Although only the former is between physiological partners, both complexes have been studied thoroughly in solution. The CCP–CY crystals were grown at 150 mM NaCl (using polyethylene glycol as precipitant) while the CCP–CH crystals contained 5 mM NaCl [with 2-methyl-2,4-pentanediol (MPD) as precipitant]. The interactions between electron transfer partners are similar in both complexes. The difference between them could arise from the differences in ionic strength of the crystallization conditions for the two crystal forms, or from specific amino acid substitutions in CY versus CH.

The CCP–CY complex is shown in *Figure 5.7a*. The interface between CCP and CY is largely hydrophobic with a surface area of about 640 Å². CCP is linked to CY by two salt bridges (Asp34 and Glu290 to Lys87 and Lys73, respectively) and by one weak hydrogen bond. Several additional acidic and basic residues form part of the interface and may add to the non-specific dipolar interactions between the proteins observed in solution (Kang *et al.*, 1977). The closest part of the CY haem group to CCP in the complex is methyl group CBC on pyrrole ring C which is in van der Waals contact with Ala173 and Ala174 of CCP. The haem planes are inclined by about 60° to each other and the iron-to-iron distance is about 26 Å. There is a predominantly covalent pathway connecting the haem groups of CCP and CY via Trp191 (the free radical site), Gly192, Ala193 and Ala194 (*Figure 5.7b*).

The structure of the CCP–CH complex differs from that of the CCP–CY complex, having a slight translation and rotation of the CH molecule relative to the CCP molecule. The interface is less hydrophobic and covers a smaller area, approximately 560 Å². The CCP and CH molecules are joined by four hydrogen bonds, two of which are mediated by charge. The remaining interactions are hydrophobic, although the interface again contains a number of additional acidic and basic residues. The hydrogen-bonding interactions between CCP and CH are located at one edge of the interface. At the other edge, the CBC methyl of pyrrole ring C is about 7 Å from Ala174 of CH and the iron-to-iron distance in the complex is about 30 Å. The pathway between the two haems now involves a considerably larger through-space jump and appears to be much less efficient than in the CCP–CY complex.

These complexes (CCP–CY and CCP–CH) are the first to allow a direct comparison between a theoretical and an observed electron transfer complex. Only one salt bridge (Asp34 of CCP to Lys87 of cytochrome *c*) agrees with the predictions of the hypothetical complex. In general, fewer hydrogen bonds and salt bridges are observed in the structures than had been predicted. The two haem groups of the hypothetical complex were predicted to be co-planar, but are observed to be tilted by about 60°. The CCP–CY structure reports an iron-to-iron distance of about 26 Å, close to the predicted value of 25 Å, but the predicted haem-to-haem distance was considerably shorter than observed in either crystal structure.

There are several questions concerning the CCP–CY complex. First, is the interface observed in the crystals the physiological one for electron transfer? Second, is there one or more than one electron transfer site on CCP for the cytochrome? Third, is the electron transfer route from cytochrome *c* to Fe(IV) the same as to the Trp191 radical? Stopped flow kinetic experiments have shown that at high ionic strength (100 mM or greater) reduction of the Trp191 radical of 'compound I' occurs first, in a rapid bimolecular reaction with cytochrome *c* (Hahm *et al.*, 1994; Nuevo *et al.*, 1993). This is followed by a second bimolecular electron transfer reaction to cytochrome *c* at a rate 5- to 10-fold lower than the first (Miller *et al.*, 1994). Mutation in CCP of either Asp34 or Glu290 (which both form salt bridges to cytochrome *c*) to Asn or Gln or of

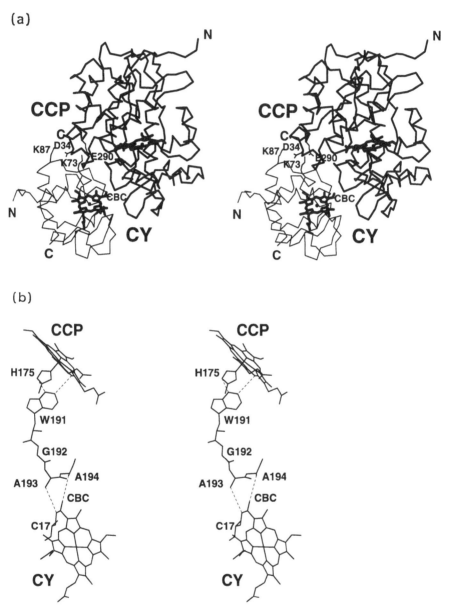

Figure 5.7. (a) The stereoview of the Cα backbone of the complex between cytochrome *c* peroxidase and cytochrome *c* (PDB entry 2PCC), both from yeast. The monomer of CCP is shown by bold lines and the cytochrome by light lines. The four residues which form salt bridges are also shown in this view. The haem cofactors from each protein are shown by bold lines. (b) Stereoviews of the pathway linking the haem groups of CCP and yeast cytochrome *c*, branching both at W191 and A193/A194 of CCP. The component, W191, is the radical site for the substrate-oxidized form of CCP. The diagrams were prepared using TURBO-FRODO (Roussel and Cambillau, 1991).

Ala173, which is in contact with the haem of cytochrome c, to Phe reduces both the fast and the slow electron transfer rate from CY or CH by 2- to 4-fold. Substitutions of amino acids in CCP which are within the interface but not directly linked to cytochrome c or are outside the interface have little or no effect on the electron transfer kinetics. Since the response of both the faster and slower electron transfer rates to the mutational changes and to variation in ionic strength are similar, it is likely that under these conditions the binding site on CCP for cytochrome c is the same for both one-electron reduction steps and that it corresponds to the binding sites observed in the crystals of the complexes.

At low ionic strength (< 100 mM), CCP binds to cytochrome c at a second site with an affinity approximately 1000-fold lower than for the first binding site (Mauk *et al.*, 1994; Stemp and Hoffman, 1993). Under these conditions, stopped flow kinetic studies (Nuevo *et al.*, 1993) indicate that 'compound I' is first reduced to the Fe(III)/Trp* form. Studies of photoinduced electron transfer quenching of metal-substituted CCP or cytochrome c (Zhou *et al.*, 1995) at low ionic strength using a redox-inactive copper-containing cytochrome as a competitive inhibitor indicate that the low affinity site for cytochrome c accommodates up to 1000-fold higher electron transfer rates than the high affinity site. Thus it may be that a second, independent site of low affinity for cytochrome c exists at low ionic strength which provides much more rapid electron transfer from the iron site of CCP than the radical site, and which has not yet been observed crystallographically.

5.5 Activity of crystals

5.5.1 Flavocytochrome b_2

The natural electron acceptor for flavocytochrome b_2 in yeast is cytochrome c. The electron transfer rates, from stopped flow studies (Capeillere-Blandin, 1982), are strongly ionic strength dependent, with bimolecular kinetics above 0.5 M ionic strength and unimolecular kinetics below, indicating that complex formation is largely mediated by ionic interactions. Crystals of FCB2 grown from 30% MPD, and their interaction with cytochrome c have been studied by single crystal polarized absorption microspectrophotometry (Tegoni *et al.*, 1983). Treatment of oxidized crystals with lactate results in simultaneous reduction of flavin and b_2 haem at a rate which is limited by diffusion of substrate through the crystal. Reduced cytochrome c will bind to the b_2 crystals up to a stoichiometry of one cytochrome c molecule per subunit of FCB2 at low ionic strength (0.01 M phosphate) but not at high ionic strength (0.1 M phosphate). The crystalline flavocytochrome b_2–cytochrome c complex could also be oxidized by ferricyanide and then reduced by lactate, and remained stable in both oxidation states.

It appears that flavocytochrome b_2 in the crystalline state is enzymatically active and is able to form a stable 1:1 complex (subunit:molecule) with cytochrome c which is functional for electron transfer, since cytochrome c cannot be reduced directly by lactate. Unfortunately it has not been possible to visualize the bound cytochrome c in crystals of the binary complex with flavocytochrome b_2 by X-ray crystallography (M. Tegoni and F.S. Mathews, unpublished results).

5.5.2 The MADH system

Enzymatic and electron transfer activities have been studied in crystals of both the binary and the ternary complexes of MADH with its redox partners (G.L. Rossi, V.L. Davidson and F.S. Mathews, unpublished results). Polarized absorption spectra were recorded from single crystals of the complexes treated with methylamine at several pH values using visible light in the wavelength range of 300–800 nm. Appearance and disappearance of oxidized semiquinone and reduced signals from TTQ and of oxidized and reduced cytochrome could be followed easily.

At pH 5.5, the crystals of the binary complex showed full reduction of the TTQ by methylamine with little or no semiquinone formation. The rate of reduction was limited by the diffusion of substrate through the crystal. At pH 7.5, signals for both semiquinone and reduced TTQ appeared in approximately equal abundance and, at pH 9.0, the TTQ was almost completely in the semiquinone form. At constant methylamine concentration in the crystal, the ratio of semiquinone to reduced TTQ could be shifted reversibly by altering the pH of the crystal bathing medium. When the ternary complex was examined, haem reduction occurred readily at pH 7.5, again at a diffusion-controlled rate, upon treatment of the crystals with methylamine. At pH 5.5, reduction could be achieved, but at a much slower rate, over a period of several hours.

These results show that the binary and ternary complexes are competent both for substrate oxidation and for electron transfer. In the case of the binary complex, formation of the TTQ semiquinone can only occur by electron transfer from the substrate-reduced TTQ to the copper. The extent of the electron transfer is dependent on pH, indicating that the relative redox potentials for the semiquinone/reduced couple of TTQ compared with the Cu^+/Cu^{2+} couple is also pH dependent. The slow rate of haem reduction in the ternary complex at pH 5.5 may reflect the unfavourable equilibrium of reduced copper available in the crystalline complex at this pH.

5.6 Complex dynamics

The crystal structures of the electron transfer complexes presented in this chapter represent static views of the interaction geometries between electron transfer partners which are probably close to the minimum energy configurations under the conditions of crystallization. Although very informative, these views do not necessarily provide the exact geometry of the proteins at physiological conditions for electron transfer and give little information on the role of dynamics in the electron transfer process.

Evidence for the need for some mobility or partial rearrangement of electron transfer partners comes from electron transfer studies within complexes stabilized by electrostatic interactions. For example, Hazzard et al. (1988) found that electron transfer within the cytochrome c–CCP complex was slower at low ionic strength (8 mM) than at high ionic strength (275 mM), implying that the low ionic strength form of the complex, presumably stabilized by strong electrostatic or hydrogen bonding interactions, was not optimal for electron transfer. The experiments also suggested that at high ionic strength ionic shielding of polar groups allowed the complex to relax to a better orientation. Similarly, studies of electron transfer between plastocyanin and cytochrome f (Meyer *et al.*, 1993) showed that the first order electron transfer rate within the electrostatically stabilized complex doubled on going from 5 to 40 mM ionic strength and then fell monotonically on going to higher ionic strength. Again, this study suggests that the electrostatically most stable configuration is not necessarily best for electron transfer.

Other suggestive evidence for the importance of mobility in electron transfer comes from crystallographic and NMR studies of complexes. When the electron density of a cytochrome c–CCP crystal [prepared under conditions where the complex was known to form and similar to those used by Pelletier and Kraut (1992)] was examined, no interpretable density for the cytochrome c molecule could be observed, even though it was known to be present in the crystal (Poulos *et al.*, 1987). This indicated a high degree of positional disorder or movement of the cytochrome molecule in the crystal. Similarly, crystals of FCB2 containing bound and functionally active cytochrome c (Tegoni *et al.*, 1983) failed to reveal ordered cytochrome c when examined by X-ray crystallography (M. Tegoni and F.S. Mathews, unpublished data). Likewise, the cytochrome b domain in FCB2 is mobile when examined by NMR relaxation (Labeyrie *et al.*, 1988) and, in part, is postionally disordered in the electron density as described earlier (Xia and Mathews, 1990).

The above examples do not prove that a given static structure of a complex observed by X-ray or other structural technique is not the optimal one for electron transfer. However, it must be borne in mind that reorientations are possible during electron transfer and that additional experiments are needed to verify the validity of any observed or, especially, theoretical model.

References

Adman ET. (1991) Copper protein structures. *Adv. Protein Chem.* **42:** 145–197.

Bhattacharyya A, Tollin G, McIntire WS, Singer TP. (1985) Laser-flash-photolysis studies of *p*-cresol methylhydroxylase. Electron transfer properties of the flavin and haem components. *Biochem. J.* **228:** 337–345.

Bosshard HR, Anni H, Yonetani T. (1991) Yeast cytochrome *c* peroxidase. In: *Peroxidases in Chemistry and Biology,* Vol. 2 (eds J Everse, K Everse, MB Grisham) CRC Press, Boca Raton, pp. 51–83.

Capeillere-Blandin C. (1982) Transient kinetics of the one-electron transfer reaction between reduced flavocytochrome b_2 and oxidized cytochrome c. *Eur. J. Biochem.* **128:** 533–542.

Causer MJ, Hopper DJ, McIntire WS, Singer TP. (1984) Azurin from *Pseudomonas putida*: an electron acceptor from *p*-cresol methylhydroxylase. *Biochem. Soc. Trans.* **12:** 1131–1132.

Chen L, Durley R, Poliks BJ, Hamada K, Chen Z, Mathews FS, Davidson VL, Satow Y, Huizinga E, Vellieux FMD, Hol WGJ. (1992) Crystal structure of an electron transfer complex between methylamine dehydrogenase and amicyanin. *Biochemistry* **31:** 4959–4964.

Chen L, Durley RC, Mathews FS, Davidson VL. (1994) Structure of an electron transfer complex: methylamine dehydrogenase, amicyanin, and cytochrome c-551$_i$. *Science* **264:** 86–90.

Chen ZW, Koh M, Vandriessche G, Van Beumen JJ, Bartsch RG, Meyer TE,

Cusanovich MA, Mathews FS. (1994) The structure of flavocytochrome c sulfide dehydrogenase from a purple phototrophic bacterium. *Science* 266: 430–432.

Cusanovich MA, Meyer TE, Bartsch RG. (1991) Flavocytochrome c. In: *Chemistry and Biochemistry of the Flavoenzymes*, Vol. 2 (ed. F Muller). CRC Press, Boca Raton, pp. 377–393.

Davidson M. (1993) Methylamine dehydrogenase. In: *Principles and Applications of Quinoproteins* (ed. V Davidson). Marcel Decker, Inc., New York, pp. 73–95.

Dolata MM, van Beeumen JJ, Ambler RP, Meyer TE, Cusanovich MA. (1993) Nucleotide sequence of the haem subunit of flavocytochrome c from the purple phototrophic bacterium, *Chromatium vinosum*. *J. Biol. Chem.* 268: 14426–14431.

Edwards SL, Poulos TL. (1990) Ligand binding and structural perturbations in cytochrome c peroxidase. *J. Biol. Chem.* 265: 2588–2595.

Farver O, Pecht I. (1992) Low activation barriers characterize intramolecular electron transfer in ascorbate oxidase. *Proc. Natl Acad. Sci. USA* 89: 8283–8287.

Farver O, Blatt Y, Pecht I. (1982) Resolution of two distinct electron transfer sites on azurin. *Biochemistry* 21: 3356–3361.

Finzel BC, Poulos TL, Kraut J. (1984) Crystal structure of yeast cytochrome c peroxidase refined at 1.7 Å resolution. *J. Biol. Chem.* 21: 13027–13036.

Gervais M, Risler J, Corazzin S. (1983) Proteolytic cleavage of *Hansenula anomala* flavocytochrome b_2 into its two functional domains. *Eur. J. Biochem.* 130: 253–259.

Hahm S, Miller MA, Geren L, Kraut J, Durham B, Millet F. (1994) Reaction of horse heart cytochrome c with the radical and the oxyferryl haem in cytochrome c peroxidase. *Biochemistry* 33: 1473–1480.

Hazzard JT, McLendon G, Cusanovich MA, Tollin G. (1988) Formation of electrostatically-stabilized complex at low ionic strength inhibits interprotein electron transfer between yeast cytochrome c and cytochrome c peroxidase. *Biochem. Biophys. Res. Commun.* 151: 429–434.

He S, Modi S, Bendall DS, Gray JC. (1991) The surface exposed Tyr83 of pea plastocyanin is involved in both binding and electron transfer reactions with cytochrome f. *EMBO J.* 10: 4011–4016.

Hopper DJ, Taylor DG. (1977) The purification and properties of p-cresol-(acceptor) oxidoreductase (hydroxylating), a flavocytochrome from *Pseudomonas putida*. *Biochem. J.* 167: 155–162.

Kang CH, Ferguson-Miller S, Margoliash E. (1977) Steady state kinetics and binding of eukaryotic cytochrome c with yeast cytochrome c peroxidase. *J. Biol. Chem.* 252: 919–926.

Kim J, Fuller JH, Cecchini G, McIntire WS. (1994) Cloning, sequencing and expression of the structural genes for the cytochrome and flavoprotein subunits of p-cresol methylhydroxylase from *Pseudomonas putida*. *J. Bacteriol.* 176: 6349–6361.

Koerber SC, Hopper DJ, McIntire WS, Singer TP. (1985) Formation and properties of flavoprotein–cytochrome hybrids by recombination of subunits from different species. *Biochem. J.* 231: 283–287.

Kraulis PJ. (1991) MOLSCRIPT: a program to produce both detailed and schematic plots of protein structures. *J. Appl. Crystallogr.* 24: 946–950.

Labeyrie F, Beloeil J-C, Thomas M-A. (1988) Evidence by NMR for mobility of the cytochrome domain within flavocytochrome b_2. *Biochim. Biophys. Acta* 953: 134–141.

Lederer F. (1991) Flavocytochrome b_2. In: *Chemistry and Biochemistry of the Flavoenzymes*, Vol. 2 (ed. F Muller). CRC Press, Boca Raton, pp. 153–242.

Mathews FS. (1985) The structure, function and evolution of cytochromes. *Prog. Biophys. Mol. Biol.* 45: 1–56.

Mathews FS, Chen Z-w, Bellamy HD, McIntire WS. (1990) The 3-dimensional structure of p-cresol methylhydroxylase (flavocytochrome c) at 3.0 Å resolution. *Biochemistry* 31: 238–247.

Mauk GR, Ferrer JC, Mauk AG. (1994) Proton linkage in formation of the cytochrome c–cytochrome c peroxidase complex: electrostatic properties of the high- and low-affinity cytochrome binding sites on the peroxidase. *Biochemistry* 33: 12609–12614.

Messerschmidt A, Ladenstein R, Huber R, Bolognesi M, Avigliano L, Petruzzelli R, Rossi A, Finazzi-Agró A. (1992) Refined crystal structure of ascorbate oxidase at 1.9 Å resolution. *J. Mol. Biol.* 224: 179–205.

Meyer TE, Marchesini A, Cusanovich MA, Tollin G. (1991) Direct measurement of intramolecular electron transfer between type-1 and type-3 copper centres in the multi-copper enzyme, ascorbate oxidase and its type-2 depleted and cyanide inhibited forms. *Biochemistry* 30: 4619–4613.

Meyer TE, Zhao AG, Cusanovich MA, Tollin, G. (1993) Transient kinetics of electron transfer from a variety of *c*-type cytochromes to plastocyanin. *Biochemistry* 32: 4552–4559.

Miles CS, Rouviere-Fourmy N, Lederer F, Mathews FS, Reid GA, Black MT, Chapman SK. (1992) Tyr-143 facilitates interdomain electron transfer in flavocytochrome b_2. *Biochem. J.* 285: 187–192.

Miller MA, Liu R-Q, Hahm S, Geren L, Hibdon S, Kraut J, Durham B, Millett F. (1994) Interaction domain for the reaction of cytochrome *c* with the radical and the oxyferryl haem in cytochrome *c* peroxidase compound I. *Biochemistry* 33: 8686–8693.

Nuevo MR, Chu HH, Vitello LB, Erman JE. (1993) Salt-dependent switch in the pathway of electron transfer from cytochrome *c* to cytochrome *c* peroxidase compound I. *J. Am. Chem. Soc.* 115: 5873–5874.

Pelletier H, Kraut J. (1992) Crystal structure of a complex between electron transfer partners, cytochrome *c* peroxidase and cytochrome *c*. *Science* 258: 1748–1755.

Poulos TL, Finzel BC. (1984) Haem enzyme structure and function. In: *Peptide and Protein Reviews,* Vol. 4 (ed. MTW Hearn). Dekker, New York, pp. 115–171.

Poulos TL, Sheriff S, Howard A. (1987) Cocrystals of yeast cytochrome *c* peroxidase and horse heart cytochrome *c*. *J. Biol. Chem.* 262: 13881–13884.

Regan JJ. (1993) *PATHWAYS II, Version 2.01.* San Diego.

Regan JJ. (1994) *GREENPATH, Version 0.97.* San Diego.

Roussel A, Cambillau C. (1991) TURBO-FRODO In: *Silicon Graphics Geometry Partners Directory.* Silicon Graphics, p. 86.

Salemme FR. (1976) A hypothetical structure for an intermolecular electron transfer complex of cytochromes *c* and b_5. *J. Mol. Biol.* 102: 563–568.

Sharp RE, White P, Chapman SK, Reid GA. (1994) Role of the interdomain hinge of flavocytochrome b_2 in intra- and inter-protein electron transfer. *Biochemistry* 33: 5225–5230.

Sivaraja M, Goodin DB, Smith M, Hoffman BM. (1989) Identification by ENDOR of Trp-191 as the free-radical site in cytochrome *c* peroxidase compound ES. *Science* 245: 738–740.

Solomon EI, Lowery MD. (1993) Electronic structure contributions to function in bioinorganic chemistry. *Science* 259: 1575–1581.

Stemp EDA, Hoffman, BM. (1993) Cytochrome *c* peroxidase binds two molecules of cytochrome *c*: evidence for a low-affinity, electron-transfer-active site on cytochrome *c* peroxidase. *Biochemistry* 32: 10848–10865.

Tegoni M, Mozzarelli A, Rossi GL. (1983) Complex formation and intermolecular electron transfer between flavocytochrome b_2 in the crystal and cytochrome *c*. *J. Biol. Chem.* 258: 5424–5427.

Theime R, Pai EF, Schirmer RH, Schulz GE. (1981) The three dimensional structure of glutathione reductase at 2 Å resolution. *J. Mol. Biol.* 152: 763–782.

Van Beeumen JJ, Demol H, Samyn B, Bartsch RG, Meyer TE, Dolata MM, Cusanovich MA. (1991) Covalent structure of the diheme cytochrome subunit and amino-terminal sequence of the flavoprotein subunit of flavocytochrome *c* from *Chromatium vinosum*. *J. Biol. Chem.* 266: 12921–12931.

van Spanning RMJ, Wansell CW, De Boer T, Hazelaar MT, Anazawa H, Harms N, Oltmann LF, Stouthamer AH. (1991) Isolation and characterization of the *MoxJ*, *MoxG*, and *MoxR* genes of *Paracoccus denitrificans*: inactivation of *moxJ*, *moxG*, and *moxR* and the resultant effect on methylotrophic growth. *J. Bacteriol.* 173: 6948–6961.

White P, Manson FDC, Brunt CE, Chapman SK, Reid GA. (1993) The importance of the interdomain hinge in intramolecular electron transfer in flavocytochrome b_2. *Biochem. J.* 291: 89–94.

Xia Z-x, Mathews FS. (1990) Molecular structure of flavocytochrome b_2 at 2.4 Å resolution. *J. Mol. Biol.* 212: 837–863.

Zhou JS, Nocek JM, DeVan ML, Hoffman BM. (1995) Inhibitor-enhanced electron transfer: copper cytochrome *c* as a redox-inert probe of ternary complexes. *Science* 269: 204–207.

Photosynthetic bacterial reaction centres

William W. Parson

6.1 Introduction

When its structure was determined by Deisenhofer, Michel and their colleagues (Deisenhofer *et al.*, 1985; Michel *et al.*, 1986), the photosynthetic reaction centre of *Rhodopseudomonas viridis* was the first integral membrane protein to yield a high resolution crystal structure and was the largest protein of any type for which such a structure had been obtained. Its four polypeptide subunits with a combined molecular weight of approximately 140 000 and its 13 pigments made the reaction centre a formidably complex system. Yet, in spite of its complexity, the bacterial reaction centre quickly became the hydrogen atom of protein electron transfer. Reasons for this are not hard to find. First, reaction centres are relatively easy to purify from several species of bacteria and are chemically stable. Their bound pigments provide a series of at least eight electron transfer reactions in a single protein of known structure, affording a wealth of data on how electron transfer dynamics depend on intermolecular distances and pigment–protein interactions. Rich optical absorption, electron paramagnetic resonance, electron–nuclear double resonance (ENDOR), Fourier transform infrared and fluorescence signals offer experimental handles on the electronic interactions among the pigments and on the energies and lifetimes of the redox states of the system. Because the electron transfer reactions can be initiated with a short flash of light, they can be examined with extremely high time resolution under a broad range of conditions. Their rate constants span 11 orders of magnitude, from approximately 10 s^{-1} to 10^{12} s^{-1}. The electron transfer reactions reveal a directional specificity whose origin is not obvious from the crystal structure, and some have other unusual features that beg for theoretical explanations, such as increases in speed with decreasing temperature. The fact that they occur on the picosecond time scale makes the first few reactions amenable to computer simulations by molecular dynamics. Finally, the biological importance of photosynthesis lends special significance to the goal of understanding how the reaction centre captures the energy of sunlight with a quantum yield of essentially 100%.

Protein Electron Transfer, Edited by D.S. Bendall

This chapter will discuss the electron transfer reactions that occur in the reaction centres of purple, non-sulphur photosynthetic bacteria. We will focus mainly on the kinetic and thermodynamic properties of the transient radical pair states that participate in the initial steps of photochemical charge separation. Reviews by Coleman and Youvan (1990), Friesner and Won (1989), Kirmaier and Holten (1987), Nitschke and Rutherford (1991), Okamura and Feher (1992), Parson (1991), Warshel and Parson (1991), and the books edited by Blankenship *et al.* (1995), Breton and Verméglio (1988, 1992), Deisenhofer and Norris (1993), Michel-Beyerle (1985, 1990, 1995) and Scheer (1992) contain articles on many aspects of reaction centre structure and function.

6.2   Reaction centre structures

High resolution crystal structures have been obtained for reaction centres of *Rp. viridis* (Deisenhofer and Michel, 1989; Deisenhofer *et al.*, 1985, 1995; Michel *et al.*, 1986) and *Rhodobacter sphaeroides* strains R-26, 2.4.1 and Y (Allen *et al.*, 1987a,b; Arnoux *et al.*, 1990; Budil *et al.*, 1987; Chang *et al.*, 1991; Chirino *et al.*, 1994; El-Kabbani *et al.*, 1991; Ermler *et al.*, 1994). Strains 2.4.1 and Y are wild-type strains; strain R-26 is a mutant that lacks the normal carotenoids. Chirino *et al.* (1994) also have described the structures in several mutant strains derived from 2.4.1. The major structural element of the reaction centres in all of these bacteria is a pair of polypeptides designated L and M, each of which has five transmembrane α-helices (*Figure 6.1*). The L and M polypeptides have homologous amino acid sequences and very similar secondary structures. About 30% of the amino acid residues in corresponding positions are identical in the two subunits and many of the others represent conservative replacements (Komiya *et al.*, 1988). A third polypeptide, H, adopts a more globular conformation and sits mainly at the cytoplasmic surface of the membrane, but also has one transmembrane α-helix. The *Rp. viridis* reaction centre has an additional, *c*-type cytochrome subunit with four haem groups, which binds tightly to L and M on the periplasmic side of the membrane.

Reaction centres from most species of purple photosynthetic bacteria contain four molecules of bacteriochlorophyll (BChl) (BChl *a* in *Rb. sphaeroides* and BChl *b* in *Rp. viridis*), two molecules of bacteriophaeophytin (BPh) (BPh *a* in *Rb. sphaeroides* and BPh *b* in *Rp. viridis*), two quinones (Q_A and Q_B) and a non-haem iron atom (*Figure 6.2*). In *Rb. sphaeroides*, Q_A and Q_B are both ubiquinone; in *Rp. viridis*, Q_B is ubiquinone and Q_A is menaquinone. Reaction centres from wild-type strains also contain one molecule of a carotenoid (neurosporene in *Rb. sphaeroides* and dihydrolycopene in *Rp. viridis*.) The non-haem Fe is replaced to a considerable extent by Mn in *Rb. sphaeroides* Y.

One of the most striking features of the reaction centre is its symmetry. The L and M polypeptides and the BChl, BPh and quinone cofactors are arranged about an axis of C2 pseudosymmetry that is oriented approximately normal to the membrane (see *Figures 6.1* and *6.2*). Two of the four BChls (P_L and P_M) sit particularly close together on either side of the symmetry axis. It is this special pair of BChls (P) that acts as the electron donor in the initial photochemical electron transfer reaction. The two BChls are about 7 Å apart centre-to-centre, and are separated by about 3.5 Å where they overlap in ring I

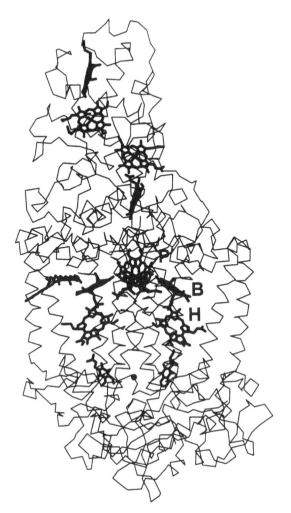

Figure 6.1. A view of the *Rp. viridis* reaction centre (Michel *et al.*, 1986) showing the pigments (heavy lines) and the α-carbon chains of the protein. The cytochrome subunit with its four haems is at the top, the H subunit at the bottom, M on the left, and L on the right. The C2 pseudosymmetry axis runs vertically through the non-haem Fe (black dot) and the special pair of BChls (P). The accessory BChl (B) and the BPh (H) associated with the L subunit are labelled. The quinones are at the bottom, Q_A on the right and Q_B on the left. The carotenoid is on the left, above B_M. The phytyl and prenyl side chains of the pigments are truncated for clarity.

of the BChl macrocycles. The two 'accessory' BChls (B_L and B_M), the BPhs (H_L and H_M) and the quinones sit in two branches stretching out from the special pair toward the intracellular face of the membrane. A rotation of 180° about the symmetry axis approximately interchanges the positions of homologous amino acid residues in the L and M polypeptides as well as the positions of corresponding pigments on either side of the axis. The carotenoid, however, is situated asymmetrically next to BChl B_M. The cytochrome subunit of the *Rp. viridis* reaction centre has an independent axis of C2 pseudosymmetry.

Reaction centres of *Rhodobacter capsulatus* and *Rhodospirillum rubrum*, two other species of purple, non-sulphur photosynthetic bacteria, probably are very similar in structure to those of *Rb. sphaeroides* and *Rp. viridis*. In *Chloroflexus aurantiacus*, a thermophilic green bacterium, the arrangement of the pigments again probably is essentially the same as that in the purple bacteria except that BChl B_M is replaced by BPh (Blankenship *et al.*, 1983; Pierson and Thornber, 1983; Shiozawa *et al.*, 1989; Shuvalov *et al.*, 1986; Vasmel *et al.*, 1983). Reaction centres from *C. aurantiacus* also lack the H subunit, although their L and M subunits are homologous to those of the purple bacteria.

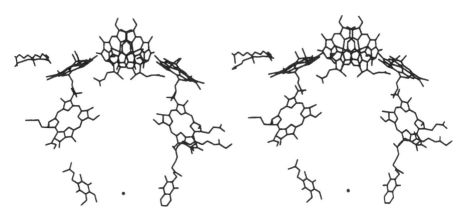

Figure 6.2. Stereo view of the electron carriers in the *Rp. viridis* reaction centre. The perspective is the same as in *Figure 6.1*. The carotenoid and the phytyl and prenyl side chains of the pigments are truncated.

The major spectroscopic signature of the reactive pair of BChls is a strong absorption band in the near-infrared (IR). This band is shifted far to the red with respect to the Q_y band of monomeric BChl; at 295 K, it typically is centred near 870 nm in *Rb. sphaeroides* and other species that contain BChl *a*, and near 960 nm in species that contain BChl *b*, such as *Rp. viridis* (*Figure 6.3*). Although each of the reaction centre's optical absorption bands probably contains contributions from all four of the BChls and both BPhs, analyses of the absorption spectrum on the basis of the crystal structure indicate that the long wavelength band is due predominantly to P (Breton, 1985; Friesner and Won, 1987; Hanson, 1988; Lathrop and Friesner, 1994; Parson and Warshel, 1987; Scherer and Fischer, 1989a; Thompson *et al.*, 1990; Thompson and Zerner, 1991). The Q_y transitions of the accessory BChls and the BPhs contribute absorption bands near 800 and 760 nm, respectively, in *Rb. sphaeroides*, and near 830 and 790 nm in *Rp. viridis* (see *Figure 6.3*). The four BChls also have Q_x bands in the region of 600 nm, and the BPhs have corresponding bands near 540 nm. Upon photooxidation of P, the long wavelength absorption band bleaches, the band in the 800- or 830-nm region shifts to shorter wavelengths by about 5 nm, and new absorption bands appear farther into the near- and mid-IR. The IR absorption bands of P^+ can be rationalized in terms of resonance interactions that delocalize electrons over the special pair of BChls (Breton *et al.*, 1992; Parson *et al.*, 1992; Plato *et al.*, 1992; Reimers and Hush, 1995a,b; Scherer and Fischer, 1992; Walker *et al.*, 1994).

The idea that the reactive complex is a BChl dimer was suggested originally by Norris *et al.* (1971), who found that the electron spin resonance (ESR) spectrum of the oxidized complex (P^+) was narrower by approximately a factor of $\sqrt{2}$ compared with the spectrum of monomeric $BChl^+$ in solution; such a narrowing would be expected if the unpaired spin is delocalized equally over two molecules. In agreement with this interpretation, the hyperfine coupling constants for the most prominent hydrogens in the ENDOR spectrum of P^+ are about half as large as those for monomeric $BChl^+$ (Norris *et al.*, 1974). More recent studies of the charge distribution in P^+ are described below.

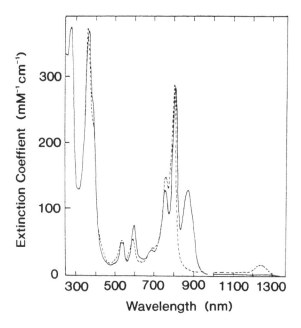

Figure 6.3. Absorption spectra of reaction centres of *Rb. sphaeroides* strain R-26 at 295 K, measured in the absence (solid curve) and presence (dotted) of continuous illumination. Under the conditions of the measurements, illumination converts most of the sample to the state $P^+Q_A^-$. Most of the absorbance changes caused by illumination are associated with oxidation of P and are similar to the changes caused by addition of an oxidant such as $K_3Fe(CN)_6$; Q_A^- contributes to the absorbance changes in the regions of 300–450 nm and (not shown) 270 nm and to the shifts of the 760- and 800-nm bands. Adapted from Parson and Cogdell (1975) The primary photochemical reaction of bacterial photosynthesis. *Biochim. Biophys. Acta* **416**: 105–149 with permission from Elsevier Science – NL.

6.3   Charge separation and competing reactions

Figure 6.4 shows a scheme of the initial electron transfer steps that occur in the reaction centres of purple bacteria. Excitation with light or resonance energy transfer from an antenna complex first promotes the special pair of BChls to an excited singlet state (P*). The excited dimer transfers an electron to one of the two BPhs (H_L) with a time constant of approximately 3 ps, generating a transient $P^+H_L^-$ radical pair that stores about 80% of the energy of the photon. The electron probably passes through the accessory BChl located between P and H_L (B_L), although the experimental identification of a $P^+B_L^-$ radical pair as a discrete kinetic intermediate is still tentative. An electron moves from H_L^- to one of the quinones (Q_A) in about 200 ps, forming $P^+Q_A^-$, and proceeds from Q_A^- to the other quinone (Q_B) in about 100 µs. Meanwhile, P^+ extracts an electron from a *c*-type cytochrome. In *Rp. viridis* the electron donor in this step is the nearest haem of the four-haem cytochrome that is bound tightly to the reaction centre; in *Rb. sphaeroides* it is cytochrome c_2, a water-soluble, single-

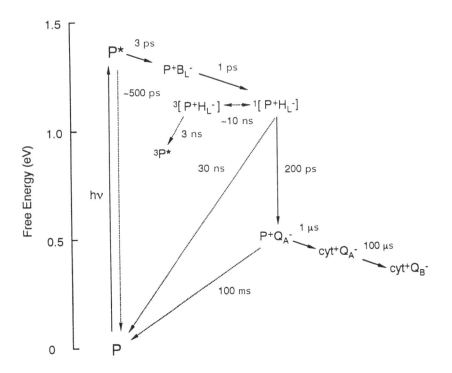

Figure 6.4. Transient excited and radical pair states in purple bacterial reaction centres. Solid arrows indicate charge separation reactions; dotted arrows, wasteful charge recombination reactions. cyt$^+$ = oxidized bound c-type cytochrome. Approximate time constants and standard partial molecular free energies are shown for *Rb. sphaeroides* reaction centres at 295 K. The decay times and free energies of most of the states were measured as described in the text; the free energy of P$^+$B$_L^-$ is not well known and the position shown is based mainly on calculations for *Rp. viridis* (Alden *et al.*, 1995b). The charge recombination reactions of ^{1}(P$^+$H$_L^-$) and ^{3}(P$^+$H$_L^-$) shown by the dotted arrows to P and ^{3}P probably are mediated by superexchange through ^{1}(P$^+$B$_L^-$) and ^{3}(P$^+$B$_L^-$); the charge recombination of P$^+$Q$_A^-$ probably involves superexchange through both P$^+$H$_L^-$ and P$^+$B$_L^-$.

haem protein that associates with the reaction centre on the periplasmic side of the membrane. Absorption of a second photon sends another electron along the same path from P* through H$_L$ and Q$_A$ to Q$_B^-$, whereupon the doubly reduced Q$_B^{2-}$ picks up protons from the intracellular solution and dissociates from the reaction centre as the quinol (Q$_B$H$_2$) into the membrane's phospholipid bilayer.

In spite of the structural symmetry of the reaction centre, the pathway of electron transfer is decidedly asymmetrical. The two BPhs can be distinguished experimentally because the Q$_x$ and Q$_y$ absorption bands of H$_L$ are shifted to the red by about 10 nm relative to the corresponding bands of H$_M$, and electron transfer from P* to H$_L$ is found to predominate by at least a factor of 10^2 over transfer to H$_M$ (Kirmaier *et al.*, 1985a,b; Kellog *et al.*, 1989; Mar and Gingras, 1990). The red shift of the absorption bands of H$_L$ can be attributed to hydrogen bonding of the ring-V keto group to a glutamic acid residue (Glu L104); the homologous residue on the M side is a valine. This supposition was tested by replacing Glu L104 by Leu or Gln by site-directed mutagenesis (Bylina and Youvan, 1988). In the Gln mutant, the Q_x absorption band of H_L shifts about halfway to the Q_x band of H_M, and in the Leu mutant the bands of the two pigments are superimposed. However, reduction of H_L is still favoured strongly in both mutants and the rate of charge separation is decreased only slightly, so the hydrogen bonding of H_L evidently is not responsible for the directionality of the electron transfer reaction. We will discuss other possible explanations for the directionality below.

The charge separation sequence can be interrupted experimentally at several stages, forcing the radical pair states to decay by alternative routes (*Figure 6.4*). In reaction centres that lack Q_B, $P^+Q_A^-$ decays by charge recombination with a time constant of about 0.1 s. If Q_A is removed or is reduced chemically before the excitation flash, $P^+H_L^-$ decays in about 20 ns, partly to the ground state (PH) and partly to an excited triplet state of P (^{3}P). Relatively weak external magnetic fields slow the decay of the radical pair and decrease the quantum yield of ^{3}P, as would be expected if $P^+H_L^-$ is created from P* in a singlet state [$^1(P^+H_L^-)$] and then evolves into a triplet radical pair [$^3(P^+H_L^-)$] that forms ^{3}P by charge recombination (Blankenship *et al.*, 1977; Boxer *et al.*, 1982, 1983; Chidsey *et al.*, 1984; Goldstein and Boxer, 1988a,b; Haberkorn and Michel-Beyerle, 1979; Hoff, 1981; Hoff *et al.*, 1977; Hunter *et al.*, 1987; Norris *et al.*, 1982; Ogrodnik *et al.*, 1982; Rademaker and Hoff, 1981; Schenck *et al.*, 1982; Werner *et al.*, 1978; Woodbury and Parson, 1984). In the absence of external magnetic fields, $^1(P^+H_L^-)$ and $^3(P^+H_L^-)$ evidently are nearly isoenergetic, and hyperfine interactions can cause oscillations between the two spin states on the nanosecond time scale. Magnetic fields in the range of 10–1000 G interfere with this process by moving two of the magnetic sublevels of $^3(P^+H_L^-)$ out of resonance with $^1(P^+H_L^-)$. At higher magnetic fields (50–150 kG) oscillations between $^3(P^+H_L^-)$ and $^1(P^+H_L^-)$ are driven by the difference between the *g* factors of the P^+ and H_L^- radicals and become considerably faster than they are at zero field (Boxer *et al.*, 1982; Goldstein and Boxer, 1988a). From the effects of magnetic fields of various strengths on the decay kinetics of the radical pair and on the quantum yield of ^{3}P, Goldstein and Boxer (1988a,b) calculated that the charge recombination reaction that generates ^{3}P from $^3(P^+H_L^-)$ has a rate constant of 4×10^8 s^{-1}, and that the rate constant for charge recombination of $^1(P^+H_L^-)$ to the ground singlet state is about one-tenth of this (3×10^7 s^{-1}).

A key point to note with regard to these alternative paths is that the forward reaction at each step in the charge separation sequence is much faster than the competing charge recombination reaction. The initial electron transfer from P* to H_L is at least 100 times faster than the decay of P* directly to the ground state by the combined processes of fluorescence and internal conversion. [Decay of P* to the ground state is undetectable in wild-type reaction centres, but occurs with time constants ranging from ~200 ps to 1 ns in several mutant strains with impaired charge separation (Breton *et al.*, 1990; Heller *et al.*, 1995; Nagarajan *et al.*, 1990, 1993].] Electron transfer from H_L^- to Q_A is faster than the decay of $^1(P^+H_L^-)$ to the ground state by a factor of about 170, and reduction of P^+ by the *c*-type cytochrome beats charge recombination of $P^+Q_A^-$ by a factor of about 10^5. Electron transfer from Q_A^- to Q_B also exceeds charge recombination of $P^+Q_A^-$ by a factor of 1000.

Favourable ratios of the charge separation and recombination reactions clearly are necessary if the reaction centre is to capture the energy of light with a high quantum yield, as it is observed to do (Wraight and Clayton, 1973). For the first two steps, $P^\star H_L \rightarrow P^+H_L^-$ and $P^+H_L^-Q_A \rightarrow P^+H_LQ_A^-$, the favourable ratios can be explained partly by the fact that the free energy change (ΔG^o) for charge recombination is much more negative than that for the forward reaction, while the reorganization energies (λ) for both reactions are relatively small. The forward reaction probably occurs near the peak of the Marcus curve, where $\Delta G^o \approx -\lambda$, whereas the charge recombination reaction falls in the 'inverted' region where $\Delta G^o \ll -\lambda$. In the classical Marcus theory (Marcus, 1956, 1965; Marcus and Sutin, 1985), the rate constant for a non-adiabatic electron transfer reaction is given by

$$k_{et} = \frac{4\pi^2}{h} \frac{|H_{AB}|^2}{\sqrt{4\pi\lambda k_B T}} \exp\left\{ -\frac{\left(\Delta G^o + \lambda\right)^2}{4\lambda k_B T} \right\} \qquad (6.1)$$

where H_{AB} is the electronic interaction matrix element between the reactant and product states, k_B and h are Boltzmann and Planck constants, respectively, and T is the temperature (see Chapter 1). The reorganization energy λ is the change in free energy necessary to raise the products from their most probable nuclear configuration to the most probable configuration of the reactants. According to Equation 6.1, the rate is maximal when $\Delta G^o \approx -\lambda$. However, quantum mechanical expressions that include nuclear tunnelling between the reactants and excited vibrational levels of the products give rate constants that are considerably less sensitive to ΔG^o than predicted by Equation 6.1 when $\Delta G^o < -\lambda$ (see, for example, Gunner and Dutton, 1989; Sarai, 1979; Warshel et al., 1989b). The smaller free energy changes thus cannot insure that the forward reactions will be much faster than the back reactions; it also is essential that the electronic interaction matrix elements for the charge separation reactions be large.

One way that the reaction centre optimizes the ratios of H_{AB} for charge separation and recombination is to break the charge separation process up into a series of small steps by using a set of electron carriers with extended π molecular orbitals. The inital electron transfer steps $P^\star B_L \rightarrow P^+B_L^-$ and $P^+B_L^-H_L \rightarrow P^+B_LH_L^-$, which must compete with the fastest back reactions, have relatively large interaction matrix elements because the reactants are essentially in contact. The edges of the π systems of P and H_L, however, are separated by about 10 Å, so that the orbital overlap between these components is relatively weak. The product of the matrix elements for the interactions of $P^\star$ with B_L and of B_L^- with H_L thus is larger than the matrix element for a direct back reaction between P^+ and H_L^-. Similarly, the product of the matrix elements for the steps that lead from $P^\star$ to $P^+Q_A^-$ is larger than the matrix element for a direct back reaction between P^+ and Q_A^-.

Efficient charge separation also requires that the decrease in free energy in each step be sufficient to defeat thermal back reactions that would return the system to an earlier radical pair or to $P^\star$. The largest drop in free energy occurs between $P^+H_L^-$ and $P^+Q_A^-$ (*Figure 6.4*). The ΔG^o of about -0.65 eV here makes the formation of $P^+Q_A^-$ almost irreversible and insures that $P^+Q_A^-$ lives long enough for P^+ to be reduced by the bound cytochrome. Reduction of P^+ locks the system in a charge-separated state that is stable for many seconds, so that even the later reactions that require diffusion of ubiquinone (UQ) or cytochromes in the membrane can occur at leisure. Although a significant fraction of the photon energy is inevitably lost as heat, the main biochemical task of the reaction centre evidently is not to store energy *per se*, but rather to provide the cytochrome bc_1 complex with UQH_2 and ferricytochrome *c*.

6.4 The free energies and enthalpies of $P^+Q_A^-$ and $P^+H_L^-$

The foregoing discussion emphasizes the importance of the free energies of the reaction centre's radical pair states for controlling the rate and reversibility of the charge separation and recombination reactions. We turn now to a description of how the free energies shown in *Figure 6.4* were obtained.

Rough estimates of the energetics of the radical pair states can be obtained from the midpoint redox potentials (E_m values) of the electron donors and acceptors. The E_m of the P/P^+ couple can be measured by titrating P with chemical oxidants (Carithers and Parson, 1985; Case and Parson, 1971; Cusanovich *et al.*, 1968; Prince *et al.*, 1976; Shuvalov *et al.*, 1976) or by electrochemical oxidation (Lin *et al.*, 1994; Moss *et al.*, 1991). The E_m is 0.50 ± 0.01 V in purified reaction centres of either *Rb. sphaeroides* or *Rp. viridis*. These numbers can be compared with the value of +0.64 V measured for monomeric BChl *a* in solution (Fajer *et al.*, 1974, 1976). Q_A can be titrated similarly, using as an assay the photooxidation of P or the *c*-type cytochrome (Case and Parson, 1971; Cogdell and Crofts, 1972; Dutton *et al.*, 1973; Jackson *et al.*, 1973; Prince *et al.*, 1976; Reed *et al.*, 1969). The apparent E_m is different in *Rp. viridis* and *Rb. sphaeroides*, as is expected from the fact that Q_A is menaquinone in the former species and ubiquinone in the latter, and the reported values are somewhat different for chromatophores and purified reaction centres. In chromatophores, the E_m becomes more negative with increasing pH. Typical values for reaction centres at pH 8 are −0.05 V in *Rb. sphaeroides* and −0.15 V in *Rp. viridis* and *C. vinosum*. Thus the free energy captured in $P^+Q_A^-$ is expected to be of the order of 0.5–0.6 eV, which is about 30–40% of the 0–0 excitation energy of P (*Figure 6.4*). This estimate neglects the electrostatic interaction between P^+ and Q_A^-, as well as differences between the energies of solvation of the $P^+Q_A^-$ radical pair and the individual radicals by the protein and surrounding medium. However, a similar number (0.52 eV for *Rb. sphaeroides*) is obtained by using the amplitude of the 'delayed' fluorescence from P* to measure the apparent equilibrium constant between P* and $P^+Q_A^-$ (Arata and Parson, 1981).

Titrations of H_L are more problematic because the BPh does not react readily with exogenous reductants. However, H_L can be forced to accumulate in the reduced state by illuminating reaction centres continuously in the presence of reductants (Dutton *et al.*, 1976; Mar and Gingras, 1990; Shuvalov and Klimov, 1976; Shuvalov *et al.*, 1976; Tiede *et al.*, 1976). With Q_A reduced, the 20 ns lifetime of $P^+H_L^-$ provides an opportunity for P^+ to extract an electron from the bound cytochrome or another reductant. Although the quantum yield of this process is only about 0.1%, repeated excitation eventually traps the reaction centre in the state PH_L^-, which then can survive for minutes or longer, depending on the ambient redox potential. Accumulation of PH_L^- can be monitored by the bleaching of the absorption bands of the BPh or by the loss of photochemical activities such as the formation of ^3P. Measurements of the extent of H_L^- accumulation in *Rp. viridis* reaction centres poised at various redox potentials have provided estimates ranging from –0.40 to –0.62 V for the E_m of the BPh (Prince *et al.*, 1976; Rutherford *et al.*, 1979; Shopes and Wraight, 1987; Shuvalov *et al.*, 1976). These values can be compared with the E_m of –0.55 for BPh *b* in solution (Fajer *et al.*, 1975). Judging from the environment of H_L in the *Rp. viridis* reaction centre, the most negative of the experimentally measured values seems likely to be the most accurate (Parson *et al.*, 1990). Combining the E_m values of +0.50 V for P/P^+ and –0.62 V for H_L^-/H_L suggests that $P^+H_L^-$ is of the order of 1.1 eV above the ground state, or about 0.2 eV below $P^\star$.

More direct measurements of the free energy of $P^+H_L^-$ have been obtained in two ways. First, the amount of the excited singlet state $P^\star$ that remains in equilibrium with $P^+H_L^-$ can be related to the amplitude of fluorescence emitted on the nanosecond time scale when electron transfer from H_L^- to Q_A is blocked by extracting or pre-reducing the quinone (Godik and Borisov, 1979; Godik *et al.*, 1982; Goldstein and Boxer, 1988a; Hörber *et al.*, 1986; Ogrodnik, 1990; Ogrodnik *et al.*, 1994; Peloquin *et al.* 1994; Rademaker and Hoff, 1981; Schenck *et al.*, 1982; Shuvalov and Klimov, 1976; Woodbury and Parson, 1984, 1986; Woodbury *et al.*, 1986). The decay of fluorescence from $P^\star$ under these conditions is multiphasic, but a major component has a 10–20 ns lifetime that matches the decay of the optical absorbance changes associated with $P^+H_L^-$. This component also has the same sensitivity to external magnetic fields as the decay of $P^+H_L^-$. The shorter lived components of the fluorescence probably do not reflect decreases in the population of $P^+H_L^-$ but, more likely, relaxations that lower the free energy of the radical pair relative to $P^\star$. They probably are not artifacts of the preparation of reaction centres because they also are seen in intact chromatophores (Woodbury and Parson, 1986). The amplitude of the long lived component indicates that, in *Rb. sphaeroides* reaction centres at 295 K, the relaxed $P^+H_L^-$ lies in the range of 0.21–0.26 eV below $P^\star$.

The second approach is to determine the equilibrium constant between $P^+H_L^-$ and ^3P (Boxer *et al.*, 1989; Chidsey *et al.*, 1985; Goldstein *et al.*, 1988; Ogrodik *et al.*, 1988; Takiff and Boxer, 1988). From the phosphorescence emission spectrum (Takiff and Boxer, 1988) and the activation energy for thermal repopulation of $P^\star$ (Shuvalov and Parson, 1981), ^3P is known to lie about 0.42 eV below $P^\star$. In reaction centres that are depleted of Q_A, ^3P evidently decays largely by thermal excitation back to $^3(P^+H_L^-)$, because the decay of ^3P is affected by external magnetic fields that (depending on the field strength) slow down or speed up interconversion of $^3(P^+H_L^-)$ and $^1(P^+H_L^-)$. The free energy and enthalpy differences between ^3P and $^3(P^+H_L^-)$ can be obtained by analysing these effects as functions of field strength and temperature. These experiments put $P^+H_L^-$ in the range of 0.25–0.26 eV below $P^\star$, again for *Rb. sphaeroides* reaction centres at 295 K, in good agreement with the fluorescence measurements at this temperature.

Although the two approaches give similar values for the free energy of $P^+H_L^-$ at 295 K, they differ on how the apparent free energy depends on temperature. Judging from the effects of magnetic fields on the decay kinetics of ^3P, the free energy of $P^+H_L^-$ varies only slightly with temperature (Goldstein et al., 1988; Ogrodnik et al., 1988). The fluorescence measurements, by contrast, suggest that the free energy difference between P^* and $P^+H_L^-$ decreases progressively with decreasing temperature, falling to about 0.05 eV by 80 K (Woodbury and Parson, 1984). Ogrodnik et al. (1994) explain that, if $P^+H_L^-$ has a distribution of energies in the population of reaction centres, the recombination fluorescence will preferentially sample radical pairs with the highest energies and this will cause the apparent free energy of $P^+H_L^-$ to change with temperature. They conclude that the mean free energy difference between P^* and $P^+H_L^-$ is 0.21–0.25 eV, independent of temperature, and that the distribution of the energies has a width of about 0.1 eV. Another possible explanation for the disagreement between the two types of measurements is that the fluorescence experiments probe the system during the 20 ns lifetime of $P^+H_L^-$, whereas the magnetic field experiments cover the 100 μs lifetime of ^3P. At low temperatures, relaxations that follow charge separation might be too slow to affect the energy of $P^+H_L^-$ on the 20 ns time scale but still affect measurements on the 100 μs time scale. Some of the absorbance changes associated with reduction of H_L appear to freeze out at temperatures between 200 and 100 K (Shuvalov and Parson, 1980; Tiede et al., 1987).

6.5 Kinetics and mechanism of the initial charge separation

The light-driven transfer of an electron from P to H_L is perhaps the most remarkable step in the charge separation sequence. Within about 3 ps, the electron moves specifically to H_L rather than H_M, traversing a distance of some 18 Å centre to centre, in a process that is almost independent of temperature. Although this might suggest that the energies and positions of the reactants and products would have to be poised delicately, the reaction has proved to be amazingly robust to structural perturbations. In this section, we discuss approaches that have been used to explore the kinetics, mechanism and specificity of the reaction.

When the reaction centre is excited with light, the excitation energy localizes on P within about 100 fs (Breton et al., 1988; Jean et al., 1988; Jia et al., 1995; Johnson et al., 1990). Formation of the excited singlet state P^* can be recognized by bleaching of the reaction centre's long wavelength absorption band and the development of stimulated emission at wavelengths to the red of this band (Woodbury, 1985). At room temperature, the stimulated emission and spontaneous fluorescence from P^* decay on the time scale of 1–10 ps (Breton et al., 1986; Du et al., 1992; Hamm et al., 1993; Holzapfel et al., 1989, 1990; Jia et al., 1995; Kirmaier and Holten, 1988a, 1990, 1991; Martin et al., 1986; Nagarajan et al. 1993; Paschenko et al., 1985; Wasielewski and Tiede, 1986; Woodbury et al., 1985, 1994). The 870 nm absorption band of P remains bleached while the emission from P^* decays, indicating that P does not return directly to the ground state. Instead, a band characteristic of P^+ develops in the region of 1250 nm. Absorbance changes in the 545-, 650- and 780-nm regions signalling the reduction of BPh H_L to H_L^- occur approximately in synchrony with this process.

Most recent observers describe the decay kinetics of P* as multiphasic. Fitting the stimulated or spontaneous emission signals to a biexponential expression typically gives a component with a time constant of approximately 2.7 ps that accounts for about 80% of the initial amplitude, and a smaller component with a time constant of about 10 ps. The multiphasic nature of the kinetics is increasingly pronounced at low temperatures in *C. aurantiacus* reaction centres (Becker *et al.*, 1991; Müller *et al.*, 1991) and in some mutant strains of *Rb. sphaeroides* (Nagarajan *et al.*, 1993; Woodbury *et al.*, 1995). The significance of the multiphasic kinetics is presently unclear. A possible explanation is that reaction centre preparations contain heterogeneous populations of molecules that differ in the rate of electron transfer (Jia *et al.*, 1993; Kirmaier and Holten, 1990; Wang *et al.*, 1993). Becker *et al.* (1991) and Gehlen *et al.* (1994) have suggested that fluctuations of the atomic positions give rise to a dynamic heterogeneity in the energy difference between P* and $P^+H_L^-$ or one of the other parameters that determine the probability of charge separation. However, results from hole-burning spectroscopy (see below) are difficult to reconcile with gross heterogeneity (Small, 1995). Nagarajan *et al.* (1993) suggest that part of the apparent complexity of the kinetics reflects shifting of the emission spectrum as the excited state relaxes prior to electron transfer. Vos *et al.* (1994c) have reported that the apparent time constant for the decay of the stimulated emission depends on the measuring wavelength, which would be consistent with the latter suggestion although it also could be explained by a correlation between kinetic and spectroscopic properties in a heterogeneous sample. Another possibility is that the slow decay component represents P* that remains in equilibrium with $P^+H_L^-$ prior to relaxations of the radical pair (Peloquin *et al.*, 1994; Woodbury *et al.*, 1994).

The rate of electron transfer from P* to H_L actually increases somewhat with decreasing temperature (Breton *et al.*, 1988; Fleming *et al.*, 1988; Nagarajan *et al.*, 1990; Woodbury *et al.*, 1985, 1994). At 10 K, the major component has a time constant of about 1.2 ps in *Rb. sphaeroides* and about 0.9 ps in *Rp. viridis*. Again, absorbance changes indicating reduction of H_L occur with similar kinetics. The observation that the kinetics are relatively insensitive to temperature suggests that the free energy change in the reaction (ΔG°) is approximately equal in magnitude, but opposite in sign, to the reorganization energy. According to Equation 6.1, the rate will increase slightly with decreasing temperature if ΔG° and λ are both independent of temperature and $\Delta G^\circ \approx -\lambda$. However, the Marcus equation describes a high temperature limit when the thermal energy (k_BT) is much greater than the quantum energies of the vibrational modes that are coupled to the reaction. Expressions that include vibrational quantization give rates that can be insensitive to temperature even if ΔG° is not equal to $-\lambda$ (see Chapter 1). The increase in the rate with decreasing temperature could be influenced by changes in ΔG° or λ (see above), or by thermal contraction, which presumably would increase the interaction matrix element H_{AB}.

As was discussed in the previous section, ΔG° for the formation of $P^+H_L^-$ from P* at 295 K is approximately -0.25 eV. If we take the observation that the reaction does not slow down with decreasing temperature to mean that $\Delta G^{\circ} \approx -\lambda$, then λ also must be of the order of 0.25 eV, which is remarkably small for a reaction that creates an electric dipole of about 70 debye. An analysis of the kinetics in a series of mutant reaction centres in which the E_m of P/P$^+$ is shifted to higher or lower values suggests an even smaller reorganization energy of about 0.10 eV (Bixon $et\ al.$, 1995). The intramolecular contribution of the electron carriers themselves to λ is expected to be small because the charges of P$^+$ and H$_L^-$ are delocalized over many atoms. In addition, the electron carriers are buried in the interior of the protein and are surrounded predominantly by hydrophobic amino acid residues. The contribution of the surroundings to λ thus is likely to be smaller than is typical of reactions at the surface of water-soluble proteins.

6.5.1 *Vibrational relaxations of P* and P⁺H$_L^-$*

The Marcus equation (Equation 6.1) and most of the other formalisms that have been developed to describe electron transfer kinetics assume that the reacting system is in thermal equilibrium with its surroundings. It is unclear whether such an equilibrium is attained on the short time scale of the initial charge separation in photosynthetic reaction centres. Several different experimental approaches have been used to address this question, and the results are somewhat discordant.

If reaction centres are excited with a pulse lasting of the order of 50 fs, the emission spectrum of P* is observed to shift to longer wavelengths with time and then to oscillate back and forth for several picoseconds (Stanley and Boxer, 1994; Streltsov $et\ al.$, 1993; Vos $et\ al.$, 1992, 1993, 1994a,b, 1995; Woodbury $et\ al.$, 1994). These oscillations probably reflect coupling of the absorption and emission of P* to a family of low frequency (15–200 cm^{-1}) delocalized vibrational modes of the protein, and possibly also to out-of-plane intramolecular vibrations of the BChls. Vibrations that modulate the resonance and excitonic interactions between the two BChls would be expected to be particularly important. Vibrational modes with similar frequencies appear in the resonance Raman spectrum of P (Palaniappan $et\ al.$, 1992; Shreve $et\ al.$, 1991). The oscillations damp out with time as a result of vibrational relaxations, electronic transitions and the 'pure' electronic dephasing caused by fluctuating interactions with the surroundings. Although a prominent oscillation with a frequency of about 120 cm^{-1} is damped with a time constant of about 0.2 ps, which is short relative to the electron transfer kinetics, other components persist for as long as 4 ps, particularly in mutant reaction centres in which electron transfer is slowed (Vos $et\ al.$, 1995). The relatively slow decay of some of the oscillations implies that vibrational equilibration of P* is incomplete on the time scale of electron transfer.

Another way to probe the dynamics of the excited state is to examine holes that are burned in the absorption spectrum by exciting reaction centres with narrow band light at low temperatures. Selective excitation of reaction centres whose 0–0 transition occurs at a particular wavelength converts these reaction centres to a metastable state such as P$^+$Q$_A^-$ and depletes the absorption spectrum at all wavelengths where these molecules absorb. The change in the absorption spectrum generally includes a relatively sharp zero-phonon hole at the 0–0 transition frequency and a progression of phonon side bands at higher frequencies. If the temperature is sufficiently low to freeze out pure dephasing, the width of the zero-phonon hole should be related reciprocally to the lifetime of the excited electronic state. (The hole cannot be broadened further by vibrational relaxations because the 0–0 transition places $P^\star$ directly in the lowest vibrational state.) Measurements of the hole width give lifetimes that agree closely with the lifetime of the fast component of the decay of stimulated or spontaneous emission from $P^\star$ (Ganago et al., 1991; Johnson et al., 1990; Lyle et al., 1993; Middendorf et al., 1991; Reddy et al., 1992; Small, 1995). This observation has several implications. First, because the time-resolved measurements of the emission from $P^\star$ are made by exciting reaction centres at a variety of frequencies above the 0–0 transition frequency, the electron transfer kinetics evidently do not depend strongly on which vibrational levels of the system are excited. The most straightforward conclusion is that $P^\star$ relaxes to its lowest vibrational level quickly, relative to the rate of electron transfer. Such rapid vibrational relaxation would be in agreement with the rapid damping of the 120 cm^{-1} oscillation in the time-resolved measurements, but is difficult to reconcile with the prolonged oscillations at other frequencies.

The agreement between the lifetimes of $P^\star$ obtained from the zero-phonon hole width and from the stimulated or spontaneous emission also indicates that the initially populated $P^\star$ probably does not relax to another electronic state prior to the electron transfer reaction (Johnson et al., 1990; Lyle et al., 1993; Middendorf et al., 1991; Reddy et al., 1992). However, Hamm and Zinth (1995) have reported that some of the IR absorption changes associated with $P^\star$ do not appear immediately after excitation, but rather develop with a rise time of about 200 fs. They suggest that $P^\star$ evolves on this time scale into a state with increasing internal charge transfer character.

The observation that the width of the hole matches only the fastest decay component of the stimulated emission, and the additional observation that the zero-phonon hole has a Lorentzian shape, argue against the notion that the electron transfer kinetics are markedly heterogeneous (Lyle et al., 1993; Small, 1995; Small et al., 1992). A subpopulation of the reaction centres that reacted slowly would give a sharp zero-phonon hole superimposed on the broad hole reflecting the rapidly reacting population. In addition, Small et al. (1992) and Lyle et al. (1993) report that the rate of electron transfer measured from the zero-phonon hole width does not change significantly as the excitation wavelength is varied across the inhomogeneous distribution of 0–0 transition wavelengths in the population of reaction centres. This means that the rate of electron transfer does not appear to be correlated with the 0–0 transition wavelength. Such a correlation might have been expected if the multiphasic kinetics of the stimulated emission reflected heterogeneities in the sample.

Modelling of the entire hole spectrum, including the broad phonon side bands, indicates that excitation of P is coupled strongly to a broad distribution of protein vibrational modes with frequencies in the region of 30 cm^{-1} and also to a characteristic mode with a frequency of about 120 cm^{-1} that could represent intermolecular motions of the two BChls (Ganago et al., 1991; Johnson et al., 1990; Lyle et al., 1993; Middendorf et al., 1991; Reddy et al.,

1992; Small, 1995). These conclusions agree well with the femtosecond spectroscopic measurements.

Theoretical studies seem not to have reached consensus on whether coherent nuclear motions in a non-thermalized system could enhance the speed and efficiency of the initial electron transfer reaction (Jean et al., 1992; Jortner, 1980; Skourtis et al., 1992). Jean et al. (1992) find that in many cases vibrational coherence would slow electron transfer. Skourtis et al. (1992) hold that vibrational coherence would make the kinetics robust to changes in ΔG° or H_{AB}.

Treutlein et al. (1992), Marchi et al. (1993) and Gehlen et al. (1994) have used molecular dynamics simulations to explore structural reorganizations that follow the formation of $P^+H_L^-$. Marchi et al. find that most of the relaxations of $P^+H_L^-$ occur within 0.1 ps, but that slower components stretching out to the picosecond time scale may contribute substantially to the total reorganization energy. Peloquin et al. (1994) and Woodbury et al. (1994, 1995) have suggested that $P^+H_L^-$ initially is very close to $P^\star$ in energy and that a dynamic equilibrium between these states is created on a subpicosecond time scale. In this adiabatic picture, temperature-dependent relaxations on the picosecond time scale lower the energy of $P^+H_L^-$ so that the equilibrium shifts progressively toward the radical pair. This model provides an attractive interpretation of the multiphasic decay kinetics of stimulated emission from $P^\star$, but requires that the interaction matrix element H_{AB} between $P^\star$ and $P^+H_L^-$ be larger than one might expect on the basis of the crystal structure (see below). Scherer and Fischer (1989a) and Kitzing and Kuhn (1990) also have suggested models in which H_{AB} is large and vibrational relaxations of $P^+H_L^-$ limit the rate of charge separation. However, accounting for the width and shape of the zero-phonon hole with any of these models is problematic (Small, 1995).

6.5.2 The role of the 'accessory' BChl (B_L)

The BChl molecule B_L situated between P and H_L undoubtedly participates in the initial electron transfer reaction because the direct, through-space electronic interaction of $P^\star$ with H_L is much too weak to account for the speed of charge separation. The weakness of this interaction is evident in the very small separation in energy between the $^1(P^+H_L^-)$ and $^3(P^+H_L^-)$ radical pairs, which is of the order of only 10^{-3} cm^{-1} (Haberkorn and Michel-Beyerle, 1979; Hoff, 1981; Marcus 1987, 1988a,b; Norris et al., 1982; Ogrodnik et al., 1982; Werner et al., 1978). It also emerges quickly in calulations of the interaction matrix element on the basis of the crystal structure (Bixon et al., 1987; Kallebring and Larsson, 1987; Michel-Beyerle et al., 1988; Parson et al., 1987; Plato et al., 1988; Scherer and Fischer, 1989a,b; Warshel et al., 1988).

Whether or not a $P^+B_L^-$ radical pair serves as a distinct kinetic intermediate in the reaction has been more difficult to establish. Zinth and colleagues (Arlt et al., 1993; Holtzapfel et al., 1989, 1990; Lauterwasser et al., 1991) and Maiti et al. (1994) have described transient absorbance changes suggesting that $P^+B_L^-$ is formed from $P^\star$ with a time constant of about 3 ps at room temperature and then gives rise to $P^+H_L^-$ with a time constant of about 1 ps. The key experimental observation is a transient bleaching in the region where B_L is expected to absorb (near 800 nm in *Rb. sphaeroides* or 820 nm in *Rp. viridis*), along with an absorbance increase in the region of 1000 nm that is consistent with the formation of a BChl anion radical. The transient bleaching near 820 nm is more pronounced in a mutant of *Rp. viridis* in which a histidine residue (L168) that is hydrogen bonded to one of the BChls of P is replaced by phenylalanine (Bibikova *et al.*, 1995). This mutation lowers the E_m of P and appears to speed up the formation of $P^+B_L^-$ relative to its decay. Judging from the magnitudes of the transient absorbance changes, the equilibrium of the reaction $P^+B_L^- \rightleftharpoons P^+H_L^-$ shifts toward $P^+B_L^-$ in reaction centres that have BPh H_L replaced by phaeophytin, which has a more negative reduction potential than BPh (Schmidt *et al.*, 1994; Shkuropatov and Shuvalov, 1993). Recent studies suggest that a similar shift in the equilibrium toward B_L^- occurs in mutant reaction centres in which the BPh is replaced by BChl (Kirmaier *et al.*, 1995a,b; Laporte *et al.*, 1995). (Monomeric BChl in solution is easier to oxidize but harder to reduce than BPh, primarily because the axial ligand of BChl interacts favourably with the cationic radical but unfavourably with the anionic radical.)   However, Kirmaier and Holten (1990, 1991) have presented conflicting experimental results on the linear dichroism of transient absorbance changes in native reaction centres, and have suggested that some of the complexity of the absorbance changes results from heterogeneity of the reaction centres. Other workers (Breton *et al.*, 1986, 1988; Du *et al.*, 1992; Fleming *et al.*, 1988; Peloquin *et al.*, 1994; Williams *et al.*, 1992; Woodbury *et al.*, 1985, 1994, 1995) also have reported finding no conclusive evidence for the formation of $P^+B_L^-$ as an intermediate.

If the $P^+B_L^-$ radical pair is not a distinct kinetic intermediate in the electron transfer reaction, it still could connect the initial and the final states (P^* and $P^+H_L^-$) by mixing quantum mechanically with both of them (Bixon *et al.*, 1987, 1988; Marcus, 1987, 1988a,b; Michel-Beyerle *et al.*, 1988; Warshel *et al.*, 1988; Woodbury *et al.*, 1985). This mechanism is referred to as 'superexchange.' The distinction between the superexchange mechanism and the two-step pathway with $P^+B_L^-$ as an intermediate hinges largely on the energy of $P^+B_L^-$. Because the rate of charge separation does not fall off at low temperatures, it cannot require formation of an intermediate that lies significantly above P^*. Superexchange, on the other hand, could continue to operate at low temperatures even if $P^+B_L^-$ does lie above P^*, as long as the mean energy gap between the two states ($\langle \Delta E_{12} \rangle$) is not much larger than the electronic interaction matrix elements that mix $P^+B_L^-$ with P^* and $P^+H_L^-$ (V_{12} and V_{23}, respectively). When $|\langle \Delta E_{12} \rangle|$ becomes large, the overall interaction matrix element for superexchange is given approximately by $2V_{12}V_{23}/|\Delta E_{12}|$.

The two-step and superexchange mechanisms are not mutually exclusive. They could occur in parallel, with superexchange perhaps playing an increasingly important role at low temperatures (Bixon *et al.*, 1990, 1991, 1995; Chan *et al.*, 1991b; Joseph *et al.*, 1991; Nagarajan *et al.*, 1993; Ogrodnik *et al.*, 1991; Tang and Norris, 1994). If $|\langle \Delta E_{12} \rangle|$ is small, the superexchange matrix element is expected to depend on both temperature and the energies of the vibrational modes that are coupled to the reaction (Kharkats *et al.*, 1995a,b; Lin, 1989).

One approach that has been used to explore whether $P^+B_L^-$ is formed in the initial electron transfer reaction is to examine the effects of applied electric fields on the kinetics. When imposed on isotropic samples, electric fields in the range of 10^6 V cm^{-1} cause an increase in the yield of fluorescence from P^* and a decrease in the quantum yields of $P^+H_L^-$ and $P^+Q_A^-$ (Boxer et al., 1989; Lao et al., 1995a,b; Lockhart et al., 1988; Moser et al., 1995; Ogrodnik, 1993; Ogrodnik et al., 1991, 1992). These effects have been interpreted as reflecting a slowing of electron transfer in reaction centres that happen to be oriented so that the vector from P to its initial electron acceptor is approximately parallel or antiparallel to the field. In the framework of the Marcus expression (Equation 6.1), the field would affect the rate by shifting the energy of the radical pair state relative to P^*. Lockhart et al. (1988) varied the angle between the field and the long wavelength (Q_y) transition dipole of P by using polarized excitation light to excite P selectively; Ogrodnik et al. (1991) probed the angle with respect to the Q_x transition dipole of H_M. Both groups concluded that the critical vector points from P to H_L, rather than to B_L. However, it is not clear that the effects of applied fields on the steady-state fluorescence yield pertain directly to the initial charge separation reaction. Time-resolved measurements suggest that these effects could involve back reactions of $P^+B_L^-$ or $P^+H_L^-$, or other processes that follow charge separation (Lao et al., 1995a,b; Ogrodnik, 1993). Moser et al. (1995) find that the effects of electric fields on the electron transfer kinetics in oriented multilayers of reaction centres are smaller than would be expected if the charge separation formed $P^+H_L^-$ directly, and are more consistent with electron transfer to B_L.

6.5.3 *Calculations of interaction matrix elements and radical pair energies*

Because no direct experimental information is available on $\langle \Delta E_{12} \rangle$, V_{12} or V_{23}, calculations of these quantities could help to clarify the reaction mechanism, provided that they are based on sufficiently realistic physical models. Several efforts have been made to calculate the electronic interaction matrix elements for electron transfer from P^* to B_L and from B_L^- to H_L. The electron donors and acceptors in these steps seem close enough together for electrons to jump directly from one carrier to the other without extensively involving the surrounding protein. To evaluate the intermolecular resonance interactions between the carriers, Kuhn (1986) used free-electron model wave functions; Parson et al. (1987) and Warshel et al. (1988) used a π-electron molecular orbital treatment (quantum consistent force field for π electrons, QCFF/π); and Plato and Winscom (1988), Scherer and Fischer (1989a,b) and Thompson and Zerner (1991) used INDO (intermediate neglect of differential overlap) treatments that also considered interactions of non-conjugated atoms. The major difficulty is that the matrix elements depend on the overlap of the molecular orbitals of the donor and acceptor at distances where the orbital amplitudes are not well characterized. Kuhn (1986), whose treatment includes scaling of the wave functions at distances beyond 1.7 Å to incorporate dielectric effects of the medium, obtained a matrix element of approximately 160 cm^{-1} for $P^* \rightarrow P^+B_L^-$ and 24 cm^{-1} for $P^+B_L^- \rightarrow P^+H_L^-$. The other groups obtained values of the order of 5–10 cm^{-1} for P* → P$^+$B$_L^-$ and 15–45 cm^{-1} for P$^+$B$_L^-$ → P$^+$H$_L^-$. Although the calculations are subject to large uncertainties, the latter values seem consistent with a two-step scheme in which the formation of P$^+$B$_L^-$ is rate limiting, and they are of roughly the right order of magnitude to account for the observed kinetics with this scheme. (If vibrational relaxations are assumed to be rapid, the necessary electronic matrix element for the rate-limiting step is ≈ 20 cm^{-1}.) Kuhn's considerably higher matrix element for the first step seems at odds with the experimental evidence for a two-step scheme in which the rate-limiting step is between P* and B$_L$ (see above). It would, however, be consistent with superexchange or with a scheme in which electron transfer from B$_L^-$ to H$_L$ is rate limiting. A scheme of the latter type has been proposed recently by Small (1995).

Calculations of the energies of the P$^+$B$_L^-$ and P$^+$H$_L^-$ radical pair states also have been described. Alden et al. (1995b), Creighton et al. (1988), Parson et al. (1990) and Warshel et al. (1994) started with the measured midpoint redox potentials of P and H$_L$, and of BChl and BPh in solution, and calculated the free energies of electrostatic interactions of the individual radicals and radical pairs in the reaction centre and in solution. Dielectric effects were treated microscopically by assigning an atomic polarizability to each atom of the protein and to a grid of points representing the surrounding solvent or membrane. In this approach, each atom or grid point acquires an induced dipole due to the fields from the charges and induced dipoles of all the other atoms and grid points. An iterative procedure is used to find a self-consistent solution for all the induced dipoles. Using a thermodynamic cycle that incorporates measured E_m values removes the need to rely on quantum calculations of gas-phase energies, and reduces the problem of calculating the energy of a state such as P$^+$B$_L^-$ largely to a calculation of the difference between solvation energies in the protein and in solution. Calculations also were done by free-energy perturbation methods and gave similar results (Alden et al., 1995b; Warshel et al., 1994).

The calculated free energy difference between P* and P$^+$H$_L^-$ in Rp. viridis agreed well with the experimental estimate of about 0.25 eV. The calculated free energy of P$^+$B$_L^-$ was slightly below that of P*, putting the P$^+$B$_L^-$ radical pair in a good position to act as an intermediate in the charge separation process (Alden et al., 1995b; Parson et al., 1990; Warshel et al., 1994). However, the estimated uncertainty of ±3 kcal mol^{-1} (12 kJ mol^{-1}) in the calculated energies was too large to allow an unambiguous choice between the two-step and superexchange mechanisms. P$^+$B$_M^-$, on the other hand, was calculated to lie about 5.5 kcal mol^{-1} (23 kJ mol^{-1}) above P*. This barrier seems sufficiently high to be an important factor in the directional specificity of the electron transfer reaction. The calculated electrostatic energies of P$^+$H$_L^-$ and P$^+$H$_M^-$ were closer together, and the difference here was deemed less significant.

Marchi et al. (1993) performed somewhat similar calculations but reached very different conclusions. They put the P$^+$B$_L^-$ radical pair about 20 kcal mol^{-1} (84 kJ mol^{-1}) above P*. However, this study used a macroscopic treatment of dielectric effects that neglected the energies that the electron carriers would have if they were separated to infinite distance in a homogeneous medium (Alden et al., 1995a,b). In addition, all ionizable residues were taken to be fully charged but were not solvated by additional water, counterions or components of the membrane. As a result, electric fields from the ionized groups played a dominant role in the calculated electrostatic energies of the radical pair states. The possible importance of such fields has been emphasized by Middendorf *et al.* (1993) and Gunner *et al.* (1996). Alden *et al.* (1995a,b), Parson *et al.* (1990) and Warshel *et al.* (1994) point out that some of the ionizable residues in the central, hydrophobic region of the reaction centre may not be ionized, and argue that most of the ionized residues in more hydrophilic regions are well shielded by solvent and electrolytes.

6.5.4 The roles of individual amino acid residues in controlling the rate, temperature dependence and directional specificity of charge separation

The calculated energy difference between $P^+B_L^-$ and $P^+B_M^-$ is a result of many small contributions from amino acid residues throughout the reaction centre. If the contributions from homologous residues in the L and M subunits are compared, the largest individual difference comes from Tyr-M210 and Phe-L181 (Parson *et al.*, 1990). (Residue numbers for *Rb. sphaeroides* are used throughout this chapter, although some of the electrostatics calculations were done with the *Rp. viridis* structure.) The side chain of Tyr-M210 is positioned so that the partial negative charge on the phenolic oxygen atom could stabilize the positive charge of P^+, while the partial positive charge of the hydrogen could stabilize a negative charge on B_L^-. Phe-L181 would not stabilize either B_M^- or P^+ in this way. Replacing Tyr-M210 by Phe, Ile or Trp by site-directed mutagenesis was found to slow the electron transfer reaction from P^* to H_L (Chan *et al.*, 1991a,b; Finkele *et al.*, 1990; Gray *et al.*, 1992; Jia *et al.*, 1993; Nagarajan *et al.*, 1990, 1993; Schochat *et al.*, 1994) and to change its temperature dependence (Nagarajan *et al.*, 1990, 1993). Instead of speeding up with decreasing temperature as it does in wild-type reaction centres, charge separation slows down in the Ile and Trp mutants and is essentially independent of temperature in the Phe mutant. These observations seem consistent with the idea that $P^+B_L^-$ is an intermediate in the electron transfer reaction. If the Ile or Trp mutation pushes $P^+B_L^-$ sufficiently high in energy, electron transfer would require thermal activation.

Norris and co-workers (Chan *et al.*, 1991a,b; Jia *et al.*, 1993) and Gray *et al.* (1992) have studied parallel mutations of Phe-L181. Remarkably, charge separation in the Phe-L181→Tyr mutant is slightly faster than in wild-type reaction centres. In the double mutant, Tyr-M210→Phe + Phe-L181→Tyr, the rate of charge separation is intermediate between the rates in the wild-type and the Tyr-M210→Phe single mutant. The photochemical reaction still leads predominantly to reduction of H_L, so it is clear that Tyr-M210 and Phe-L181 do not, by themselves, determine the directionality of the reaction. However, Gray *et al.* (1992) found increased phototrapping of H_M^- in both the double mutant and the Tyr-M210→Phe single mutant, indicating that these residues probably do contribute to the directional specificity. In addition, Tyr-M210 evidently plays a role in stabilizing P^+. Replacing the tyrosine by a less polar residue raises the E_m of P/P^+, whereas replacing Phe-L181 by Tyr lowers the E_m (Jia *et al.*, 1993; Nagarajan *et al.*, 1993).

The hydrogen bonding of the ring-V keto group of H_L to Glu-L104 would be expected to stabilize H_L^- and thus to favour electron transfer in the observed direction. However, the calculated effects of the hydrogen bond on the energy of $P^+H_L^-$ are relatively small (Hanson, 1988; Parson *et al.*, 1990) and, as mentioned above, replacing the Glu residue by Gln or Leu has little or no effect on the measured directionality of charge separation (Bylina *et al.*, 1988).

Heller *et al.* (1995) appear to have succeeded in redirecting the initial electron transfer reaction partly to the M branch of the pigments in *Rb. capsulatus* reaction centres by a combination of mutations that raise the energies of the radical pair states on the L branch. They began by substituting His for Leu-M212, which is located near the centre of one face of H_L. As had been found previously with other similar mutations in this region (Kirmaier *et al.*, 1991, 1995a,b), the availability of the His imidazole as an axial ligand resulted in incorporation of BChl in place of BPh H_L. The BChl at the H_L position is referred to as 'β.' Previous studies also had shown that P^* initiates charge separation to give $P^+β^-$ with kinetics only slightly slower those seen in wild-type reaction centres (Kirmaier *et al.*, 1991, 1995a,b). However, $P^+β^-$ is higher in energy than $P^+H_L^-$ and it decays to the ground state more rapidly, probably by way of $P^+B_L^-$. Heller *et al.* (1995) combined the β mutation with the substitution Gly-M201→Asp, which introduces a potentially ionizable carboxyl group close to ring V of B_L. If the Asp is indeed ionized, the negatively charged carboxyl group would be expected to elevate the free energy of $P^+B_L^-$. In the double mutant, electron transfer to β is slowed ($\tau \approx 21$ ps) and charge separation appears to occur partly along the normally inactive branch of the pigments to give $P^+H_M^-$ with a time constant of about 100 ps and a quantum yield of about 15%. These results fit the idea that the directional specificy of charge separation in favour of the L branch in wild-type reaction centres results largely from an energy difference between $P^+B_L^-$ and $P^+B_M^-$.

6.5.5 *Contributions to the directional specificity from orbital overlap, internal charge transfer states of P and dielectric asymmetry*

In addition to an energy difference between $P^+B_L^-$ and $P^+B_M^-$, several other factors could contribute to the directionality of the photochemical electron transfer reaction. One possibly important factor is an asymmetric admixture of internal charge transfer (CT) states in the excited state P^*. Molecular orbital analyses of the crystal structure indicate that such CT states would mix strongly with the exciton states of the BChls and that this mixing could account for much of the red shift of the long wavelength absorption band of P relative to the Q_y band of monomeric BChl (Lathrop and Friesner, 1994; Parson and Warshel, 1987; Scherer and Fischer, 1989a; Thompson and Zerner, 1991; Thompson *et al.*, 1990; Warshel and Parson, 1987). Mixing with asymmetric CT states would cause the excited state to have a dipolar and polarizable character, which is qualitatively in agreement with the observation of large Stark effects on the absorption band (Lockhart and Boxer, 1987, 1988; Lösche *et al.*, 1987; Middendorf *et al.*, 1993) and with an unusually strong coupling to low frequency vibrational modes of the protein (Johnson *et al.*, 1990; Lathrop and Friesner, 1994; Lyle *et al.*, 1993).

Analyses of the crystal structure indicate that one of the two BChls of P (P_M) has greater orbital overlap with B_L than with B_M, while the other (P_L) has better overlap with B_M (Plato et al., 1988). A CT state in which an electron moves to from P_L to P_M ($P_L^+P_M^-$) thus could favour photoreduction of B_L, whereas a CT state in which an electron moved in the opposite direction ($P_M^+P_L^-$) could promote electron transfer to B_M. Differences in structural distortions of the two BChls or in electrostatic interactions with the protein could favour one of these CT states over the other so that it made the predominant contribution to P*. In addition, Plato et al. (1988) found that the calculated overlap integrals for P_M and B_L, which would figure prominently in electron transfer from P* to or through B_L, were about three times larger than those for P_L and B_M. They suggested that the CT character of P* comes predominantly from $P_L^+P_M^-$, and that the larger orbital overlap of P_M^- with B_L is a major factor in directing electron transfer to B_L and H_L.

Experimental information on the asymmetry of electrostatic interactions between P and the protein has come from resonance Raman and mid-IR spectroscopy of reaction centres in solution (Breton et al., 1992; Mattioli et al., 1991, 1994) and from ENDOR studies of oriented crystals (Lendzian et al., 1993). The ENDOR measurements show that hyperfine interactions of the unpaired electron of P$^+$ generally are stronger with hydrogen atoms of P_L than with the corresponding hydrogens of P_M, which means that the electron and the positive charge tend to localize preferentially on P_L. The apparent ratio of P_L^+ to P_M^+ is of the order of 2:1. This asymmetry implies that the $P_L^+P_M^-$ CT state is likely to be favoured in P*, which is consistent with the proposal that an asymmetric charge distribution in P* plays a role in the directionality of the electron transfer reaction. However, the significance of the electrostatic asymmetry of P$^+$ is called into question by the observation that the asymmetry is less pronounced in chromatophores and depends on the nature of the detergent used to purify reaction centres (Rautter et al., 1994).

The charge distribution in P$^+$ can be shifted strongly in either direction by mutations that result in the substitution of BPh for one of the two BChls of P. In wild-type reaction centres, the imidazole groups of His-M202 and -L173 provide axial ligands to the Mg atoms of P_L and P_M, respectively. If His-M202 is replaced by Leu, the reaction centre incorporates BPh in place of BChl P_L (Bylina and Youvan, 1988; Bylina et al., 1990; Hammes et al., 1990). Replacing His-L173 by Leu results in incorporation of BPh in place of PM (McDowell et al., 1991). The spectroscopic properties of the 'heterodimer' reaction centres indicate that replacing one of the BChls by BPh lowers the energy of a CT state in which an electron is transferred to the BPh from the remaining BChl. The ENDOR, Raman and IR spectra of P$^+$ in heterodimer reaction centres are consistent with the view that the positive charge now localizes almost exclusively on the BChl (Breton et al., 1992; Huber et al., 1990; Rautter et al., 1995). Yet the photochemical electron transfer reaction still favours reduction of H_L strongly over H_M in heterodimer mutants of either His-M202 or -L173, and the rates of charge separation are not very different in mutants at the two positions (Kirmaier et al., 1989; McDowell et al., 1990, 1991). This observation indicates that the directionality of charge separation in wild-type reaction centres probably does not arise from an asymmetric CT state of P, and also is unlikely to depend on the different orbital overlaps of the two BChls of P with B_L and B_M.

Steffen *et al.* (1994) have suggested another possible explanation for the directionality of charge separation. They propose that the effective dielectric constant in the region surrounding H_L is larger than that in the region of H_M. Such dielectric asymmetry could stabilize the $P^+H_L^-$ radical pair and increase the reorganization energy associated with its formation, which in principle could either facilitate or retard the formation of $P^+H_L^-$ relative to $P^+H_M^-$, depending on the relative magnitudes of λ and ΔG°. Steffen *et al.* (1994) found that the electric fields from P^+ and Q_A^- in the $P^+Q_A^-$ radical pair appear to have a larger Stark effect on the Q_y absorption band of H_M than they have on the corresponding band of H_L. This is surprising because Q_A is much closer to H_L than to H_M (see *Figure 6.2*). There are, however, several possible explanations for the observations other than dielectric asymmetry. For example, the hydrogen bond formed by the ring-V keto group of H_L could rotate the vector for the change in permanent dipole moment associated with the Q_y transition, which would alter the observed response to an oriented electric field. Analysis of the results also may be complicated by exciton interactions of H_L and H_M with the neighbouring BChls (B_L and B_M), and by changes in the electrical fields as ionizable amino acid side chains near the quinones bind or release protons. Although anisotropic electrostatic screening factors are to be expected for charged groups in the interior of any protein and are seen in electrostatics calculations on the reaction centre (Alden *et al.*, 1995b), additional work seems needed to show whether such effects have much bearing on the directionality of charge separation.

6.5.6 *Computer simulations*

Several groups have described molecular dynamics simulations of electron transfer from $P^\star$ to H_L, with or without $P^+B_L^-$ as an intermediate (Creighton *et al.*, 1988; Gehlen *et al.*, 1994; Marchi *et al.*, 1993; Schulten and Tesch, 1991; Treutlein *et al.*, 1992; Warshel and Parson, 1991; Xu and Schulten, 1994). In the 'semi-classical trajectory' or 'surface hopping' model (Warshel, 1982; Warshel and Hwang, 1986), fluctuations of the energy gap between the reactant and product states are followed during a molecular dynamics trajectory. Opportunities for electron transfer are considered to occur whenever the gap passes through zero. Creighton *et al.* (1988) and Warshel and Parson (1991) evaluated transitions from $P^\star$ to $P^+B_L^-$ and from $P^+B_L^-$ to $P^+H_L^-$ in separate trajectories, and also included superexchange-mediated jumps from $P^\star$ to $P^+H_L^-$. The simulations agreed reasonably well with the experimental observations: $P^+H_L^-$ rose with a time constant of about 3.5 ps, with the build-up of a small amount of $P^+B_L^-$ during the first 1 ps. With the particular values that were used for the mean energy gaps and electronic interaction matrix elements, superexchange played a relatively minor role in the overall kinetics. However, similar kinetics can be obtained in models that invoke a larger mean energy gap between $P^\star$ and $P^+B_L^-$ ($\langle\Delta E_{12}\rangle$) so that the reaction depends entirely on superexchange, provided that the matrix elements V_{12} and V_{23} are made sufficiently large (Gehlen *et al.*, 1994; Marchi *et al.*, 1993).

Because the semi-classical trajectory model neglects nuclear tunnelling, it is not directly applicable to low temperatures. Molecular dynamics simulations at room temperature can, however, be used to obtain information on how strongly nuclear motions of various frequencies are coupled to the electron transfer reaction, and this information can be incorporated into quantum mechanical models that should continue to hold at low temperatures. In the 'dispersed polaron' approach (Warshel and Hwang, 1986; Warshel et $al.$, 1989a,b), a Fourier transform of the fluctuations of the time-dependent energy gap during a molecular dynamics simulation is used to relate the complex, anharmonic motions of the protein to the motions of an equivalent multidimensional harmonic system. Each vibrational mode contributes a Fourier transform peak at its characteristic frequency, with an amplitude proportional to the displacement of the coordinates between the reactant and product states. The set of frequencies and displacements can be used in a quantum mechanical expression for electron transfer in a multidimensional harmonic system. The similar 'spin boson' model (Bader et $al.$, 1990; Marchi et $al.$, 1993; Xu and Schulten, 1994) uses a Fourier transform of the autocorrelation function of the energy gap. Either of these treatments can reproduce the observed increase in the rate of electron transfer reaction with decreasing temperature, again provided that the mean energy gaps are chosen judiciously. Although the predictive value of such simulations presently is limited by the uncertainties in the energy gaps and the interaction matrix elements, it is encouraging that treatments based on the actual structure of the reaction centre can capture both the extremely high rate and the novel temperature dependence of the reaction.

6.6 Electron transfer from bacteriophaeophytin H_L^- to Q_A

The kinetics of electron transfer from H_L^- to Q_A can be measured from the disappearance of the 650-nm absorption band of H_L^- or recovery of the 545- and 760-nm absorption bands of H_L. The time constant at 295 K is about 200 ps in reaction centres of both $Rb.$ $sphaeroides$ and $Rp.$ $viridis$ (Holten et $al.$, 1978; Kaufmann et $al.$, 1975; Kirmaier and Holten, 1987; Kirmaier et $al.$, 1985a; Rockley et $al.$, 1975).

In $Rb.$ $sphaeroides$ reaction centres, electron transfer from H_L^- to Q_A speeds up with decreasing temperature; it is about 2.5 times faster at 4 K than at 295 K (Kirmaier and Holten, 1988b; Kirmaier et $al.$, 1985a). In $Rp.$ $viridis$, the rate is essentially independent of temperature over this range (Kirmaier and Holten, 1988b). As was discussed above for the initial electron transfer step, this observation suggests that ΔG^o is approximately equal in magnitude, but opposite in sign, to λ. However, an extensive study by Gunner and Dutton (1989) showed that changing ΔG^o by substituting various anthraquinones or naphthoquinones for the native Q_A (ubiquinone) had relatively little effect on the temperature dependence of the kinetics, although the absolute value of the rate constant decreased as the reaction was made less exothermic. ΔG^o for the reaction in native reaction centres is about −0.65 eV at 295 K (Woodbury et $al.$, 1986). Gunner and Dutton (1989) found that the rate began to drop off significantly when ΔG^o was reduced in magnitude below about −0.50 eV. They fitted the rate constant to quantum mechanical expressions that included coupling to one or more vibrational modes with energies in the range of 120 cm^{-1}. They concluded that higher energy intramolecular modes of H_L and Q_A also could be important, but that there was little coupling to modes with energies of less than about 8 cm^{-1}. Kirmaier *et al.* (1985a), who used a multimode quantum mechanical expression to fit the temperature dependence of the kinetics in native reaction centres, also emphasized the importance of high energy modes of the electron carriers.

The insensitivity of the kinetics to temperature also can be reproduced by dispersed-polaron simulations that use a distribution of vibrational modes (Warshel *et al.*, 1989b). As discussed above, this approach has the merit that the vibrational frequencies and displacements are obtained from molecular dynamics simulations of the actual system, rather than being adjusted phenomenologically to fit the kinetics. However, the method still requires rather arbitrary adjustments of ΔG^0, in addition to the assumption that ΔG^0 is independent of temperature. The validity of this last assumption is in question because measurements of delayed fluorescence suggest that the free energy of $P^+H_L^-$ may increase with decreasing temperatures (see above).

The rate of electron transfer from H_L^- to Q_A decreases by about a factor of 20 if the reaction centre's non-haem Fe atom is removed (Kirmaier *et al.*, 1986). Reconstitution of Fe-depleted reaction centres with either Fe^{2+} or Zn^{2+} restores the original kinetics. Although Q_A is not bound directly to the Fe, the two are linked through a histidine residue that is hydrogen bonded to the quinone and liganded to the Fe. Movements of the Fe and its other ligands thus are likely to contribute to the reorganization energy of the electron transfer reaction. In addition, the positive charge of the Fe presumably contributes to ΔG^0.

The side chain of Trp-M252 sits directly between H_L and Q_A and seems likely to participate in the path of electron transfer. The rate of the reaction at room temperature decreases by a factor of 3–4 if the tryptophan is replaced by tyrosine or phenylalanine (Stilz *et al.*, 1994). However, these mutations also loosen the binding of the quinone, so it is possible that the slower kinetics result more from a repositioning of the quinone than from differences in the capacities of the tryptophan, tyrosine and phenylalanine side chains to mediate superexchange.

Acknowledgements

I am indebted to R. Alden, V. Nagarajan and A. Warshel for stimulating discussions and the National Science Foundation grant DMB-9111599 for support.

References

Alden RG, Parson WW, Chu ZT, Warshel A. (1995a) Macroscopic and microscopic estimates of the energetics of charge separation in bacterial reaction centers. In: *Reaction Centres of Photosynthetic Bacteria. Structure and Dynamics* (ed. ME Michel-Beyerle). Springer-Verlag, Berlin (in press).
Alden RG, Parson WW, Chu ZT, Warshel A. (1995b) Calculations of electrostatic energies in photosynthetic reaction centers. *J. Am. Chem. Soc.* **117**: 12 284–12 298.
Allen JP, Feher G, Yeates TO, Komiya H, Rees DC. (1987a) Structure of the reaction center from *Rhodobacter sphaeroides* R-26: the protein subunits. *Proc. Natl Acad. Sci. USA* **84**: 6162–6166.

Allen JP, Feher G, Yeates TO, Komiya H, Rees DC. (1987b) Structure of the reaction center from *Rhodobacter sphaeroides* R-26: the cofactors. *Proc. Natl Acad. Sci. USA* **84:** 5730–5734.

Arata H, Parson WW. (1981) Delayed fluorescence from *Rhodopseudomonas sphaeroides* reaction centers. Enthalpy and free energy changes accompanying electron transfer from P-870 to quinones. *Biochim. Biophys. Acta* **638:** 201–209.

Arlt T, Schmidt S, Kaiser W, Lauterwasser C, Meyer M, Scheer H, Zinth W. (1993) The accessory bacteriochlorophyll: a real electron carrier in primary photosynthesis. *Proc. Natl Acad. Sci. USA* **90:** 11757–11762.

Arnoux B, Ducruix A, Astier C, Picaud M, Roth M, Reiss-Husson F. (1990) Towards the understanding of the function of *Rb. sphaeroides* Y wild type reaction center: gene cloning, protein and detergent structures in the three-dimensional crystals. *Biochimie* **72:** 525–530.

Bader JS, Kuharski RA, Chandler D. (1990) Role of nuclear tunneling in aqueous ferrous–ferric electron transfer. *J. Chem. Phys.* **93:** 230–236.

Becker M, Nagarajan V, Middendorf D, Parson WW, Martin JE, Blankenship RE. (1991) Temperature dependence of the initial electron-transfer kinetics in photosynthetic reaction centers of *Chloroflexus aurantiacus*. *Biochim. Biophys. Acta* **1057:** 299–312.

Bibikova M, Arlt T, Zinth W, Oesterhelt D. (1995) Site-specific mutagenesis around the special pair of the reaction center in *Rhodopseudomonas viridis*. *Proceedings of the Xth International Congress on Photosynthesis* (ed. P Mathis). Kluwer Academic Publishers, Dordrecht, in press.

Bixon M, Jortner J, Michel-Beyerle ME, Ogrodnik A, Lersch W. (1987) The role of the accessory bacteriochlorophyll in reaction centers of photosynthetic bacteria: intermediate acceptor in the primary electron transfer? *Chem. Phys. Lett.* **140:** 626–630.

Bixon M, Michel-Beyerle ME, Jortner J. (1988) Formation dynamics, decay kinetics, and singlet-triplet splitting of the (bacteriochlorophyll dimer)$^+$ (bacteriopheophytin)$^-$ radical pair in bacterial photosynthesis. *Isr. J. Chem.* **28:** 155–168.

Bixon M, Jortner J, Michel-Beyerle ME. (1990) On the primary charge separation in bacterial photosynthesis. In: *Reaction Centres of Photosynthetic Bacteria* (ed. ME Michel-Beyerle). Springer-Verlag, Berlin, pp. 389–400.

Bixon M, Jortner J, Michel-Beyerle ME. (1991) On the mechanism of the primary charge separation in bacterial photosynthesis. *Biochim. Biophys. Acta* **1056:** 301–315.

Bixon M, Jortner J, Michel-Beyerle ME. (1995) A kinetic analysis of the primary charge separation in bacterial photosynthesis. *Chem. Phys.* **197:** 389–404.

Blankenship RE, Schaafsma TJ, Parson WW. (1977) Magnetic field effects on radical pair intermediates in bacterial photosynthesis. *Biochim. Biophys. Acta* **461:** 297–305.

Blankenship RE, Feick R, Bruce BD, Kirmaier C, Holten D, Fuller RC. (1983) Primary photochemistry in the facultative green photosynthetic bacterium *Chloroflexus aurantiacus*. *J. Cell Biochem.* **22:** 251–261.

Blankenship RE, Madigan MT, Bauer CE, eds. (1995) *Anoxygenic Photosynthetic Bacteria*. Kluwer Academic Publishers, Dordrecht (in press).

Boxer SG, Chidsey CED, Roelofs MG. (1982) Anisotropic magnetic interactions in the primary radical ion-pair of photosynthetic reaction centers. *Proc. Natl Acad. Sci. USA* **79:** 4632–4636.

Boxer SG, Chidsey CED, Roelofs MG. (1983) Magnetic field effects on reaction yields in the solid state: an example from photosynthetic reaction centers. *Annu. Rev. Phys. Chem.* **34:** 389–417.

Boxer SG, Goldstein RA, Lockhart DJ, Middendorf TR, Takiff L. (1989) Excited states, electron-transfer reactions, and intermediates in bacterial photosynthetic reaction centers. *J. Phys. Chem.* **93:** 8280–8294.

Breton J. (1985) Orientation of the chromophores in the reaction center of *Rhodopseudomonas viridis*. Comparison of low-temperature linear dichroism spectra with a model derived from X-ray crystallography. *Biochim. Biophys. Acta* **810:** 235–245.

Breton J, Verméglio A, eds. (1988) *The Photosynthetic Bacterial Reaction Center*. Plenum Press, New York.

Breton J, Verméglio A, eds. (1992) *The Photosynthetic Reaction Centre II: Structure, Spectroscopy and Dynamics*. Plenum Press, New York.

Breton J, Martin J-L, Migus A, Antonetti A, Orszag A. (1986) Femtosecond spectroscopy of excitation electron transfer and initial charge separation in the reaction center of the photosynthetic bacterium *Rhodopseudomonas viridis*. *Proc. Natl Acad. Sci. USA* **83**: 5121–5175.

Breton J, Martin J-L, Fleming GR, Lambry J-C. (1988) Low temperature femtosecond spectroscopy of the initial step of electron transfer in reaction centers from photosynthetic purple bacteria. *Biochemistry* **27**: 8276–8284.

Breton J, Martin J-L, Lambry J-C, Robles SJ, Youvan DC. (1990) Ground state and femtosecond transient absorption spectroscopy of a mutant of *Rhodobacter capsulatus* which lacks the initial electron acceptor bacteriopheophytin. In: *Reaction Centers of Photosynthetic Bacteria* (ed. ME Michel-Beyerle). Springer-Verlag, Berlin, pp. 293–302.

Breton J, Nabedryk E, Parson WW. (1992) A new infrared electronic transition of the oxidized primary electron donor in bacterial reaction centers: a way to assess resonance interactions between the bacteriochlorophylls. *Biochemistry* **31**: 7503–7510.

Budil DE, Gast P, Chang CH, Schiffer, M, Norris JR. (1987) Three-dimensional X-ray crystallography of membrane proteins: insights into electron transfer. *Annu. Rev. Phys. Chem.* **38**: 561–583.

Bylina EJ, Youvan DC. (1988) Directed mutations affecting spectroscopic and electron transfer properties of the primary donor in the photosynthetic reaction center. *Proc. Natl Acad. Sci. USA* **85**: 7226–7230.

Bylina EJ, Kirmaier C, McDowell L, Holten D, Youvan DC. (1988) Influence of an amino-acid on the optical properties and electron transfer dynamics of a photosynthetic reaction centre complex. *Nature* **336**: 182–184.

Bylina EJ, Kolaczkowski SV, Norris JR, Youvan DC. (1990) EPR characterization of genetically modified reaction centers of *Rhodobacter capsulatus*. *Biochemistry* **29**: 6203–6210.

Carithers RP, Parson WW. (1975) Delayed fluorescence from *Rhodopseudomonas viridis* following single flashes. *Biochim. Biophys Acta* **387**: 194–211.

Case GD, Parson WW. (1971) Thermodynamics of the primary and secondary photochemical reactions in *Chromatium*. *Biochim. Biophys. Acta* **253**: 187–202.

Chan C-K, Chen LX-Q, DiMagno TJ, Hanson DK, Nance SL, Schiffer M, Norris JR, Fleming GR. (1991a) Initial electron transfer in photosynthetic reaction centers of *Rhodobacter capsulatus* mutants. *Chem. Phys. Lett.* **176**: 366–372.

Chan C-K, DiMagno TJ, Chen LX-Q, Norris JR, Fleming GR. (1991b) Mechanism of the initial charge separation in photosynthetic bacterial reaction centers. *Proc. Natl Acad. Sci. USA* **88**: 11202–11206.

Chang CH, El-Kabbani O, Tiede D, Norris J, Schiffer M. (1991) Structure of the membrane-bound protein photosynthetic reaction center from *Rhodobacter sphaeroides*. *Biochemistry* **30**: 5352–5360.

Chidsey CED, Kirmaier C, Holten D, Boxer SG. (1984) Magnetic field dependence of radical-pair decay kinetics and molecular triplet quantum yield in quinone-depleted reaction centers. *Biochim. Biophys. Acta* **766**: 424–437.

Chidsey CED, Takiff L, Goldstein RA, Boxer SG. (1985) Effect of magnetic fields on the triplet state lifetime in photosynthetic reaction centers: evidence for thermal repopulation of the initial radical pair. *Proc. Natl Acad. Sci. USA* **82**: 6850–6854.

Chirino AJ, Lous EJ, Huber M, Allen JP, Schenck CC, Paddock ML, Feher G, Rees DC. (1994) Crystallographic analyses of site-directed mutants of the photosynthetic reaction center from *Rhodobacter sphaeroides*. *Biochemistry* **33**: 4584–4593.

Cogdell RJ, Crofts AR. (1972) Some observations on the primary acceptor of *Rhodopseudomonas viridis*. *FEBS Lett.* **27**: 176–178.

Coleman WJ, Youvan DC. (1990) Spectroscopic analysis of genetically modified photosynthetic reaction centers. *Annu. Rev. Biophys. Biophys. Chem.* **19**: 333–367.

Creighton S, Hwang J-K, Warshel A, Parson WW, Norris J. (1988) Simulating the dynamics of the primary charge separation process in bacterial photosynthesis. *Biochemistry* **27**: 774–781.

Cusanovich MA, Bartsch RG, Kamen MD. (1968) Light-induced electron transport in *Chromatium* strain D. II. Light-induced absorbance changes in *Chromatium* chromatophores. *Biochim. Biophys. Acta* **153**: 397–417.

Deisenhofer J, Michel H. (1989) The photosynthetic reaction center from the purple bacterium *Rhodopseudomonas viridis*. *Science* **245**: 1463–1473.

Deisenhofer J, Norris JR, eds. (1993) *The Photosynthetic Reaction Centre*. Academic Press, New York.

Deisenhofer J, Michel H. (1995) Crystallographic refinement at 2.3 Å resolution and refined model of the photosynthetic reaction center from *Rhodopseudomonas viridis*. *J. Mol. Biol.* **246:** 429–457.

Deisenhofer J, Epp O, Miki K, Huber R, Michel H. (1985) Structure of the protein subunits in the photosynthetic reaction center of *Rhodopseudomonas viridis* at 3 Å resolution. *Nature* **318:** 618–624.

Deisenhofer J, Epp O, Sinning I, Michel H. (1995) Crystallographic refinement at 2.3 Å resolution and refined model of the photosynthetic reaction center from *Rhodopseudomonas viridis*. *J. Mol. Biol.* **246:** 429–457.

Du M, Rosenthal SJ, Xie X, DiMagno TJ, Schmidt M, Hanson DK, Schiffer M, Norris JR, Fleming GR. (1992) Femtosecond spontaneous emission studies of reaction centers from photosynthetic bacteria. *Proc. Natl Acad. Sci. USA* **89:** 8517–8521.

Dutton PL, Leigh JS, Wraight CA. (1973) Direct measurement of the midpoint potential of the primary electron acceptor in *Rhodopseudomonas spheroides in situ* and in the isolated state: some relationships with pH and *o*-phenanthroline. *FEBS Lett.* **36:** 169–173.

Dutton PL, Prince RC, Tiede DM, Petty KM, Kaufmann KJ, Netzel TL, Rentzepis PM. (1976) Electron transfer in the photosynthetic reaction center. *Brookhaven Symp. Biol.* **28:** 213–236.

El-Kabbani O, Chang CH, Tiede D, Norris J, Schiffer M. (1991) Comparison of reaction centers from *Rhodobacter sphaeroides* and *Rhodopseudomonas viridis*: overall architecture and protein–pigment interactions. *Biochemistry* **30:** 5361–5369.

Ermler U, Fritzsch G, Buchanan SK, Michel H. (1994) Structure of the photosynthetic reaction center from *Rhodobacter sphaeroides* at 2.65 Å resolution: cofactors and protein–cofactor interactions. *Structure* **2:** 925–936.

Fajer J, Borg DC, Forman A, Felton RH, Dolphin D, Vegh L. (1974) The cation radicals of free base and zinc bacteriochlorin, bacteriochlorophyll, and bacteriopheophytin. *Proc. Natl Acad. Sci. USA* **71:** 994–998.

Fajer J, Brune DC, Davis MS, Forman A, Spaulding LD. (1975) Primary charge separation in bacterial photosynthesis: oxidized chlorophylls and reduced pheophytin. *Proc. Natl Acad. Sci. USA* **72:** 4956–4960.

Fajer J, Davis MS, Brune DC, Spaulding LD, Borg DC, Forman A. (1976) Chlorophyll radicals and primary events. *Brookhaven Symp. Biol.* **28:** 74–103.

Finkele U, Lauterwasser C, Zinth W, Gray KA, Oesterhelt D. (1990) Role of tyrosine M210 in the initial charge separation of reaction centers of *Rhodobacter sphaeroides*. *Biochemistry* **29:** 8517–8521.

Fleming GR, Martin J-L, Breton J. (1988) Rates of primary electron transfer in photosynthetic reaction centers and their mechanistic implications. *Nature* **33:** 190–192.

Friesner RA, Won, Y. (1989) Spectroscopy and electron transfer dynamics of the bacterial photosynthetic reaction center. *Biochim. Biophys. Acta* **977:** 99–122.

Ganago AO, Shkuropatov AYa, Shuvalov VA. (1991) Sub-picosecond dynamics of excited state of primary electron donor in reaction centers of *Rhodopseudomonas viridis* as revealed by hole burning at 1.7 K broad and narrow holes. *FEBS Lett.* **284:** 199–202.

Gehlen JN, Marchi M, Chandler M. (1994) Dynamics affecting the primary charge transfer in photosynthesis. *Science* **263:** 499–502.

Godik VI, Borisov AY. (1979) Short-lived delayed luminescence of photosynthetic organisms. I. Nanosecond afterglows in purple bacteria at low redox potentials. *Biochim. Biophys. Acta* **548:** 296–308.

Godik VI, Kotova EA, Borisov AY. (1982) Nanosecond recombination luminescence of purple bacteria. The lifetime temperature dependence in *Rhodospirillum rubrum* chromatophores. *Photochem. Photophys.* **4:** 219–226.

Goldstein RA, Boxer SG. (1988a) The effect of very high magnetic fields on the reaction dynamics in bacterial reaction centers: implications for the reaction mechanism. *Biochim. Biophys. Acta* **977:** 70–77.

Goldstein RA, Boxer SG. (1988b) The effect of very high magnetic fields on the delayed fluorescence from oriented bacterial reaction centers. *Biochim. Biophys. Acta* **977:** 78–86.

Goldstein RA, Takiff L, Boxer SG. (1988) Energetics of initial charge separation in bacterial photosynthesis: the triplet decay rate in very high magnetic fields. *Biochim. Biophys. Acta* **934**: 253–263.

Gray KA, Wachtveitl J, Oesterhelt D. (1992) Photochemical trapping of a bacteriopheophytin anion in site-specific reaction-center mutants from the photosynthetic bacterium *Rhodobacter sphaeroides*. *Eur. J. Biochem.* **207**: 723–731.

Gunner MR, Dutton PL. (1989) Temperature and $-\Delta G^0$ dependence of the electron transfer from BPh⁻ to Q_A in reaction center protein from *Rhodobacter sphaervides* with different quinones as Q_A. *J. Am. Chem. Soc.* **111**: 3400–3412.

Gunner MR, Nicholls A, Honig B. (1996) Electrostatic potentials in *Rhodopseudomonas viridis* reaction centers: implications for the driving force and directionality of electron transfer. *J. Phys. Chem.* **100**: 4277–4291.

Haberkorn R, Michel-Beyerle ME. (1979) On the mechanism of magnetic field effects in bacterial photosynthesis. *Biophys. J.* **26**: 489–498.

Hamm P, Zinth W. (1995) Ultrafast initial reaction in bacterial photosynthesis revealed by femtosecond infrared spectroscopy. *J. Phys. Chem.* **99**: 13537– 13544.

Hamm P, Gray KA, Oesterhelt D, Feik R, Scheer H, Zinth W. (1993) Subpicosecond emission studies of bacterial reaction centers. *Biochim. Biophys. Acta* **1142**: 99–105.

Hammes SL, Mazzola L, Boxer SG, Gaul DF, Schenck CC. (1990) Stark spectroscopy of the *Rhodobacter sphaeroides* reaction center heterodimer mutant. *Proc. Natl Acad. Sci. USA* **87**: 5682–5686.

Hanson LK. (1988) Theoretical calculations of reaction centers. *Photochem. Photobiol.* **47**: 903–921.

Heller BA, Holten D, Kirmaier C. (1995) Control of electron transfer between the L- and M-sides of the photosynthetic reaction center. *Science* **269**: 940–945.

Hoff AJ. (1981) Magnetic field effects on photosynthetic reaction centers. *Q. Rev. Biophys.* **14**: 599–665.

Hoff AJ, Rademaker H, Van Grondelle R, Duysens LNM. (1977) On the magnetic field dependence of the yield of the triplet state in reaction centers of photosynthetic bacteria. *Biochim. Biophys. Acta* **460**: 547–554.

Holten D, Windsor MW, Parson WW, Thornber JP. (1978) Primary photochemical processes in isolated reaction centers of *Rhodopseudomonas viridis*. *Biochim. Biophys. Acta* **501**: 112–126.

Holzapfel W, Finkele U, Kaiser W, Oesterhelt D, Scheer H, Stilz HU, Zinth W. (1989) Observation of a bacteriochlorophyll anion radical during the primary charge separation in a reaction center. *Chem. Phys. Lett.* **160**: 1–7.

Holzapfel W, Finkele U, Kaiser W, Oesterhelt D, Scheer H, Stilz HU, Zinth W. (1990) Initial electron transfer in the reaction center from *Rhodobacter sphaeroides*. *Proc. Natl Acad. Sci. USA* **87**: 5168–5172.

Hörber JKH, Göbel W, Ogrodnik A, Michel-Beyerle ME, Cogdell RJ. (1986) Time-resolved measurements of fluorescence from reaction centers of *Rhodopseudomonas sphaeroides*. *FEBS Lett.* **198**: 273–278.

Huber M, Lous EJ, Isaacson RA, Feher G, Gaul D, Schenck CC. (1990) EPR and ENDOR studies of the oxidized donor in reaction centers of *Rhodobacter sphaeroides* strain R-26 and two heterodimer mutants in which histidine M202 or L173 was replaced by leucine. In: *Reaction Centers of Photosynthetic Bacteria* (ed. ME Michel-Beyerle). Springer-Verlag, Berlin, pp. 219–228.

Hunter DA, Hoff AJ, Hore PJ. (1987) Theoretical calculations of RYDMAR effects in photosynthetic bacteria. *Chem. Phys. Lett.* **134**: 6–11.

Jackson JB, Cogdell RJ, Crofts AR. (1973) Some effects of *o*-phenanthroline on electron transport in chromatophores from photosynthetic bacteria. *Biochim. Biophys. Acta* **292**: 218–225.

Jean JM, Chan C-K, Fleming GR. (1988) Electronic energy transfer in photosynthetic bacterial reaction centers. *Isr. J. Chem.* **28**: 169–175.

Jean JM, Friesner RA, Fleming GR. (1992) Application of a multilevel Redfield theory to electron transfer in condensed phases. *J. Chem. Phys.* **96**: 5827–5842.

Jia Y, DiMagno TJ, Chan C-K, Wang Z, Du M, Hanson DK, Schiffer M, Norris JR, Fleming GR, Popov MS. (1993) Primary charge separation in mutant reaction centers of *Rhodobacter capsulatus*. *J. Phys. Chem.* **9**: 13180–13191.

Jia Y, Jonas DM, Joo T, Nagasawa Y, Lang MJ, Fleming GR. (1995) Observations of ultrafast energy transfer from the accessory bacteriochlorophylls to the special pair in photosynthetic reaction centers. *J. Phys. Chem.* **99**: 6263–6266.

Johnson SG, Tang D, Jankowiak R, Hayes JM, Small GJ, Tiede DM. (1990) Primary donor state model structure and energy transfer in bacterial reaction centers. *J. Phys. Chem.* **94**: 5849–5855.

Jortner J. (1980) Dynamics of the primary events in bacterial photosynthesis. *J. Am. Chem. Soc.* **102**: 6676–6686.

Joseph J, Bruno W, Bialek W. (1991) Bleaching of the bacteriochlorophyll monomer band: can absorption kinetics distinguish virtual from two-step electron transfer in bacterial photosynthesis? *J. Phys. Chem.* **95**: 6242–6247.

Kallebring B, Larsson S. (1987) Singlet spectra of chromophores in the reaction center of *Rhodopseudomonas viridis* as calculated by the CNDO/S method. *Chem. Phys. Lett.* **138**: 76–82.

Kaufmann KJ, Petty KM, Dutton PL, Rentzepis PM. (1975) Picosecond kinetics of events leading to reaction center bacteriochlorophyll oxidation. *Science* **188**: 1301–1304.

Kharkats YI, Kuznetsov AM, Ulstrup J. (1995a) Coherence, friction and electric field effects in primary charge separation of bacterial photosynthesis. *J. Phys. Chem.* **99**: 13545–13554.

Kharkats YI, Kuznetsov AM, Ulstrup J. (1995b) Dynamics of fast optical band-shape time evolution of three-center electron transfer systems. *J. Phys. Chem.* **99**: 13555–13559.

Kellog EC, Kolaczkowski S, Wasielewski MR, Tiede DM. (1989) Measurement of the extent of electron transfer to the bacteriopheophytin in the M-subunit in reaction centers of *Rhodopseudomonas viridis*. *Photosynth. Res.* **22**: 47–59.

Kirmaier C, Holten D. (1987) Primary photochemistry of reaction centers from the photosynthetic purple bacteria. *Photosynth. Res.* **13**: 225–260.

Kirmaier C, Holten D. (1988a) Subpicosecond spectroscopy of charge separation in *Rhodobacter capsulatus* reaction centers. *Isr. J. Chem.* **28**: 79–85.

Kirmaier C, Holten D. (1988b) Temperature effects on the ground state absorption spectra and electron transfer kinetics of bacterial reaction centers. In: *The Photosynthetic Bacterial Reaction Center: Structure and Dynamics* (eds J Breton, A Verméglio). Plenum Press, New York, pp. 219–228.

Kirmaier C, Holten D. (1990) Evidence that a distribution of bacterial reaction centers underlies the temperature and detection-wavelength dependence of the rates of the primary electron-transfer reactions. *Proc. Natl Acad. Sci. USA* **87**: 3552–3556.

Kirmaier C, Holten D. (1991) An assessment of the mechanism of initial electron transfer in bacterial reaction centers. *Biochemistry* **30**: 609–613.

Kirmaier C, Holten D, Parson WW. (1985a) Temperature and detection-wavelength dependence of the picosecond electron-transfer kinetics measured in *Rhodopseudomonas sphaeroides* reaction centers. Resolution of new spectral and kinetic components in the primary charge-separation process. *Biochim. Biophys. Acta* **810**: 33–48.

Kirmaier C, Holten D, Parson WW. (1985b) Picosecond-photodichroism studies of the transient states in *Rhodopseudomonas sphaeroides* reaction centers at 5K. Effects of electron transfer on the six bacteriochlorin pigments. *Biochim. Biophys. Acta* **810**: 49–61.

Kirmaier C, Holten D, Debus RJ, Feher G, Okamura MY. (1986) Primary photochemistry of iron-depleted and zinc-reconstituted reaction centers from *Rhodopseudomonas sphaeroides*. *Proc. Natl Acad. Sci. USA* **83**: 6407–6411.

Kirmaier C, Bylina EJ, Youvan DC, Holten D. (1989) Subpicosecond formation of the intradimer charge transfer state $[BChl_{LP}{}^{+}BPh_{MP}{}^{-}]$ in reaction centers from the $HIS^{M200}{\rightarrow}LEU$ mutant of *Rhodobacter capsulatus*. *Chem. Phys. Lett.* **159**: 251–257.

Kirmaier C, Gaul D, DeBey R, Holten D, Schenck CC. (1991) Charge separation in a reaction center incorporating bacteriochlorophyll for photoactive bacteriopheophytin. *Science* **251**: 922–927.

Kirmaier C, Laporte L, Schenck CC, Holten D. (1995a) The nature and dynamics of the charge-separated intermediate in reaction centers in which bacteriochlorophyll replaces the photoactive bacteriopheophytin. 1. Spectral characterization of the transient state. *J. Phys. Chem.* **99**: 8903–8909.

Kirmaier C, Laporte L, Schenck CC, Holten D. (1995b) The nature and dynamics of the charge-separated intermediate in reaction centers in which bacteriochlorophyll replaces the photoactive bacteriopheophytin. 2. The rates and yields of charge separation and recombination. *J. Phys. Chem.* **99**: 8910–8917.

Kitzing EV, Kuhn H. (1990) Primary electron transfer in photosynthetic reaction centers. *J. Phys. Chem.* **94**: 1699–1702.

Komiya H, Yeates TO, Rees DC, Allen JP, Feher G. (1988) Structure of the reaction center from *Rhodobacter sphaeroides* R-26 and 2.4.1: symmetry relations and sequence comparisons between different species. *Proc. Natl Acad. Sci. USA* **85**: 9012–9016.

Kuhn H. (1986) Electron transfer mechanism in the reaction center of photosynthetic bacteria. *Phys. Rev. A.* **34**: 3409–3425.

Lathrop EJP, Friesner RA. (1994) Simulation of optical spectra from the reaction center of *Rb. sphaeroides*. Effects of an internal charge-separated state of the special pair. *J. Phys. Chem.* **98**: 3056–3066.

Lao K, Franzen S, Steffen M, Lambright D, Stanley R, Boxer SG. (1995a) Effects of applied electric fields on the quantum yields for the initial electron transfer steps in bacterial photosynthesis. 1. Quantum yield failure. *J. Phys. Chem.* **97**: 13165–13171.

Lao K, Franzen S, Steffen M, Lambright D, Stanley R, Boxer SG. (1995b) Effects of applied electric fields on the quantum yields for the initial electron transfer steps in bacterial photosynthesis. II. Dynamic Stark effect. *Chem. Phys.* **197**: 259–275.

Laporte L, Kirmaier C, Schenck CC, Holten D. (1995) Free-energy dependence of the rate of electron transfer to the primary quinone in beta-type reaction centers. *Phys. Chem.* **197**: 225–237.

Lauterwasser C, Finkele U, Scheer H, Zinth W. (1991) Temperature dependence of the primary electron transfer in photosynthetic reaction centers from *Rhodobacter sphaeroides*. *Chem. Phys. Lett.* **183**: 471–477.

Lendzian F, Huber M, Isaacson RA, Endeward B, Plato M, Bönigk B, Möbius K, Lubitz W, Feher G. (1993) The electronic structure of the primary donor cation radical in *Rhodobacter sphaeroides* R-26: ENDOR and TRIPLE resonance studies in single crystals of reaction centers. *Biochim. Biophys. Acta* **1183**: 139–160.

Lin SH. (1989) Theory of photoinduced intramolecular electron transfer in condensed media. *J. Chem. Phys.* **90**: 7103–7113.

Lin X, Murchison HA, Nagarajan V, Parson WW, Allen JP, Williams JC. (1994) Specific alteration of the oxidation potential of the electron donor in reaction centers from *Rhodobacter sphaeroides*. *Proc. Natl Acad. Sci. USA* **91**: 10265–10270.

Lockhart DJ, Boxer SG. (1987) Magnitude and direction of the change in dipole moment associated with excitation of the primary electron donor in *Rhodopseudomonas sphaeroides* reaction centers. *Biochemistry* **26**: 664–668.

Lockhart DJ, Boxer SG. (1988) Stark effect spectroscopy of *Rhodobacter sphaeroides* and *Rhodopseudomonas viridis* reaction centers. *Proc. Natl Acad. Sci. USA* **85**: 107–111.

Lockhart DJ, Goldstein RF, Boxer SG. (1988) Structure-based analysis of the initial electron transfer step in bacterial photosynthesis: electric-field induced fluorescence anisotropy. *J. Chem. Phys.* **89**: 1408–1415.

Lösche M, Feher G, Okamura MY. (1987) The Stark effect in reaction centers from *Rhodobacter sphaeroides* R-26 and *Rhodopseudomonas viridis* reaction centers. *Proc. Natl Acad. Sci. USA* **84**: 7537–7541.

Lyle PA, Kolaczkowski SV, Small GJ. (1993) Photochemical hole-burned spectra of protonated and deuterated reaction centers of *Rhodobacter sphaeroides*. *J. Phys. Chem.* **97**: 6924–6933.

Maiti S, Walker GC, Cowen BR, Pippenger R, Moser CC, Dutton PL, Hochstrasser RM. (1994) Femtosecond coherent transient infrared spectroscopy of reaction centers from *Rhodobacter sphaeroides*. *Proc. Natl Acad. Sci. USA* **91**: 10360–10364.

Mar T, Gingras G. (1990) Relative phototrapping rates of the two bacteriopheophytins in the photoreaction center of *Ectothiorhodospira* sp. *Biochim. Biophys. Acta* **1017**: 112–117.

Marchi M, Gehlen JN, Chandler D, Newton M. (1993) Diabatic surfaces and the pathway for primary electron transfer in a photosynthetic reaction center. *J. Am. Chem. Soc.* **115**: 4178–4190.

Marcus RA. (1956) On the theory of oxidation–reduction reactions involving electron transfer. I. *J. Chem. Phys.* **24**: 966–978.

Marcus RA. (1965) On the theory of oxidation–reduction reactions involving electron transfer. VI. Unified treatment for homogeneous and electrode reactions. *J. Chem. Phys.* **43**: 679–701.

Marcus RA. (1987) Superexchange versus an intermediate BChl$^-$ mechanism in reaction centers of photosynthetic bacteria. *Chem. Phys. Lett.* **133**: 471–477.

Marcus RA. (1988a) An internal consistency test and its implications for the initial steps in bacterial photosynthesis. *Chem. Phys. Lett.* **146**: 13–22.

Marcus RA. (1988b) Mechanisms of the early steps in bacterial photosynthesis and their implications for experiment. *Isr. J. Chem.* **28**: 205–213.

Marcus RA, Sutin N. (1985) Electron transfers in chemistry and biology. *Biochim. Biophys. Acta* **811**: 265–322.

Martin J-L, Breton J, Hoff AJ, Migus A, Antonetti A. (1986) Femtosecond spectroscopy of electron transfer in the reaction center of the photosynthetic bacterium *Rhodopseudomonas sphaeroides* R-26: direct electron transfer from the dimeric bacteriochlorophyll primary donor to the bacteriopheophytin acceptor with a time constant of 2.8 ± 0.2 psec. *Proc. Natl Acad. Sci. USA* **83**: 957–961.

Mattioli TA, Hoffmann A, Robert B, Schrader B, Lutz M. (1991) Primary donor structure and interactions in bacterial reaction centers from near-infrared Fourier transform resonance Raman spectroscopy. *Biochemistry* **30**: 4648–4654.

Mattioli TA, Williams JC, Allen JP, Robert B. (1994) Changes in primary donor hydrogen bonding interactions in mutant reaction centers from *Rhodobacter sphaeroides*: identification of the vibrational frequencies of all the conjugated carbonyl groups. *Biochemistry* **33**: 1636–1643.

McDowell LM, Gaul D, Kirmaier C, Holten D, Schenck CC. (1991) Investigation into the source of electron transfer asymmetry in bacterial reaction centers. *Biochemistry* **30**: 8315–8322.

McDowell LM, Kirmaier C, Holten D. (1990) Charge transfer and charge resonance states of the primary electron donor in wild-type and mutant bacterial reaction centers. *Biochim. Biophys. Acta* **1020**: 239–246.

Michel H, Epp O, Deisenhofer J. (1986) Pigment–protein interactions in the photosynthetic reaction center from *Rhodobacter sphaeroides*. *EMBO J.* **5**: 2445–2451.

Michel-Beyerle ME, ed. (1985) *Antennas and Reaction Centers of Photosynthetic Bacteria.* Springer-Verlag, New York.

Michel-Beyerle ME, ed. (1990) *Reaction Centers of Photosynthetic Bacteria.* Springer-Verlag, New York.

Michel-Beyerle ME, ed. (1995) *Reaction Centers of Photosynthetic Bacteria. Structure and Dynamics.* Springer-Verlag, New York.

Michel-Beyerle ME, Plato M, Deisenhofer J, Michel H, Bixon M, Jortner J. (1988) Unidirectionality of charge separation in reaction centers of photosynthetic bacteria. *Biochim. Biophys. Acta* **932**: 52–70.

Middendorf TR, Mazzola LT, Gaul DF, Schenck CC, Boxer SG. (1991) Photochemical hole-burning spectroscopy of a photosynthetic reaction center mutant with altered charge separation kinetics: properties and decay of the initially excited state. *J. Phys. Chem.* **95**: 10142–10151.

Middendorf TR, Mazzola LT, Lao K, Steffen MA, Boxer SG. (1993) Stark effect (electrostriction) spectroscopy of photosynthetic reaction centers at 1.5 K: evidence that the special pair has a large excited-state polarizability. *Biochim. Biophys. Acta* **1143**: 223–234.

Moser CC, Sension RJ, Szarka AZ, Repinec ST, Hochstrasser RM, Dutton PL. (1995) Initial charge separation kinetics of bacterial photosynthetic reaction centers in oriented Langmuir–Blodgett films in an applied electric field. *Chem. Phys.* **197**: 343–354.

Moss DA, Leonhard M, Bauscher M, Mäntele W. (1991) Electrochemical redox titration of cofactors in the reaction center from *Rhodobacter sphaeroides*. *FEBS Lett.* **283**: 33–36.

Müller MG, Griebenow K, Holzwarth AR. (1991) Primary processes in isolated photosynthetic bacterial reaction centers from *Chloroflexus aurantiacus* studied by picosecond fluorescence spectroscopy. *Biochim. Biophys. Acta* **1098**: 1–12.

Nagarajan V, Parson WW, Gaul D, Schenck C. (1990) Effect of specific mutations of tyrosine-(M)210 on the primary photosynthetic electron-transfer process in *Rhodobacter sphaeroides*. *Proc. Natl Acad. Sci. USA* **87**: 7888–7892.

Nagarajan V, Parson WW, Davis D, Schenck CC. (1993) Kinetics and free energy gaps of electron-transfer reactions in *Rhodobacter sphaeroides* reaction centers. *Biochemistry* **32**: 12324–12336.

Nitschke W, Rutherford AW. (1991) Photosynthetic reaction centres: variations on a common structural theme? *Trends Biochem. Sci.* **16:** 241–245.

Norris JR, Uphaus RA, Crespi HL, Katz JJ. (1971) Electron spin resonance of chlorophyll and the origin of signal I in photosynthesis. *Proc. Natl Acad. Sci. USA* **68:** 625–628.

Norris JR, Scheer H, Druyan ME, Katz JJ. (1974) An electron–nuclear double resonance (endor) study of the special pair model for photo-reactive chlorophyll in photosynthesis. *Proc. Natl Acad. Sci. USA* **71:** 4897–4900.

Norris JR, Bowman MK, Budil DE, Tang J, Wraight CA, Closs GL. (1982) Magnetic characterization of the primary state of bacterial photosynthesis. *Proc. Natl Acad. Sci. USA* **79:** 5532–5536.

Ogrodnik A. (1990) The free energy difference between the excited primary donor $^1P^*$ and the radical pair state P^+H^- in reaction centers of *Rhodobacter sphaeroides*. *Biochim. Biophys. Acta* **1020:** 65–71.

Ogrodnik A. (1993) Electric field effects on steady state and time-resolved fluorescence from photosynthetic reaction centers. *Mol. Crystals Liquid Crystals* **230:** 35–56.

Ogrodnik, A, Kruger HW, Orthuber H, Haberkorn R, Michel-Beyerle ME, Scheer H. (1982) Recombination dynamics in bacterial photosynthetic reaction centers. *Biophys. J.* **39:** 91–100.

Ogrodnik A, Volk M, Letterer R, Feik R, Michel-Beyerle ME. (1988) Determination of free energies in reaction centers of *Rb. sphaeroides*. *Biochim. Biophys. Acta* **936:** 361–371.

Ogrodnik A, Eberl U, Heckman R, Kappl M, Feik R, Michel-Beyerle ME. (1991) Excitation dichroism of electric field modulated fluorescence yield for the identification of primary electron acceptor in photosynthetic reaction center. *J. Phys. Chem.* **95:** 2036–2041.

Ogrodnik A, Langenbacher T, Bieser G, Siegl J, Eberl U, Michel-Beyerle ME. (1992) Electric field-induced decrease of quantum yield of charge separation in reaction centers. *Chem. Phys. Lett.* **198:** 653–658.

Ogrodnik A, Keupp W, Volk M, Auermeier G, Michel-Beyerle ME. (1994) Inhomogeneity of radical pair energies in photosynthetic reaction centers revealed by differences in recombination dynamics of $P^+H_A^-$ when detected in delayed emission and in absorption. *J. Phys. Chem.* **98:** 3432–3439.

Okamura MY, Feher G. (1992) Proton transfer in reaction centers from photosysnthetic bacteria. *Annu. Rev. Biochem.* **61:** 861–896.

Palaniappan V, Aldema MA, Frank HA, Bocian DF. (1992) Q_y-Excitation resonance Raman scattering from the special pair in *Rhodobacter sphaeroides* reaction centers. Implications for primary charge separation. *Biochemistry* **31:** 11050–11058.

Parson WW. (1991) Reaction centers. In: *Chlorophylls* (ed. H Scheer). CRC Press, Boca Raton, FL, pp. 1153–1180.

Parson WW, Cogdell RJ. (1975) The primary photochemical reaction of bacterial photosynthesis. *Biochim. Biophys. Acta* **416:** 105–149.

Parson WW, Warshel A. (1987) Spectroscopic properties of photosynthetic reaction centers. 2. Application of the theory to *Rhodopseudomonas viridis*. *J. Am. Chem. Soc.* **109:** 6152–6163.

Parson WW, Creighton S, Warshel A. (1987) Calculations of spectroscopic properties and electron transfer kinetics of photosynthetic bacterial reaction centers. In: *Primary Reactions of Photobiology* (ed. T Kobayashi). Springer-Verlag, New York, pp. 43–51.

Parson WW, Chu Z-T, Warshel A. (1990) Electrostatic control of charge separation in bacterial photosynthesis. *Biochim. Biophys. Acta* **1017:** 251–272.

Parson WW, Nabedryk E, Breton J. (1992) Mid- and near-IR electronic transitions of P^+: new probes of resonance interactions and structural asymmetry in reaction centers. In: *The Photosynthetic Bacterial Reaction Center II* (eds J Breton, A Verméglio). Plenum Press, New York, pp. 79–88.

Paschenko VZ, Korvatovskii BN, Kononenko AA, Chamorovsky SK, Rubin AB. (1985) Estimation of the rate of photochemical charge separation in *Rhodopseudomonas sphaeroides* reaction centers by fluorescence and absorption picosecond spectroscopy. *FEBS Lett.* **191:** 245–248.

Peloquin JM, Williams JC, Lin X, Alden RG, Taguchi AKW, Allen JP, Woodbury NW. (1994) Time-dependent thermodynamics during early electron transfer in reaction centers from *Rhodobacter sphaeroides*. *Biochemistry* **33**: 8089–8100.

Pierson BK, Thornber JP. (1983) Isolation and characterization of photochemical reaction centers from the thermophilic green bacterium *Chloroflexus aurantiacus* strain J-10-f1. *Proc. Natl Acad. Sci. USA* **80**: 80–84.

Plato M, Winscom CJ. (1988) A configuration interaction (CI) description of vectorial electron transfer in bacterial reaction centers. In: *The Photosynthetic Bacterial Reaction Center* (eds J Breton, A Verméglio). Plenum Press, New York, pp. 421–424.

Plato M, Möbius K, Michel-Beyerle ME, Bixon M, Jortner J. (1988) Intermolecular electronic interactions in bacterial photosynthesis. *J. Am. Chem. Soc.* **110**: 7279–7285.

Plato M, Lendzian F, Lubitz W, Möbius K. (1992) Molecular orbital study of electronic asymmetry in primary donors of bacterial reaction centers. In: *The Photosynthetic Bacterial Reaction Center II* (eds J Breton, A Verméglio). Plenum Press, New York, pp. 109–118.

Prince RC, Leigh JS, Dutton PL. (1976) Thermodynamic properties of the reaction center of *Rhodopseudomonas viridis*. *In vivo* measurement of the reaction center bacteriochlorophyll–primary acceptor intermediary electron carrier. *Biochim. Biophys. Acta* **440**: 622–636.

Rademaker J, Hoff AJ. (1981) The balance between primary forward and back reaction in bacterial photosynthesis. *Biophys. J.* **34**: 325–344.

Rautter J, Lendzian F, Wang S, Allen JP, Lubitz W. (1994) Comparative study of reaction centers from photosynthetic purple bacteria: electron paramagnetic resonance and electron nuclear double resonance spectroscopy. *Biochemistry* **33**: 12077–12084.

Rautter J, Lendzian F, Schulz C, Fetsch A, Kuhn M, Lin X, William JC, Allen JP, Lubitz W. (1995) ENDOR studies of the primary donor cation radical in mutant reaction centers of *Rhodobacter sphaeroides* with altered hydrogen-bond interactions. *Biochemistry* **34**: 8130–8143.

Reddy NRS, Lyle PA, Small GJ. (1992) Applications of spectral hole burning spectroscopies to antenna and reaction center complexes. *Photosynth. Res.* **31**: 167–194.

Reed DW, Zankel KL, Clayton RK. (1969) The effect of redox potential on P870 fluorescence in reaction centers from *Rhodopseudomonas spheroides*. *Proc. Natl Acad. Sci. USA* **63**: 42–46.

Reimers JR, Hush NS. (1995a) The nature of the near-infrared electronic absorption at 1250 nm in the spectra of the radical cations of the special pairs in the photosynthetic reaction centers of *Rhodobacter sphaeroides* and *Rhodopseudomonas viridis*. *J. Am. Chem. Soc.* **117**: 1302–1308.

Reimers JR, Hush NS. (1995b) Nature of the ground and first excited states of the radical cations of photosynthetic reaction centers. *Chem. Phys.* **197**: 323–332.

Rockley MG, Windsor MW, Cogdell RJ, Parson WW. (1975) Picosecond detection of an intermediate in the photochemical reaction of bacterial photosynthesis. *Proc. Natl Acad. Sci. USA* **72**: 2251–2255.

Rutherford AW, Heathcote P, Evans MCW. (1979) Electron-paramagnetic-resonance measurements of the electron transfer components in the reaction center of *Rhodopseudomonas viridis*. *Biochem. J.* **182**: 515–523.

Sarai A. (1979) Energy and temperature dependence of electron and excitation transfer in biological systems. *Chem. Phys. Lett.* **63**: 360–366.

Scheer H, ed. (1992) *Chlorophylls.* CRC Press, Boca Raton, FL.

Schenck CC, Blankenship RE, Parson WW. (1982) Radical-pair decay kinetics, triplet yields and delayed fluorescence from bacterial reaction centers. *Biochim. Biophys. Acta* **680**: 44–59.

Scherer POJ, Fischer SF. (1989a) Quantum treatment of the optical spectra and the initial electron transfer process within the reaction center of *Rhodopseudomonas viridis*. *Chem. Phys.* **131**: 115–127.

Scherer POJ, Fischer SF. (1989b) Long range electron transfer via energy tuning of molecular orbitals within the hexamer of the photosynthetic reaction center of *Rhodopseudomonas viridis*. *J. Phys. Chem.* **93**: 1633–1637.

Scherer POJ, Fischer SF. (1992) Theoretical studies on the electronical structure of the special pair dimer and the charge separation process for the reaction center *Rhodopseudomonas viridis*. In: *The Photosynthetic Bacterial Reaction Center II* (eds J Breton, A Verméglio). Plenum Press, New York, pp. 193–207.

Schmidt S, Arlt T, Hamm P, Huber H, Naegele T, Wachtveitl J, Meyer M, Scheer H, Zinth W. (1994) Energetics of the primary electron transfer reaction revealed by ultrafast spectroscopy on modified bacterial reaction centers. *Chem. Phys. Lett.* **223**: 116–120.

Schulten K, Tesch M. (1991) Coupling of protein motion to electron transfer: molecular dynamics and stochastic quantum mechanics study of photosynthetic reaction centers. *Chem. Phys.* **158**: 421–446.

Shiozawa JA, Lottspeich F, Oesterhelt D, Feick R. (1989) The primary structure of the *Chloroflexus aurantiacus* reaction center polypeptides. *Eur. J. Biochem.* **180**: 75–84.

Shkuropatov AY, Shuvalov VA. (1993) Electron transfer in pheophytin *a*-modified reaction centers from *Rhodobacter sphaeroides* (R-26). *FEBS. Lett.* **322**: 168–172.

Shochat S, Arlt T, Francke C, Gast P, van Noort PI, Otte SCM, Schelvis HPM, Schmidt S, Vigjengoom E, Vriezel J, Zinth W, Hoff AJ. (1994) Spectroscopic characterization of reaction centers of the (M)Y210W mutant of the photosynthetic bacterium *Rhodobacter sphaeroides*. *Photosynth. Res.* **40**: 55–66.

Shopes RJ, Wraight CA. (1985) The acceptor quinone complex of *Rhodopseudomonas viridis* reaction centers. *Biochim. Biophys. Acta* **806**: 348–356.

Shreve AP, Cherepy NJ, Franzen S, Boxer SG, Mathies RA. (1991) Rapid-flow resonance Raman spectroscopy of bacterial photosynthetic reaction centers. *Proc. Natl Acad. Sci. USA* **88**: 11207–11211.

Shuvalov VA, Klimov VV. (1976) The primary photoreactions in the complex cytochrome-P-890•P-760 (bacteriopheophytin$_{760}$) of *Chromatium minutissimum* at low redox potentials. *Biochim. Biophys. Acta* **440**: 587–599.

Shuvalov VA, Parson WW. (1980) Energies and kinetics of radical pairs involving bacteriochlorophyll and bacteriopheophytin in bacterial reaction centers. *Proc. Natl Acad. Sci. USA* **78**: 957–961.

Shuvalov VA, Parson WW. (1981) Triplet states of monomeric bacteriochlorophyll *in vitro* and of bacteriochlorophyll dimers in antenna and reaction center complexes. *Biochim. Biophys. Acta* **638**: 50–59.

Shuvalov VA, Krakhmaleva IN, Klimov VV. (1976) Photooxidation of P960 and photoreduction of P800 (bacteriopheophytin *b*-800) in reaction centers from *Rhodopseudomonas viridis*. *Biochim. Biophys. Acta* **449**: 597–601.

Shuvalov VA, Shkuropatov AYa, Kulakova SM, Ismailov MA, Shkuropatova VA. (1986) Photoreactions of bacteriopheophytins and bacteriochlorophylls in reaction centers of *Rhodopseudomonas sphaeroides* and *Chloroflexus aurantiacus*. *Biochim. Biophys. Acta* **849**: 337–346.

Skourtis SS, da Silva AJR, Bialek W, Onuchic JN. (1992) A new look at the primary charge separation in bacterial photosynthesis. *J. Phys. Chem.* **96**: 8034–8041.

Small GJ. (1995) On the validity of the standard model for primary charge separation in bacterial reaction centers. *Chem. Phys.* **197**: 239–257.

Small GJ, Hayes JM, Silbey RJ. (1992) The question of dispersive kinetics for the initial phase of charge separation in bacterial reaction centers. *J. Phys. Chem.* **96**: 7499–7501.

Stanley RJ, Boxer SG. (1995) Oscillations in spontaneous fluorescence from photosynthetic reaction centers. *J. Phys. Chem.* **99**: 859–863.

Steffen MA, Lao K, Boxer SG. (1994) Dielectric asymmetry in the photosynthetic reaction center. *Science* **264**: 810–816.

Stilz HU, Finkele U, Holzapfel W, Lauterwasser C, Zinth W, Oesterhelt D. (1994) Influence of M subunit Thr222 and Trp252 on quinone binding and electron transfer in *Rhodobacter sphaeroides* reaction centers. *Eur. J. Biochem.* **223**: 233–242.

Streltsov AM, Vakovlev AG, Shkuropatov AY, Shuvalov VA. (1993) Dynamic hole burning within special pair absorption band of *Rhodobacter sphaeroides* (R-26) reaction center at room temperature. *FEBS Lett.* **357**: 239–241.

Takiff L, Boxer SG. (1988) Phosphorescence from the primary electron donor in *Rhodobacter sphaeroides* and *Rhodopseudomonas viridis* reaction centers. *Biochim. Biophys. Acta* **932**: 325–334.

Tang J, Norris JR. (1994) On superexchange electron-transfer reactions involving three paraboloidal potential surfaces in a two-dimensional reaction coordinate. *J. Chem. Phys.* **101**: 5615–5622.

Thompson MA, Zerner MC. (1991) A theoretical examination of the electronic structure and spectroscopy of the photosynthetic reaction center from *Rhodopseudomonas viridis*. *J. Am. Chem. Soc.* **113**: 8210–8215.

Thompson MA, Zerner MC, Fajer JJ. (1990) Electronic structure of bacteriochlorophyll dimers. 1. Bacteriochloroin. *J. Phys. Chem.* **94**: 3820–3828.

Tiede DM, Prince RC, Dutton PL. (1976) EPR and optical spectroscopic properties of the electron carrier intermediate between the reaction center bacteriochlorophylls and the primary acceptor in *Chromatium vinosum. Biochim. Biophys. Acta* **449**: 447–469.

Tiede DM, Kellog E, Breton J. (1987) Conformational changes following reduction of the bacteriopheophytin electron acceptor in reaction centers of *Rhodopseudomonas viridis. Biochim. Biophys. Acta* **892**: 294–302.

Treutlein H, Schulten K, Brunger AT, Karplus M, Deisenhofer J, Michel H. (1992) Chromophore–protein interactions and the function of the photosynthetic reaction center: a molecular dynamics study. *Proc. Natl Acad. Sci. USA* **89**: 75–79.

Vasmel H, Meiburg RF, Kramer HJM, DeVos LJ, Amesz J. (1983) Optical properties of the photosynthetic reaction center of *Chloroflexus aurantiacus* at low temperature. *Biochim. Biophys. Acta* **724**: 333–339.

Vos MH, Lambry J-C, Robles SJ, Youvan DC, Breton J, Martin J-L. (1992) Femtosecond spectral evolution of the excited state of bacterial reaction centers at 10 K. *Proc. Natl Acad. Sci. USA* **89**: 613–617.

Vos MH, Rappoport R, Lambry J-C, Breton J, Martin J-L. (1993) Visualization of coherent nuclear motion in a membrane protein by femtosecond spectroscopy. *Nature* **363**: 320–325.

Vos MH, Jones MR, Hunter CN, Breton J, Lambry J-C, Martin J-L. (1994a) Coherent dynamics during the primary electron-transfer reaction in membrane-bound reaction centers of *Rhodobacter sphaeroides. Biochemistry* **33**: 6750–6757.

Vos MH, Jones MR, Hunter CN, Breton J, Lambry J-C, Martin J-L. (1994b) Influence of the membrane environment on vibrational motions in reaction centers of *Rhodobacter sphaeroides. Biochim. Biophys. Acta* **1186**: 117–122.

Vos MH, Jones MR, Hunter CN, Breton J, Martin J-L. (1994c) Coherent nuclear dynamics at room temperature in bacterial reaction centers. *Proc. Natl Acad. Sci. USA* **91**: 12701–12705.

Vos MH, Jones MR, Hunter CN, Breton J, Lambry J-C, Martin J-L. (1995) Femtosecond spectroscopy and vibrational coherence of membrane-bound RCs of *Rhodobacter sphaeroides* genetically modified at positions M210 and L181. In: *Reaction Centers of Photosynthetic Bacteria. Structure and Dynamics* (ed. ME Michel-Beyerle). Springer-Verlag, Berlin (in press).

Walker GC, Maiti S, Cowen BR, Moser CC, Dutton PL, Hochstrasser RM. (1994) Time resolution of electronic transitions of photosynthetic reaction centers in the infrared. *J. Phys. Chem.* **98**: 5778–5783.

Wang Z, Pearlstein RM, Jia Y, Fleming GR, Norris JR. (1993) Inhomogeneous electron transfer kinetics in reaction centers of bacterial photosyntheisis. *Chem. Phys.* **176**: 421–425.

Warshel A. (1982) Dynamics of reactions in polar solvents. Semiclassical trajectory studies of electron-transfer and proton-transfer reactions. *J. Phys. Chem.* **86**: 2218–2224.

Warshel A, Hwang J-K. (1986) Simulation of the dynamics of electron transfer reactions in polar solvents: semiclassical trajectories and dispersed polaron approaches. *J. Chem. Phys.* **84**: 4938–4957.

Warshel A, Parson WW. (1987) Spectroscopic properties of photosynthetic reaction centers. *J. Am. Chem. Soc.* **109**: 6143–6152.

Warshel A, Parson WW. (1991) Computer simulations of electron transfer reactions in solution and in photosynthetic reaction centers. *Annu. Rev. Phys. Chem.* **42**: 279–309.

Warshel A, Creighton S, Parson WW. (1988) Electron-transfer pathways in the primary event of bacterial photosynthesis. *J. Phys. Chem.* **92**: 2696–2701.

Warshel A, Chu Z-T, Parson WW. (1989a) Microscopic simulation of quantum dynamics and nuclear tunneling in bacterial reaction centers. *Photosynth. Res.* **22**: 39–46.

Warshel A, Chu Z-T, Parson WW. (1989b) Dispersed polaron simulations of electron transfer in photosynthetic reaction centers. *Science* **246**: 112–116.

Warshel A, Chu ZT, Parson WW. (1994) On the energetics of the primary electron-transfer process in bacterial reaction centers. *Photochem. Photobiol. A. Chem.* **82**: 123–128.

Wasielewski MR, Tiede DM. (1986) Sub-picosecond measurements of primary electron transfer in *Rhodopseudomonas viridis* reaction centers using near-infrared excitation. *FEBS Lett.* **204**: 368–372.

Werner H, Schulten K, Weller A. (1978) Electron transfer and spin exchange contributing to the magnetic field dependence of the primary photochemical reaction of bacterial photosynthesis. *Biochim. Biophys. Acta* **502**: 255–268.

Williams JC, Alden RG, Murchison HA, Peloquin JM, Woodbury NW, Allen JP. (1992) Effects of mutations near the bacteriochlorophylls in reaction centers from *Rhodobacter sphaeroides*. *Biochemistry* **31**: 11029–11037.

Woodbury NWT, Parson WW. (1984) Nanosecond fluorescence from isolated photosynthetic reaction centers of *Rhodopseudomonas sphaeroides*. *Biochim. Biophys. Acta* **767**: 345–361.

Woodbury NWT, Parson WW. (1986) Nanosecond fluorescence from chromatophores of *Rhodopseudomonas sphaeroides* and *Rhodospirillum rubrum*. *Biochim. Biophys. Acta* **850**: 197–210.

Woodbury NW, Becker M, Middendorf D, Parson WW. (1985) Picosecond kinetics of the initial photochemical electron transfer reaction in bacterial photosynthetic reaction centers. *Biochemistry* **24**: 7516–7521.

Woodbury NWT, Parson WW, Gunner MR, Prince RC, Dutton PL. (1986) Radical-pair energetics and decay mechanisms in reaction centers containing anthraquinones, naphthoquinones or benzoquinones in place of ubiquinone. *Biochim. Biophys. Acta* **851**: 6–22.

Woodbury NW, Peloquin JM, Alden RG, Lin X, Taguchi AKW, Williams JC, Allen JP. (1994) Relationship between thermodynamics and mechanism during photoinduced charge separation in reaction centers from *Rhodobacter sphaeroides*. *Biochemistry* **33**: 8101–8112.

Woodbury NW, Lin S, Lin X, Peloquin JM, Taguchi AKW, Williams JC, Allen JP. (1995) The role of reaction center excited state evolution during charge separation in a *Rb. sphaeroides* mutant with an initial electron donor midpoint potential 260 mV above wild type. *Chem. Phys.* **197**: 405–421.

Wraight CA, Clayton RK. (1973) The absolute quantum efficiency of bacteriochlorophyll photooxidation in reaction centers of *Rhodopseudomonas sphaeroides*. *Biochim. Biophys. Acta* **333**: 246–260.

Xu D, Schulten K. (1994) Coupling of protein motion to electron transfer in a photosynthetic reaction center: investigating the low temperature behavior in the framework of the spin-boson model. *Chem. Phys.* **182**: 91–117.

Copper proteins

Ole Farver

7.1 Introduction

Many metal ions play decisive roles in living organisms ranging from bacteria to mammals. The copper ion is thus essential in several biological systems and, with the exception of the large multisubunit copper protein, haemocyanin, which transports molecular oxygen in a variety of invertebrates, all copper proteins are in one way or the other involved in electron transfer (ET), particularly those processes associated with bioenergetics and substrate oxidation (Karlin and Tyeklar, 1993). In excess, copper is highly toxic, and the subtle balance of copper homeostasis is controlled by rather complicated transport and bioregulatory processes. The understanding of the biochemistry of copper has developed considerably, particularly during the last few decades, thanks to intensive studies of structure–reactivity relationships among copper proteins (Adman, 1985, 1991).

Several features make the biochemistry of copper unique. The Cu(II) ion has the highest affinity among the divalent metal ions for coordination to organic molecules, just as the Cu(I) ion is the most effective monovalent ion. This property makes the Cu(II)/Cu(I) couple exceedingly useful as a redox couple in biological oxidation–reduction processes. Further, Cu(I) is a strong π-electron donor which may coordinate ligands such as O_2 and CO (Cotton and Wilkinson, 1988).

One important group of copper-containing electron transfer proteins is the so-called 'blue' copper proteins which constitute a rather large family widespread in nature (Chapman, 1991; Farver and Pecht, 1984). These proteins have relatively small molecular weights, 10–20 kDa, and exchange electrons between large immobilized, membrane-bound proteins. A second group, the copper-containing oxidases, maintains the blue copper centre in multicopper enzymes where electrons are passed in sequential one-electron steps from the reducing substrate to a dioxygen-binding polynuclear copper centre where the oxidizing substrate, O_2, eventually becomes reduced to two water molecules (Messerschmidt, 1993). In both types of proteins, intramolecular ET becomes the fundamental physical event through which specificity and control is exerted (Farver and Pecht, 1994). Another, less colourful group of copper proteins catalyses the oxidation of a variety of organic

Protein Electron Transfer, Edited by D.S. Bendall
© 1996 BIOS Scientific Publishers Ltd, Oxford

substrates with the concomitant two-electron reduction of O_2 to H_2O_2. These enzymes, which all contain a single copper ion as part of the prosthetic group, include galactose oxidase and amine oxidases, and also the copper–zinc superoxide dismutase is found in this group (Karlin and Tyeklar, 1993). The main theme of this chapter is long range electron transfer (LRET) in the above types of copper proteins.

7.2   Ligand structure

A considerable amount of structural information is now available on copper proteins, spanning from the single blue copper proteins (cupredoxins) and single copper oxidases to the large multicopper enzymes like ascorbate oxidase (AO) as well as the dioxygen transport protein, haemocyanin. The first cupredoxins for which the three-dimensional structure was solved were plastocyanin (Pc) from the poplar tree (*Populus nigra*) (Colman *et al.*, 1978) and azurin (Az) from *Pseudomonas aeruginosa* (Adman *et al.*, 1978) (*Figure 7.1*). High resolution structural information is now available for both oxidized [Cu(II)] and reduced [Cu(I)] forms of these proteins and at different pH values (Baker, 1988; Guss and Freeman, 1983; Nar *et al.*, 1991a; Shephard *et al.*, 1990). Later, the structures of other single blue copper proteins, like the cucumber blue protein (Guss *et al.*, 1988) and amicyanin (Chen *et al.*, 1992; Kalverda *et al.*, 1994) were determined. The former is interesting in its lack of methionine which otherwise constitutes one of the copper ligands. The

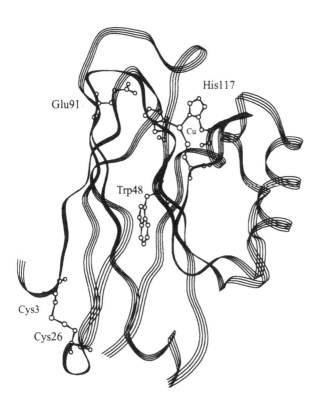

Figure 7.1. Three-dimensional structure of the main chain polypeptide of *Pseudomonas aeruginosa* azurin (Nar *et al.*, 1991a). The copper centre is shown together with some other amino acid residues of particular interest.

amicyanin structure is particularly noteworthy since this copper protein has been crystallized with its physiological partners (Chen *et al.*, 1992, 1994; see Chapter 5).

X-ray structures have been reported for different metal ion-substituted cupredoxins (Adman, 1991), and recently many Az mutant structures have been examined as well (Murphy *et al.*, 1993; Nar *et al.*, 1991b; Romero *et al.*, 1993; Tsai *et al.*, 1995), and several important conclusions have emerged. The folding topology of all cupredoxins studied so far is very similar, consisting of an eight-stranded β-sandwich structure with a highly hydrophobic interior (see *Figure 7.1*. The 'type-1' (T1) copper site is found at one end of the protein (commonly referred to as the 'northern region') but the metal ion is isolated from direct contact with the surrounding medium by the ligands and a high density of surrounding hydrophobic residues. This arrangement is important in determining the specificity and control of the physiological function in this group of proteins.

The blue copper centre seems to be coordination saturated and shows no tendency to bind substrates or other exogenous ligands. Three of the copper ligands consist of two imidazoles (His) and one thiolate group (Cys) providing a trigonal planar ligand arrangement for the copper ion (*Figure 7.2*) which stabilizes the Cu(I) state relative to the Cu(II) state. It is charge transfer from the cysteine thiolate to Cu(II) that gives rise to the intense blue colour of this group of proteins, (see Section 7.3).

The small differences that do exist in the structure of the copper centre in cupredoxins are seen mainly in the strength and geometry of the weaker axial

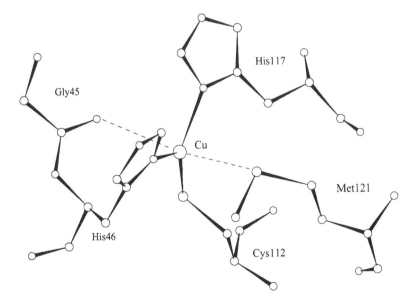

Figure 7.2. The blue copper centre in azurin (Nar *et al.*, 1991a). The three close ligands are shown attached to the copper ion by solid lines, while the two distal ligands are connected to the metal centre by broken lines.

interactions. Further, it is noteworthy that the configuration of the copper site is only slightly affected by the oxidation state of the metal ion (Guss *et al.*, 1986; Romero *et al.*, 1993; Shephard *et al.*, 1990). The site geometry has thus evolved to assist ET, which is in line with the concept of a rack-induced binding where the structure is determined solely by the protein (Malmström, 1994). Even removal of the metal ion does not change the conformation of the protein in any significant way (Garrett *et al.*, 1984; Nar *et al.*, 1992).

As opposed to T1 electron transfer activity, the non-blue or type-2 (T2) Cu(II) centres are involved in chemical activity, and are found not only in enzymes like amine- and galactose oxidase and in superoxide dismutase (SOD), but also contribute to the prosthetic dioxygen binding site in the blue copper oxidases. In AO, the T2 copper ion has two His ligands and one water or hydroxyl ligand in a trigonal geometry (Messerschmidt and Huber, 1990; Messerschmidt *et al.*, 1989), while in SOD the geometry of the copper site is a distorted square pyramid with four histidines in the plane and an exchangeable water molecule in the apical position comprising the ligand sphere (*Figure 7.3*) (Tainer *et al.*, 1982). This centre is thus reminiscent of the smaller inorganic copper complexes.

The type-3 (T3) copper serves as a binding centre for O_2 and is characterized by a pair of copper ions which in the oxidized Cu(II) state is strongly anti-ferromagnetically coupled (Gray and Solomon, 1981). In some proteins, like the dioxygen-carrying haemocyanin, it is the only copper type present, while in the blue oxidases the T3 site together with T2 copper constitute a trinuclear copper centre. In deoxy-haemocyanin, the copper ions are 0.35 nm apart and each metal ion is connected by N^ε-nitrogens from two tightly bound histidines, while a third histidine is found at a more distant axial position (see *Figure 7.4*) (Volbeda and Hol, 1989). In oxy-haemocyanin the two copper ions are connected by O_2 in a μ–η^2–η^2 configuration (Magnus *et al.*,

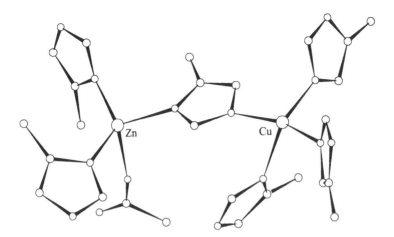

Figure 7.3. The distorted tetrahedral copper and zinc centres in bovine superoxide dismutase (SOD) (Tainer *et al.*, 1982, 1983). The aspartate and the six histidine ligands are shown, while a loosely bound water molecule positioned 0.3 nm from the copper ion is not included.

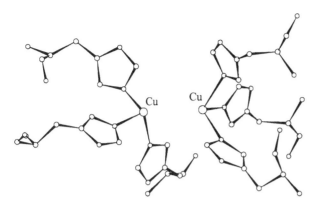

Figure 7.4. The copper sites in deoxy-haemocyanin from *Panulirus interruptus* (Volbeda and Hol, 1989). In the oxygenated form, the two copper ions are connected by a $\mu-\eta^2-\eta^2$ peroxo bridge (Magnus *et al.*, 1993).

1993). In AO, the copper ions have quite a different geometry, being coordinated to six histidines arranged in a trigonal position and with the ion pair bridged by OH$^-$ (Messerschmidt and Huber, 1990; Messerschmidt *et al.*, 1989, 1992) (see below).

The multicopper oxidases have a minimum of four copper atoms per molecule, with one of each of the above-mentioned types, T1, T2 and T3. These enzymes all catalyse the oxidation of organic substrates by O_2 with a concomitant four-electron reduction of dioxygen to water (Farver and Pecht, 1984). The blue oxidase, AO, is a 140-kDa dimer, shown to consist of two identical subunits (Messerschmidt and Huber, 1990; Messerchmidt *et al.*, 1989, 1992). Each monomer is constructed of three domains with T1 copper bound inside domain-3, while three copper ions constitute a novel trimeric arrangement held together symmetrically by four histidines from domain-1 and four histidines from domain-3 (*Figure 7.5*). Of the three copper ions, two have similar coordination to three histidines reminiscent of the haemocyanin structure, and are by analogy called T3, while the remaining copper ion would then, by exclusion, be T2. None of the histidines are involved in bridging of the copper ions. The T2 site is the one most accessible to exogenous ligands and may bind anions such as F$^-$, N$_3^-$ and CN$^-$.

Since the electrons are picked up from the reducing substrate one at a time by the T1 Cu(II) ion while O_2 is coordinated to the trinuclear centre during the catalytic reduction–oxidation process, intramolecular ET must be an important part of the enzymatic function. It is therefore quite noteworthy that adjacent amino acid residues bind to T1 and T3 (Messerschmidt and Huber, 1990; Messerschmidt *et al.*, 1989, 1992). In this sequence, His–Cys–His, the cysteine thiolate binds to T1 copper and the two histidines coordinate two separate copper ions of the trinuclear cluster (see Section 7.9).

7.3 Redox potentials and spectroscopic properties

In the blue proteins, the copper ion is approximately co-planar with the three strong ligands, leading to an electronic ground state which is highly covalent,

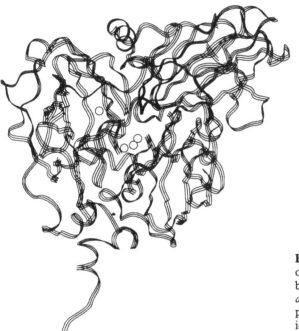

Figure 7.5. Ribbon drawing of the ascorbate oxidase backbone (Messerschmidt *et al.*, 1989, 1992). The positions of the four copper ions are indicated by white circles.

and with a particularly strong π bonding interaction between the thiolate sulphur and Cu(II) (Lowery and Solomon, 1992). The structure of the T1 site, shown in *Figure 7.2*, demonstrates neither the preferred geometry of a Cu^{2+} complex which is tetragonal nor the tetrahedral structure preferred by Cu^+ complexes. Rather, the geometry is distorted tetrahedral, that is a compromise between the two. The unique structure of the copper site in the blue proteins, however, is not necessary for imposing the unusual spectroscopic properties. The intense colour is simply due to a $\pi S \rightarrow Cu(d_{x^2-y^2})$ ligand to metal charge transfer involving a cysteine ligand as the electron donor (Lowery *et al.*, 1993) and with a molar extinction coefficient of 2000–6000 M^{-1} cm^{-1}. This is several hundred times larger than that found for simple Cu(II) complexes. A second characteristic property associated with T1 copper(II) is an electron paramagnetic resonance (EPR) spectrum displaying a hyperfine splitting in the $g_{\parallel}$ region, due to interaction of the copper nuclear spin with the electron spin, which is unusually narrow (~0.008 cm^{-1}) or approximately 50% smaller when compared with those of smaller copper complexes (Boas, 1984). This is attributed to delocalization of the unpaired $Cu(d_{x^2-y^2})$ electron onto the Cys(S)pπ orbital, so reducing the nuclear–electron interaction (Gray and Solomon, 1981). The unique spectroscopic properties are believed to be related to the function of blue copper proteins, and the relationship between the unusual structure and ET reactivity is a central issue in studies of biological ET.

The reduction potentials of the Cu(II)/Cu(I) centre in the blue proteins are higher than generally observed for copper complexes. The protein-forced deformation of the copper site away from a square planar towards a distorted tetrahedral geometry would explain the stabilization of the Cu(I) state relative to Cu(II). Structure-imposed Cu→L π backbonding has been proposed to further account for stabilization of the cuprous state, since strong π interaction with the d_π orbitals results in an increase in the ligand field strength (Gray and Malmström, 1983).

It seems quite perplexing that, although the spectroscopic properties of the blue copper proteins are relatively constant, the midpoint reduction potentials span a region from 180 to 800 mV at pH 7. Site-directed mutagenesis of the blue copper protein, Az, has been very useful in revealing the structural control of the reduction potentials. While the axial ligands are not required for maintaining the blue copper site, variations in the potential seems to be related to the nature of these ligands (Pascher et al., 1993). In Az, Met121 provides a sulphur atom as copper ligand at a distance of about 0.3 nm, and the midpoint potential of the wild-type Az is 304 nm at 298 K and pH 7.0 (Goldberg and Pecht, 1976). Producing a series of Met121 mutants showed that large hydrophobic residues would raise the potential whereas negatively charged residues lower it (Pascher et al., 1993). The total electrode potential span from Met121Glu to Met121Leu Az is 300 mV or about half the range of the potentials for the naturally occurring blue copper proteins. The large increases in reduction potentials caused by substitutions involving bulky hydrophobic side chains at position 121 were explained by the ability of such groups to exclude water from the metal site or simply to provide a low dielectric environment for the copper ion. Opposite, negatively charged hydrophilic residues or water are expected to stabilize the copper ion in the +2 state and thus lower the potential, as is indeed observed (Pascher et al., 1993). Smaller effects on the reduction potentials were observed when mutations were made in positions other than 121, so further effects which may induce changes in solvation or subtle geometric perturbations could also be involved. It should, however, be emphasized that a thiolate copper ligand is not a prerequisite for high reduction potentials per se.

The reduction potential of Az shows a 60–70 mV decrease upon increasing the pH from 5 to 10 (Pascher et al., 1993; StClair et al., 1992; van de Kamp et al., 1993). The latter group has examined how much deprotonation of the two titratable histidines, His35 and His83, would decrease the potential, and have found that the contributions from these two residues are 50 and 13 mV, respectively. However, the H35K Az mutant displays essentially the same pH dependence (Pascher et al., 1993), which seems to argue against this hypothesis, at least if the same mechanism should determine the pH dependence of the reduction potential in all Azs. Indirect pH effects on the protein conformation could, instead, give rise to subtle geometric changes in the copper centre.

Although T2 copper with tetragonally arranged ligands shows the expected lower reduction potentials characteristic of small copper complexes, the T3 copper pair generally have redox potentials as high as that of the T1 copper site, although the ligand sphere consists solely of imidazoles and hydroxide (Messerschmidt, 1993). Distortion of the copper site towards a more tetrahedral symmetry stabilizing Cu(I) relative to Cu(II) is again the reason for the high potentials. In AO, the six histidine ligands (five N^ε and one N^δ) are arranged in a trigonal prism (Messerschmidt and Huber, 1990), while in haemocyanin the arrangement of the T3 site is quite different (Magnus *et al.*, 1991; Volbeda and Hol, 1989), as noted above (*Figure 7.4*). This disparity in geometry is not surprising when the different chemistry of the two proteins is taken into consideration: AO catalyses the irreversible four-electron reduction of dioxygen to water while in haemocyanin the copper centre serves as a reversible O_2 binding site. The reduction potentials of the T1 and T3 sites in AO are both 350 mV, at pH 7.0 (Marchesini and Kroneck, 1979). It is of interest that the similarity of T1 and T3 potentials is conserved in the other blue oxidases, and it is quite intriguing that in fungal laccase both potentials are very high (~780 mV) (Fee and Malmström, 1968). The concordance between the T1 and T3 potentials is still an unanswered puzzle. Considering the reversibility of the

$$Cu^+-Cu^+ + O_2 \rightleftharpoons Cu^{2+}-O_2^{2-}-Cu^{2+} \tag{7.1}$$

reaction in haemocyanin, this would place the Cu(II)/Cu(I) reduction potential at around 800 mV. This again illustrates that high potentials are determined by the three-dimensional structure rather than by particular ligands such as a thiolate.

7.4   Role of ligand geometry on ET

Unlike haem iron and iron–sulphur proteins, the active copper site in electron transfer proteins has no extended coordination sphere. Rather, the copper ion chelates directly to the protein, which then defines the active site structure. As noted in Section 7.2, the proteins provide a geometry which is a compromise between the preferred tetragonal structure for Cu(II) and the tetrahedral structure for Cu(I) and, besides a slight lengthening of the copper–ligand bonds in the reduced proteins, the structures are remarkably similar in the two oxidation states (Guss *et al.*, 1986; Romero *et al.*, 1993; Shephard *et al.*, 1990). As most copper proteins are involved in fast ET reactions, this is not surprising. The structural changes occurring upon reduction of the Cu(II) ion should be insignificant, and the sterically imposed compromise, the 'enthatic state', has been considered to be the main reason for rapid outer sphere ET, since the reorganization energy associated with this process becomes minimal. Thus, the reorganization enthalpy has been determined to be about 30 kJ mol^{-1} (~7 kcal mol^{-1}) (Margalit *et al.*, 1984), which again confirms that the blue copper centre is constructed optimally for ET. Besides minimal changes in copper–ligand bonds, the metal ion is imbedded in the hydrophobic interior of the protein, so solvation contributions to outer sphere reorganization enthalpies also become negligible. It should, however, be noted that entropic factors associated with conformational changes and hydrophobic interactions may also be influential. Comparing the self-exchange rate constants for the two blue copper proteins, plant Pc ($k_{ESE} \leq 10^4$ M^{-1} s^{-1}) (da Silva *et al.*, 1992) and Az ($k_{ESE} \sim 10^6$ M^{-1} s^{-1}) (Groeneveld and Canters, 1988), it is obvious that minimal reorganization energies cannot be the only determining factor for effective ET.

Extensive electronic overlap between the electron donor and acceptor orbitals is a prerequisite for providing a strong electronic coupling. Spectroscopic studies of single crystals of Pc have been combined with self-consistent field calculations and indicate an orientation of the half-filled $d_{x^2-y^2}$ orbital bisecting the Cu–Cys(S$^\gamma$) bond which leads to a high degree of anisotropic covalency favouring ET through the cysteine by a factor of more than 1000 compared with the two histidine ligands (Lowery et al., 1993). In AO where intramolecular ET is expected to take place from T1 copper(I) via a thiolate (see Section 7.9), it is tempting to suggest a similar extent of anisotropic covalency. However, in Az with a similar very intense S(Cys)pπ→Cu^{2+} ($d_{x^2-y^2}$) transition, the thiolate is directed towards the interior of the protein while the physiological ET is supposed to take place via a histidine ligand (His117; see Section 7.6). It remains to be explained if and how apparently conformationally identical copper sites may nonetheless have distinct electronic structures.

7.5   ET pathway calculations

Several excellent reviews can be found on the ET theory (Broo and Larsson, 1990; Jortner and Bixon, 1993; Marcus and Sutin, 1985; van de Kamp and Canters, 1992; Winkler and Gray, 1992), and the reader is referred to Chapters 1 and 2 for a comprehensive and up to date discussion of this subject. LRET in proteins is characterized by a weak interaction between electron donor and acceptor, and in the non-adiabatic limit the rate constant is proportional to the square of the electronic coupling between the electronic states of reactant and product, represented in the form of a matrix element, H_{AB}. For intramolecular ET, the rate constant is given by Fermi's golden rule (Marcus and Sutin, 1985):

$$k = \frac{2\pi}{\hbar} H_{AB}^2 (FC) \tag{7.2}$$

The Franck–Condon factor (FC) for nuclear movements can, for relatively low vibrational frequencies, where $kT > \hbar v$, be treated classically, a condition which often applies to biological ET. The electronic motion, however, requires a quantum mechanical approach, and the semi-classical Marcus equation may be expressed as (Marcus and Sutin, 1985):

$$k = \frac{2\pi}{\hbar} \frac{H_{AB}^2}{(4\pi\lambda RT)^{1/2}} e^{-(\Delta G^\circ + \lambda)^2/4\lambda RT} \tag{7.3}$$

Since wave functions decay exponentially with distance, the tunnelling matrix element, H_{AB} will decrease with the distance, $(r - r_o)$, as:

$$H_{AB} = H_{AB}^\circ \, e^{-\beta(r-r_o)/2} \tag{7.4}$$

H_{AB}° is the electronic coupling at direct contact between electron donor and acceptor (where $r = r_o$) and the decay rate of electronic coupling with distance is determined by the coefficient, β. ET pathways may be identified by analysing the bonding interactions which maximize H_{AB} (Beratan et al., 1991, 1992) (see Chapter 2).

7.6   Pathways in protein–protein ET

Electron transfer copper proteins contain the metal ion as part of the prosthetic group, and resolution of the pathways through which electrons migrate to and from these centres becomes one of the major challenges in biophysics (Farver and Pecht, 1991). In all the copper-containing ET proteins, the metal is relatively inaccessible for external reactants or solvent molecules, which then raises the fundamental question as to how electrons proceed to and from the copper ion. One obvious starting point in such an endeavour is identification of reaction loci on the surface of the redox proteins. Information about the detailed three-dimensional structures of redox proteins has become available during the last two decades and has made resolution of ET pathways possible, thereby allowing a meaningful discussion of structure–function correlation for ET proteins.

Although active sites on ET proteins represent only a small fraction of the total protein, the surrounding parts obviously still play important roles. The relative orientations of the reaction partners should be optimized in order to ensure effective electronic coupling and thereby to control ET probability between the redox centres.

One attempt at identifying active sites on ET proteins has been by affinity labelling, using Cr(II) as electron donor (Farver and Pecht, 1989a). The ET reaction between Cr(II) aqua ions and Cu(II) in several blue copper proteins (P) proceeds as: Cr(II) + P[Cu(II)] $\rightarrow$ Cr(III)–P[Cu(I)]. Cr(III) complexes are generally substitution inert, in marked contrast to the labile Cr(II) complexes, so the essential feature of the above procedure is based on the notion that the coordination sphere of Cr(II) in the transition state is retained in the substitution-inert Cr(III) product. Thus, the Cr(III) ion will remain in the very same coordination sphere as during the ET from Cr(II) to the Cu(II) centre. In this way, redox active sites on the blue copper proteins, Az (Farver and Pecht, 1981a), Pc (Farver and Pecht, 1981b) and stellacyanin (Farver et al., 1987; Morpurgo and Pecht, 1982) were identified. In all three proteins, the Cr(III) label was found in one particular site only, in spite of the fact that a number of potentially complexing chromium ligands are available. The uniqueness of the Cr(III) binding centres was demonstrated by the large effects observed on the reduction kinetics of Cr(III)-labelled proteins (Skov et al., 1993). With identification of redox active sites on blue copper proteins, the question naturally arises as to whether the Cr(III) label attached to redox active sites on the surface perturbs the physiological function of the proteins.

7.6.1 *Azurin*

In Az, Cr(III) was found attached to the carboxylate group of Glu91 at a direct distance of 1.3 nm from the copper centre (see *Figure 7.1*) (Farver and Pecht, 1981a). Pathway calculations (Section 7.5) show a very efficient electronic coupling from O^{ε} of Glu91 to N^{δ} of His46, one of the Cu ligands. Further, it is noteworthy that the imidazole ring of His35 is placed between the Cr(III) binding site and the imidazole ring of the copper ligand, His46. The two rings are virtually parallel, with an interplane distance of 0.38 nm. This structural arrangement in Az has, therefore, been suggested to be of significance in the physiological ET reactions of this protein (Farver and Pecht, 1981a).

Azs are found in many bacteria where they serve as mobile electron mediators (Adman, 1985), most probably between cytochrome *c*-551 and nitrite reductase (also called bacterial cytochrome oxidase). The interesting question now arises as to whether Az uses the same site for interaction with the two partners. Analysis of the complex kinetics of the Az–cytochrome c-551 ET process revealed that the chromium label attenuates the rate of the direct bimolecular ET from Az[Cu(I)] to cytochrome c-551[Fe(III)] by a factor of 2, and it was suggested that the Cr(III) binding site is perturbing the optimal docking of the two proteins by steric or electrostatic effects, and that His35 plays a critical role in the ET process by providing an enhanced electronic coupling by π interaction with the Cu-ligating His46 (Farver et al., 1982a). This assignment was challenged by kinetic studies of the reduction of single site-mutated Az[Cu(II)] by Cyt[Fe(II)] in which Az residues His35 and Glu91 were exchanged (Pascher et al., 1989; van de Kamp et al., 1990). Substitution of His35 by Lys had a very limited effect on the reduction potential of Az, but nevertheless the ET rate increased by 50% rather than showing an expected decrease in rate due to the lack of π interaction here (Pascher et al., 1989). Also when Glu91 was exchanged with Gln, a relatively large increase in rate was observed, so the negative charge on Glu91 seems to be unimportant for effective ET from cytochrome c-551 to Az (Pascher et al., 1989). In another study, His35 was substituted by other amino acid residues, both aromatic and non-aromatic, with small decreases in ET rates as a result, but this could be explained by a similar decrease in reduction potential of the Az mutants, lowering the driving force of the process (van de Kamp et al., 1990). Changes in kinetics, however, do point towards a functional role for Glu91 as part of the binding interface for cytochrome c-551. The studies with the Az mutants, however, indicate that ET between Az and cytochrome c-551 probably takes place via another pathway, most likely taking advantage of the partially exposed imidazole of the copper ligand, His117.

The other half of the physiological role assigned to Az involves ET to nitrite reductase. The kinetic parameters for this reaction did not change upon Cr(III) modification of Az, which led to the conclusion that ET from Az takes place via another functional site, most probably His117 placed in a partially exposed position in the hydrophobic northern end of the protein (Figure 7.1) (Farver et al., 1982a). Site-specific mutants of Az involving both the hydrophobic patch near His117 and the site close to His35 have been used in kinetic studies with nitrite reductase, and have further supported this notion (van de Kamp et al., 1990). With native Az, the rate constant is independent of ionic strength, but substitution of a methionine (Met44) on the north pole with lysine increases the rate with decreasing ionic strength, implicating interaction between the protonated lysyl residue and a negatively charged surface area on nitrite reductase.

In summary, mutant studies on Az show that ET between Az and nitrite reductase takes place via the hydrophobic patch, with the copper ligand His117 serving as electron conductor. The region around His35 and Glu91 may play a role by forming a precursor ET complex in the cytochrome c-551–Az reaction, but the significance of the conservative His35 site in Az ET is still an unanswered puzzle.

Studies of the electron self exchange (ESE) in Az by nuclear magnetic resonance (NMR) line broadening (Groeneveld and Canters, 1988) and rapid-freeze EPR (Groeneveld *et al.*, 1987) indicate that the hydrophobic surface patch surrounding the copper-ligating His117 is involved. The reaction shows minimal dependence on pH and ionic strength and the rate constant ($\sim 10^6$ M^{-1} s^{-1}) is determined by a very favourable entropy of activation which is expected for the hydrophobic surface interaction leading to a precursor protein–protein complex. The Cu–Cu distance in this arrangement is 1.5 nm and ET is assumed to take place via the two His117 ligands. The high self-exchange rate constant in Az is in marked contrast to the slow rate in plant Pc ($<10^4$ M^{-1} s^{-1}) (da Silva *et al.*, 1992) where the high net negative charge apparently prevents the formation of an association complex with a favourable electronic coupling between the copper ions. In *P. aeruginosa* Az crystals, two Az molecules are associated via their hydrophobic north poles, and the dimer packing shows two water molecules connecting the monomers by hydrogen bonding (Nar *et al.*, 1991a). This results in a continuous electron density bridge between the two Cu ions. The electronic coupling for this structurally well defined system has been determined using extended Hückel calculations, and the theoretical rate constant is in good agreement with that determined experimentally ($k_{ESE} = \sim 10^6$ M^{-1} s^{-1}) (Mikkelsen *et al.*, 1993). It is very interesting that the calculations demonstrate that the electronic coupling and thus the probability for ET strongly depends on the orientation of the water molecules bridging the system. Water molecules intercalated between electron donor and acceptor may well play a key role as switches for biological ET (Mikkelsen *et al.*, 1993).

7.6.2 *Plastocyanin*

Pc is found in all higher plants and many algae where it functions as an electron carrier between the membrane-bound cytochrome *bf* complex of photosystem II (PSII) and the chlorophyll-containing centre of photosystem I (PSI) (Chapman, 1991; Gross, 1993). NMR studies using negatively and positively charged inorganic complexes have demonstrated two different sites of interaction in Pc (Cookson *et al.*, 1980). The former complexes interact via the partially exposed His87, which is one of the copper ligands, while the latter interacts with a peptide patch of acidic residues. The chromium labelling experiments discussed above show that Cr(III) forms a substitution-inert complex with this peptide which contains four carboxylate side chains, Glu42, Asp43, Glu44 and Asp45, defining a conservative negative patch, adjacent to a solvent-exposed Tyr83 aromatic residue (Farver and Pecht, 1981b). The distance from this binding centre to the copper ion is 1.2 nm. In agreement with the earlier NMR studies, it was proposed, therefore, that the acidic patch serves as one specific redox partner binding site. This had already been suggested by Freeman and co-workers based on their high resolution three-dimensional structure of poplar Pc (Colman *et al.*, 1978), and implies that Pc may utilize two different sites on the protein surface for interaction with its ET partners. The question thus arises as to whether these two sites may serve as entry and exit ports for electrons in the physiological ET processes between PSI and PSII.

Solubilized cytochrome *f* reduces Pc at a rate 40% slower upon Cr(III) labelling (Beoku-Betts *et al.*, 1983, 1985). As the interaction between Pc and cytochrome *f* is essentially electrostatic (Qin and Kostic, 1992), this further substantiates the notion that acidic residues on Pc are involved in formation of the Pc–cytochrome f interprotein complex. Also, chemical modification of carboxyl groups on reduced Pc with ethylenediamine produced a large degree of inhibition of cytochrome f oxidation (Anderson et al., 1987), which further supports the notion that the acidic patch forms a recognition site for cytochrome f.

Site-directed mutagenesis of Pc has led to further progress in the understanding of the complex mechanism of protein–protein ET reactions and binding (Gross, 1993). He et al. (1991) have studied ET from cytochrome f to Pc in which Tyr83 has been substituted by Phe and Leu, and they have shown very convincingly that the surface-exposed Tyr83 plays a decisive role both in the association and the ET reactions with cytochrome f. It is interesting that, while replacing Tyr83 with Phe mainly affects the binding constant, substitution with Leu causes a major decrease in the ET rate, as this lends support to a definite role for an aromatic side chain in ET (He et al., 1991). In another study using Pc mutants (Modi et al., 1992), the results seem less clear, however. Mutation of Leu12 on the 'northern' hydrophobic surface increased the binding constant for Pc with cytochrome f significantly, while substituting one of the carboxylates, Asp42, with Asn showed no change in binding of ET reactivity with cytochrome f as would have been expected from the reduced negative charge of the acidic patch. It seems likely, therefore, that both the northern and the eastern patches could be involved in formation of the interprotein complex.

Most evidence suggests that Tyr83 is involved in ET to the Cu(II) centre in Pc, and all Pcs from green plants have an aromatic residue at this position. Tyr83 is found next to the copper-ligating Cys84 and this configuration would define a direct through-bond pathway (*Figure 7.6*). The question arises, however, as to why Pc should choose this 'remote' pathway, involving 12 covalent bonds from Tyr83 to copper, when a shorter route including only four bonds from His87 is available. This may be caused by the anisotropic covalency discussed in Section 7.4, which enhances ET by a strong overlap between the half-filled $Cu(d_{x^2-y^2})$ orbital and the $Cys(S)p\pi$ orbital (Lowery et al., 1993). The high degree of covalent anisotropic interaction thus makes the 'remote' pathway highly effective.

The interaction of Pc with PSI is less well understood (see Haehnel et al., 1994). Although the same mechanism applies for photo-oxidation of native and Cr(III)-labelled Pc[Cu(I)] by P700$^+$, the modified Pc was oxidized at a rate significantly slower than that of the native Pc ($k_{OX} = 6.9 \times 10^7$ M^{-1} s^{-1} for Cr(III)–Pc[Cu(I)] and $k_{OX} = 1.8 \times 10^8$ M^{-1} s^{-1} for wild-type Pc) (Farver et al., 1982b). This indicates that the region around the acidic patch indeed plays an important role in either docking or protein–protein ET. Substitution of Leu12 on the 'northern' hydrophobic patch by a negatively charged Glu decreased the binding of Pc to PSI and no intracomplex ET was observed here (Nordling et al., 1991), which suggests that this surface is important for the process and seems to contradict the former observation. However, on replacement of Tyr83 with His, the rate of ET decreased by a factor of 3, but the reduction potential of the mutant was 35 mV higher than for the wild-type protein, so this effect on the ET rate could be a reflection of the lower driving force of the reaction. The

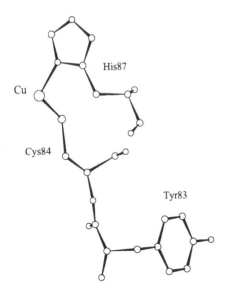

Figure 7.6. Electron transfer pathways in plastocyanin. The 'direct' pathway via His87 and the 'remote' pathway via Tyr83 and Cys84 to the copper ion are shown. The drawing is based on the coordinates from poplar plastocyanin (Guss and Freeman, 1983).

rate was found to be slowed down upon protonation of the His83 residue (Sigfridsson *et al.*, 1995) which was interpreted as a combination of destabilization of the intraprotein complex and the change in driving force. This agrees with the earlier observation that attachment of a Cr^{3+} ion near Tyr83 has influence on the Pc–P700 interaction.

The kinetics of the interaction between Pc mutants and PSI has been studied further by laser flash techniques (Haehnel *et al.*, 1994). Just as for the interaction between Pc and cytochrome *f*, the process involves an association step followed by ET in the protein–protein complex. Interestingly, ET from Pc, where Tyr83 has been substituted with Leu, to $P700^+$ is faster than with the native protein, which seems to exclude a role for this aromatic residue in ET to PSI. Two other mutants were produced where hydrophobic amino acids on the north pole, Gly10 and Ala90, were replaced with the more bulky Leu residue. These mutants completely prevented the formation of a complex between the two proteins, which indicates that this area is involved in the interprotein complex formation. Further, the second order rate constant for ET from Pc[Cu(I)] to $P700^+$ decreases by a factor of 3–6. However, the reduction potentials of these Pc mutants have not been assessed, so it is not clear to what degree a change in ET driving force may affect the rate.

The crystal structure of PSI from cyanobacterium *Synechococcus elongatus* has been determined at a resolution of 0.6 nm (Krauss *et al.*, 1993) and, based on the three-dimensional model, possible docking sites for Pc have been proposed (Fromme *et al.*, 1994). According to the electron density map, the most feasible position for Pc is with the north pole directed towards the chorophylls of P700 and with the copper ion at a distance of approximately 2.0 nm.

From the above studies, it emerges that two distinct binding sites exist on Pc, one near the copper-ligating His87 at the northern hydrophobic patch and the other close to Tyr83 at the eastern acidic region. Most evidence now indicates that cytochrome f binds to the negatively charged carboxylates and donates electrons via a pathway consisting of Tyr83–Cys84. PSI, on the other hand, binds to the hydrophobic northern patch of Pc which represents a decisive structural element for efficient ET from Cu(I) via His87 to P700$^+$.

7.7   Intramolecular LRET in type-1 copper proteins

7.7.1 *Azurin*

Although Azs from different bacteria have a similar three-dimensional structure, a marked difference is seen in their amino acid composition, thus providing proteins with distinct reactivities and reduction potentials (see Section 7.3). All known Azs contain a disulphide bridge at one end of the β-sandwich structure separated from the copper centre in the opposite end of the protein at a distance of 2.6 nm, (see *Figure 7.1*). With high resolution, three-dimensional structures of several wild-type and single site-mutated Azs available (see Section 7.2), this supplies us with a highly interesting system for studies of the different factors which determine the rate of intramolecular LRET (Farver and Pecht, 1994).

As indicated in the reaction scheme below, besides a direct reduction of the Cu^{2+} ion (i), pulse radiolytically produced CO$_2^-$ reacts with the disulphide bridge in oxidized Cu(II)–Az to yield the radical anion RSSR$^-$ (ii) which exhibits a strong absorption band around 410 nm (ε_{410} = 10 000 M^{-1} cm^{-1}). This transient was found to disappear again (iii) concomitantly with a decrease in the characteristic Cu(II) absorption at 625 nm (ε_{625} = 5700 M^{-1} cm^{-1}) with a rate constant independent of the protein concentration (Farver and Pecht, 1989b, 1992a; Farver *et al.*, 1992, 1993, 1996a,b).

$$\text{RSSR-Az[Cu(II)]} + \text{CO}_2^- \rightarrow \text{RSSR-Az[Cu(I)]} + \text{CO}_2 \qquad \text{(i)}$$
$$\text{RSSR-Az[Cu(II)]} + \text{CO}_2^- \rightarrow \text{RSSR}^-\text{-Az[Cu(II)]} + \text{CO}_2 \qquad \text{(ii)}$$
$$\text{RSSR-Az[Cu(II)]} \rightarrow \text{RSSR-Az[Cu(I)]} \qquad \text{(iii)}$$

Scheme 7.1

The observed process (iii) must therefore be intramolecular ET from RSSR$^-$ to Cu(II). The specific rates at pH 7.0 and 298 K for a large number of different Azs, both wild-type and single site mutants, are summarized in *Table 7.1*, together with the Cu(II)/Cu(I) reduction potentials and activation parameters.

Pathway calculations for the above intramolecular ET in Az (reaction iii) using high resolution three-dimensional structures of wild-type Azs and, when available, structures of *Pseudomonas* mutants (Nar *et al.*, 1991b) predict similar ET routes in all the native and the mutated proteins studied so far. Two main pathways were found with only small differences in the electronic coupling factors (*Figure 7.7*). One pathway proceeds along the polypeptide backbone from Cys3 to Asn10, which is connected to the copper-ligating His46 by a hydrogen bond between the carbonyl group and N$^\varepsilon$H of the imidazole group. The alternative pathway goes directly from Cys3 to Thr30 via a hydrogen bond and then from Val31 by a through-space jump of 0.4 nm to Trp48 (or the mutated residues #48). Val49 is connected to Phe111 by another hydrogen bond which may lead the electron further to the thiolate of Cys112 (Farver and Pecht, 1994).

Table 7.1. Kinetic and thermodynamic data for the intramolecular reduction of Cu(II) by RSSR⁻

Azurin	k_{298} (s⁻¹)	E' (mV)	$-\Delta G°$ (kJ mol⁻¹)	$\Delta H^{\neq}$ (kJ mol⁻¹)	$\Delta S^{\neq}$ (J K⁻¹ mol⁻¹)
Wild-type					
P. aeruginosa [a]	44 ± 7	304	68.9	47.5 ± 4.0	−56.5 ± 7.0
P. fluorescens [b]	22 ± 3	347	73.0	36.3 ± 1.2	−97.7 ± 5.0
Alc. species [a]	28 ± 1.5	260	64.6	16.7 ± 1.5	−171 ± 18
Alc. faecalis [b]	11 ± 2	266	65.2	54.5 ± 1.4	−43.9 ± 9.5
Mutant					
W48L [c]	40 ± 4	323	70.7	48.3 ± 0.9	−51.5 ± 5.7
W48M [c]	33 ± 5	312	69.7	48.4 ± 1.3	−50.9 ± 7.4
F114A [c]	72 ± 14	358	74.1	52.1 ± 1.3	−36.1 ± 8.2
M121L [c]	38 ± 7	412	79.3	45.2 ± 1.3	−61.5 ± 7.2
M44K [d]	134 ± 12	370	75.3	47.2 ± 0.7	−46.4 ± 4.4
H35Q [d]	53 ± 11	268	65.4	37.3 ± 1.3	−86.5 ± 5.8
D23A [e]	15 ± 3	311	69.6	47.8 ± 1.4	−61.4 ± 6.3
F110S [f]	38 ± 10	314	69.9	55.5 ± 5.0	−28.7 ± 4.5
I7S [f]	42 ± 8	301	68.6	56.6 ± 4.1	−21.5 ± 4.2
M64E [f]	55 ± 8	278	66.4	46.3 ± 6.2	−56.2 ± 7.2

[a]Farver and Pecht (1989b); [b]Farver and Pecht (1992a); [c]Farver *et al.* (1993); [d]Farver *et al.* (1992); [e]Farver *et al.*. (1996b); [f]Farver *et al.* (1996a). *Alc., Alcaligenes.*

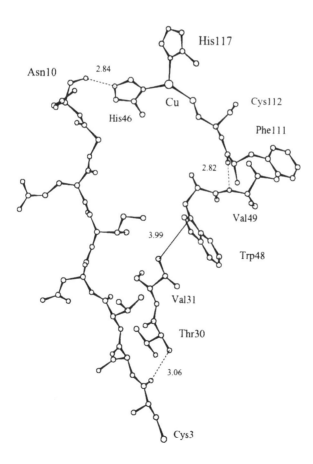

Figure 7.7. Two prominent calculated ET pathways from disulphide to copper in azurin. Some of the interconnecting distances are shown (in Å). Hydrogen bonds are shown as broken lines and the through-space jump is indicated by a thin line. The structure coordinates were taken from Nar *et al.*, (1991a) and the pathway calculations were performed using the methodology developed by Beratan and Onuchic (Beratan *et al.*, 1991).

Since the same LRET pathway from RSSR$^-$ to Cu(II) applies to all Azs studied so far, it was possible, from the kinetic data and activation parameters (see Section 7.5), to calculate the reorganization energy, $\lambda = 99 \pm 5$ kJ mol^{-1} (24 ± 1 kcal mol^{-1}), and the electronic decay factor, $\beta = 6.7 \pm 0.3$ nm^{-1} (Farver et al., 1996a). The latter value is in excellent agreement with both the experimentally determined decay factor for electron tunnelling in modified cytochrome c (7.1 nm^{-1}) (Wuttke et al., 1992) and the theoretical value calculated for tunnelling through a saturated -(CH$_2$)$_n$ chain (6.5 nm^{-1}) (Broo and Larsson, 1990). In these calculations, no distinction is made between systems involving σ and π orbitals. Another theoretical study, however, applying extended Hückel calculations, implies that aromatic residues may enhance the electronic coupling (Broo and Larsson, 1991). In systems which have been selected by evolution for efficient ET, aromatic residues may be found in positions that enlarge the electronic coupling. Examples of this are found in the tryptophan-mediated reduction of quinone in the photosynthetic reaction centre (Plato et al., 1989) as well as in the cytochrome c peroxidase–cytochrome c complex (Pelletier and Kraut, 1992). Other examples are Tyr83 in Pc and the aromatic Cu ligands, His87 and His117 in Pc and Az, respectively, discussed above in Section 7.6. Recent studies using an Az mutant with a Trp substitution in position #31 (V31W) shows a 10-fold increase in intramolecular ET rate from RSSR$^-$ to Cu(II) which cannot be explained in terms of driving force or reorganization energy (unpublished results).

The control of the protein matrix on LRET is still a subject of intense discussion. There is substantial evidence for superexchange through-bond ET, but Dutton and co-workers (see Chapter 1) have analysed data from intramolecular ET processes in a large number of biological systems and found a good correlation between the free energy optimized rate constant, k_{MAX} (i.e. for $-\Delta G^\circ = \lambda$) and the edge-to-edge distance between electron donor and acceptor, and determined a decay factor of $\beta = 14$ nm^{-1} (Moser et al., 1992). Using a value for $(r - r_0)$ between Cys3(S$^\gamma$) and Cys112(S$^\gamma$) in P. aeruginosa Az of 2.46 nm, which defines the shortest ET distance, this yields $\beta = 9$ nm^{-1}. The difference between the two calculated values for the exponential decay seems too large to be accounted for in terms of experimental error. Also, using $\beta = 14$ nm^{-1} gives a calculated maximum rate constant which is two orders smaller than that estimated from the kinetic parameters determined for Az ($k_{MAX} = 10^3$ s^{-1}). The experimental results for LRET in the Azs, therefore, do not fit an exponential decay correlation with a direct through-space distance. Interestingly, the above $k_{MAX} = 10^3$ s^{-1} fits perfectly on a linear plot of log k_{MAX} versus the through-bond distance for ruthenium-substituted cytochrome c mutants drawn with a slope of 7.1 nm^{-1} (Wuttke et al., 1992).

Ruthenium ammines are very attractive reagents for modification of redox proteins as they may be attached to histidine side chains on the protein surface forming substitutionally inert complexes (Gray, 1986). ET between ruthenium(II) and the native redox centre in structurally well characterized proteins can be studied and, by coordination to different surface histidines, the ET distance and nature of the intervening medium can be modified in a controlled manner. Further, changing the ligands of the Ru label permits studies of the driving force effect on the rate of intramolecular ET.

Az possesses a surface-exposed His83 which can be derivatized by aquapentaammine-ruthenium(II) (Margalit *et al.*, 1984), and intramolecular ET from $[(NH_3)_5Ru(His83)]^{2+}$ to the blue copper centre has been studied by flash photolysis. The rate constant is quite small ($1.9 \ s^{-1}$) and independent of temperature (Kostic *et al.*, 1983). Interestingly, this rate is more than one order of magnitude slower than in $[(NH_3)_5Ru(His33)]$cytochrome *c* (Nocera *et al.*, 1984), although the direct distance between the redox centres is the same (1.2 nm), and the driving force is slightly larger in the copper protein (23 vs. 17 kJ mol^{-1}; 5.5 vs. 4.1 kcal mol^{-1}). The difference in rates can be explained in terms of a through-bond mechanism, since the pathway in modified Az involves twice as many peptide bonds as that of Ru-modified cytochrome *c*. Intramolecular ET in Az with much higher driving forces has also been studied by using other Ru(II) complexes attached to His83. The $(bipy)_2Ru(II)His83$ to Cu(II) system has a driving force of 73 kJ mol^{-1} (17.4 kcal mol^{-1})and the rate constant is 10^6 s^{-1} (Wuttke and Gray, 1993). This dramatic increase in rate is not unreasonable since the large driving force is sufficient to overcome the reorganization energy demand for the reaction. The only barrier to the ET process then becomes the distance between electron donor and acceptor.

7.7.2 *Other blue copper proteins*

Stellacyanin has two free histidine residues which can be labelled with ruthenium, and intramolecular ET between Ru(II) and Cu(II) has been observed with a rate constant of $0.05 \ s^{-1}$ at $18°C$ (Farver and Pecht, 1990). The direct distance is approximately 2.0 nm, but the absence of a known three-dimensional structure prevents a more detailed pathway analysis. Another striking example of an unusually slow ET process involves Pc modified with $(NH_3)_5Ru^{2+}$ at His59. Although the edge-to-edge distance between the two redox centres here is only 1.2 nm, the ET rate constant is less than $0.3 \ s^{-1}$ (Jackman *et al.*, 1988). The slow rate is most likely due to a very poor Ru–Cu electronic coupling caused by a very long through-bond ET pathway. Interestingly, extended Hückel calculations have been performed on five different potential ET pathways in Pc which have demonstrated that intramolecular through-bond ET can be electronically favourable even when compared with shorter outer sphere ET where the wave functions decay much faster than by superexchange through the protein matrix (Christensen *et al.*, 1990, 1992).

7.8   ET in type-2 copper proteins

The 'normal' or T2 copper–zinc SOD is one of the best characterized T2 copper proteins (Fridovich, 1989). The enzyme is found in all eukaryotic cells and is a dimer with two identical subunits of 16 kDa molecular weight, each of which contains one copper ion and one zinc ion. As the name indicates, it catalyses the dismutation of the superoxide radical anion into dioxygen and peroxide. The three-dimensional structure has been determined and the overall protein structure bears similarities to the blue copper proteins discussed in Section 7.2 (Tainer *et al.*, 1982, 1983). The metal coordinating site is shown in *Figure 7.3* and consists of a Cu(II) ion ligated to four histidines and one exchangeable water molecule in a distorted tetragonal pyramidal arrangement and a Zn(II) ion ligated to three histidines and one aspartate in a distorted tetrahedral geometry. One of the histidines forms a bridge between the two metal ions. The Cu(II) ion is found in the bottom of a crevice where anions such as cyanide, azide as well as the substrate molecule, the superoxide radical, O_2^-, may enter and replace the water molecule thus initiating the enzymatic dismutation reaction. In the first step, the superoxide anion coordinates to the copper centre and reduces Cu(II) to Cu(I) concomitantly with the liberation of O_2, and then the reduced metal ion becomes oxidized by another superoxide ion producing and releasing one molecule of HO_2^- or H_2O_2. Thus, direct ET takes place within the primary coordination sphere.

The Zn(II) ion is not accessible to solvent and seems to serve as a structural component only, and there is no evidence that exogenous ligands, such as for example, water, can bind to this site. A number of metal ions like Ag(I), Co(II), Cd(II) and others can be substituted for the zinc centre with little or no effect on the SOD activity (Valentine and Freitas, 1985; Valentine and Pantoliano, 1981).

Another example of direct ET between the T2 copper centre and a substrate ligand is found in galactose oxidase which contains a single Cu centre in a square pyramidal geometry with two histidines, a cysteine–tyrosine derivative and the alcohol substrate in equatorial positions, while another tyrosine occupies the axial position (Ito et al., 1991). The enzyme catalyses the oxidation of primary alcohols to the corresponding aldehyde concomitantly with the reduction of dioxygen to hydrogen peroxide. Here the special equatorial tyrosine ligand is apparently undergoing a redox change which, in combination with the Cu(II)/Cu(I) cycle, provides the two redox equivalents required for the two-electron transfer (Ito et al., 1991).

Organic cofactors are also found in the T2 copper-containing amine oxidases where primary amines are oxidatively deaminated, and the two oxidation equivalents are again provided from the $O_2 \rightarrow H_2O_2$ reduction. The cofactor here is a tyrosine-derived quinone (topaquinone) not directly coordinated to the copper centre (Duine, 1991), but both redox centres are required for the amine oxidase activity (Mu et al., 1992). Intramolecular ET between the reduced quinone and the Cu(II) ion has been observed and was suggested to be part of the catalytic cycle (Dooley et al., 1991). The three-dimensional structure of this T2 copper enzyme has not yet been determined.

7.9   Intramolecular LRET in blue oxidases

The blue copper oxidases catalyse specific one-electron substrate oxidation by O_2 which eventually becomes reduced to water (Farver and Pecht, 1984; Messerschmidt, 1993). The minimal catalytic unit consists of four copper ions bound to distinct sites in the protein, called T1, T2 and T3, which are described above in Section 7.2. Earlier studies of the blue oxidases, laccase and ceruloplasmin have focused mainly on the kinetics of T1 [Cu(II)] reduction and have contributed to the notion that the T1 site acts as the electron uptake site in these enzymes. The breakthrough in the studies of the kinetics of this important group of enzymes occurred with the determination of the high resolution three-dimensional structure of AO (Messerschmidt and Huber, 1990; Messerschmidt *et al.*, 1989, 1992), since this provided insight into the detailed structure and spatial relationship among the copper coordinating sites (see *Figure 7.5*).

AO exists as a dimer of identical 70 kDa subunits each containing a catalytic unit of one T1, one T2 and one T3 copper site. The reduction potentials of the T1 and T3 Cu(II)/Cu(I) couples are identical, 350 mV at pH 7.0 (Marchesini and Kroneck, 1979) while the T2 potential is considerably lower (<300 mV). The catalytic cycle of this protein proceeds by a sequential mechanism where single electrons are transferred from the organic reducing substrates to T1[Cu(II)] while the trinuclear T2–T3 centre serves as the dioxygen binding and reducing site. Intramolecular ET from T1[Cu(I)] to T3[Cu(II)] therefore seems to be the essential step required for the transfer of the four electrons (and four protons) necessary for reduction of O_2 to two H_2O molecules.

Three independent groups have studied the intramolecular ET from T1[Cu(I)] to T3[Cu(II)] in AO under anaerobic conditions and have reported similar rates (Farver and Pecht, 1992b; Hazzard *et al.*, 1994; Kyritsis *et al.*, 1993; Meyer *et al.*, 1991; Tollin *et al.*, 1993). Photochemically produced lumiflavin semiquinone reduces T1[Cu(II)] in a fast second order process ($k = 2.7 \times 10^7$ M^{-1} s^{-1} at pH 7.0) followed by partially reoxidation of the T1[Cu(I)] site with a rate constant of 160 s^{-1} (Meyer *et al.*, 1991). The flavin absorption in the near-UV region prevented monitoring the changes at 330 nm where T3[Cu(II)] absorbs, but the latter process was interpreted as intramolecular ET from T1 to T3. This idea gained further support using AO where T2 copper was removed specifically, without altering the rate of intramolecular ET (Meyer *et al.*, 1991). Specific rates at both 610 nm (T1) and 330 nm (T3) were observed in another study (Farver and Pecht, 1992b) using pulse radiolytically produced CO_2^- as primary electron donor in the bimolecular, diffusion-controlled T1[Cu(II)] reduction (1.2×10^9 M^{-1} s^{-1} at pH 7.0). The ensuing processes of T1[Cu(I)] reoxidation and T3[Cu(II)] reduction occurred with identical rate constants, which were concentration independent, confirming their assignment as intramolecular ET from T1 to T3. Several phases were observed with rate constants of 200 s^{-1} for the fastest phase and 2 s^{-1} for the slowest. Similar ET rates were determined in another pulse radiolysis study using different organic radicals and monitoring the 610 nm chromophore (Kyritsis *et al.*, 1993). Interestingly, using pulse radiolytically produced reducing radicals (CO_2^- and O_2^-), an intramolecular ET from T1 to T3 with a rate constant of 2 s^{-1} was observed in the related blue copper oxidase, laccase (Farver and Pecht, 1991). However, in this protein, no faster intramolecular ET process occurred.

The structure of the blue oxidases has most probably been optimized by evolution for catalytic dioxygen reduction, and AO therefore represents a very interesting system for studies of intraprotein ET. The following important questions then arise. Is there some kind of control of the intramolecular ET in AO during the multielectron reduction and oxidation? Does the internal ET rate depend on the number of reduction equivalents taken up by the molecule? How does the rate of ET relate to the conformational changes which were resolved by the structure determination? Does the presence of substrates, organic reductants or dioxygen affect the internal ET rates? These questions gain more significance when it is noted that steady-state kinetic measurements of AO yield turnover numbers of 12 000–15 000 s^{-1} (Kroneck et al., 1982; Marchesini and Kroneck, 1979) which are considerably faster than the above observed values for intramolecular T1 to T3 ET.

No difference in the internal ET rate constants was observed with enzymes 'activated' or 'pulsed' by turning over 1 mM ascorbate in the presence of 0.25 mM O$_2$ prior to the determination of the intramolecular ET processes (Farver and Pecht, 1992b). However, in contrast to the above experiments which were all performed under anaerobic conditions, when AO solutions containing small and controlled concentrations of O$_2$ (15–65 µM) were employed, quite conspicuous differences were observed (Farver et al., 1994). A new fast intramolecular ET was discovered which depended on the presence of dioxygen. The rate constant of this phase, monitored at both 610 nm (T1) and 330 nm (T3), was 1100 s^{-1} (293 K, pH 5.8) and was maintained at this value as long as O$_2$ remained in the solution. Large spectral changes occurred around 330 nm, demonstrating interaction between the trinuclear site and dioxygen following the intramolecular ET from T1[Cu(I)] to T3[Cu(II)]. Oxygen intermediates coordinated to the trinuclear centre will increase the driving force significantly, enhancing the rate of intramolecular ET. Calculations show that a 100 mV increase in the reduction potential of the T3 centre would lead to the observed rise in ET rate. Similar effects have been observed in cytochrome oxidase in its catalytic reactions with O$_2$ (Brunori et al., 1992).

From determination of the activation parameters for the intramolecular ET processes in AO and applying the equations presented in Section 7.5, the reorganization energy, λ, could be calculated (Farver and Pecht, 1992b). Using a through-bond distance between the T1 Cu-ligating cysteine and one of the histidines coordinated to the T3 copper pair of 1.34 nm, together with the previously determined electronic decay factor, $\beta = 6.7$ nm^{-1} for ET in proteins (Section 7.7.1), λ becomes 153 kJ mol^{-1} (37 kcal mol^{-1}). This value for the reorganization energy in AO is significantly larger than that found for intramolecular ET in Az ($\lambda = 99$ kJ mol^{-1}, 24 kcal mol^{-1}, see Section 7.8). This observation can be rationalized in terms of the three-dimensional structure of different forms of the enzyme. Upon reduction of the T3 site, the copper–copper distance increases from 0.37 to 0.51 nm (Messerschmidt et al., 1993) as the anti-ferromagnetic coupling between the copper pair is disrupted causing considerable local conformational change around the trinuclear site.

An observed maximal rate constant for intramolecular ET in AO of 1100 s^{-1} is still considerably smaller than the turnover number of approximately 14 000 s^{-1}. Thus, dioxygen coordination to the trinuclear site is not sufficient to ensure maximal enzymatic activity. Under optimal conditions, the concentration of reducing substrate (e.g. ascorbate) is sufficient to maintain a steady state of fully reduced copper sites. Thus, an antithetical approach was taken by Tollin and co-workers studying the reoxidation of fully reduced AO by a laser-generated triplet state of 5-deazariboflavin (Hazzard et al., 1994). Subsequent to the one-electron oxidation of the reduced T2–T3 cluster, a rapid biphasic intramolecular ET occurs from T1[Cu(I)] (and presumably) to the oxidized trinuclear centre. The faster of the two observed rate constants (9500 and 1400 s^{-1}, respectively) is comparable with the turnover number for AO under steady-state conditions, and renders it very likely that this is the rate-limiting step in catalysis.

In the three-dimensional model of AO, one of the T1 copper ligands is the thiolate Cys507(S^γ), while imidazoles of the two neighbouring histidines (His506 and His508) coordinate to the T3 copper ions. This led to the proposal that the shortest ET pathway from T1[Cu(I)] to T3[Cu(II)] takes place via Cys507 and either His506 or His508 (Messerschmidt et al., 1989) (see *Figure 7.8*). Both pathways consist of nine covalent bonds yielding a total distance of 1.34 nm. Performing the pathway calculations developed by Beratan and Onuchic (see Chapter 2) confirms this notion and further demonstrates an alternative path via a hydrogen bond between the carbonyl oxygen of Cys507 and His506(N^δ). The electronic coupling for the two covalent pathways is 0.010, while for the latter the coupling is 0.014. Another noteworthy point is the similarity between the structural arrangement here and the one observed in Pc. In both cases, T1 copper utilizes the cysteine ligand in its ET. As discussed above (Section 7.4), Solomon and co-workers have calculated a high degree of anisotropic covalency for the copper centre in Pc, resulting in a very favourable electronic coupling via the cysteine ligand (Lowery et al., 1993). It would be highly interesting to see whether the same anisotropic covalent interaction exists in the structural arrangement of T1–T3 in AO.

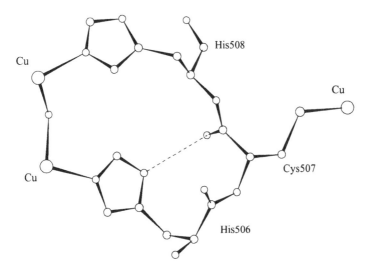

Figure 7.8. The suggested pathways for intramolecular ET from T1 Cu(I) (right) to T3 Cu(II) (left) in ascorbate oxidase. The hydrogen bond is shown as a broken line. Pathway calculations were based on the three-dimensional structure coordinates of AO (Messerschmidt et al., 1992).

7.10 ET pathways in proteins: conclusion

The great challenge in biophysical ET chemistry is trying to understand how ET rates depend on protein structure, and how structural concepts may determine particular pathways. Decay of the electronic wave function through the protein can be described either as a through-space or through-bond coupling interaction (see Chapters 1 and 2). Recent theoretical and experimental studies have demonstrated that the protein structure must be included in an adequate description of the electronic coupling between electron donor and acceptor (Regan *et al.*, 1993; Siddarth and Marcus, 1993; Skourtis *et al.*, 1994). Most results, including the systems discussed here in Sections 7.7–7.9, reflect that the protein constitutes an anisotropic medium for LRET. In conclusion, through-bond pathways seem to be more efficient than through-space pathways, at least in proteins which are mainly of the β-sandwich type like the copper ET proteins, and LRET in these proteins is thus determined by the detailed protein structure.

Acknowledgement

The author wishes to thank Dr Lars K. Skov for producing the figures of the protein structures.

References

Adman ET. (1985) Structure and function of small blue copper proteins, In: *Topics in Molecular and Structural Biology: Metalloproteins.* (ed. PM Harrison). Chemie Verlag, pp. 1–42.

Adman ET. (1991) Copper protein structures. *Adv. Protein Chem.* **42**: 145–197.

Adman ET, Stenkamp RE, Sieker LC, Jensen LH. (1978) A crystallographic model for azurin at 3 Å resolution. *J. Mol. Biol.* **123**: 35–47.

Anderson GP, Sanderson DG, Lee CH, Durell S, Anderson LB, Gross EL. (1987) The effect of ethylenediamine chemical modification of plastocyanin on the rate of cytochrome *f* oxidation and P700⁺ reduction. *Biochim. Biophys. Acta* **894**: 386–398.

Baker EN. (1988) Structure of azurin from *Alcaligenes denitrificans*. *J. Mol. Biol.* **203**: 1071–1095.

Beoku-Betts D, Chapman SK, Knox CV, Sykes AG. (1983) Concerning the binding site on plastocyanin for its natural redox partner cytochrome *f*. *J. Chem. Soc. Chem. Commun.* 1150–1152.

Beoku-Betts D, Chapman SK, Knox CV, Sykes AG. (1985) Kinetic studies on 1:1 electron transfer reactions involving blue copper proteins. 1. Effects of pH, competitive inhibition, and chromium(III) modification on the reaction of plastocyanin with cytochrome *f*. *Inorg. Chem.* **24**: 1677–1681.

Beratan DN, Betts JN, Onuchic, JN. (1991) Protein electron transfer rates set by the bridging secondary and tertiary structure. *Science* **252**: 1285–1288.

Beratan DN, Onuchic JN, Winkler JR, Gray HB. (1992) Electron-tunneling pathways in proteins. *Science* **258**: 1740–1741.

Boas JF. (1984) Electron paramagnetic resonance of copper proteins. In: *Copper Proteins and Copper Enzymes*, Vol. 1 (ed. R Lontie). CRC, Boca Raton, FL, pp. 5–62.

Broo A, Larsson S. (1990) Electron transfer due to through-bond interactions: study of aliphatic chains. *Chem. Phys.* **148**: 103–115.

Broo A, Larsson S. (1991) Electron transfer in azurin and the role of aromatic side groups of the protein. *J. Phys. Chem.* **95**: 4925–4928.

Brunori M, Antonini G, Malatesta F, Sarti P, Wilson MT. (1992) The oxygen reactive species of cytochrome-*c*-oxidase: an alternative view. *FEBS Lett.* **314**: 191–194.

Chapman SK. (1991) Blue copper proteins. *Perspect. Bioinorg. Chem.* **1**: 95–140.

Chen L, Durley R, Poliks BJ, Hamada K, Chen Z, Mathews FS, Davidson VL, Satow Y, Huizinga E, Vellieux FMD, Hol WGJ. (1992) Crystal structure of an electron-transfer complex between methylamine dehydrogenase and amicyanin. *Biochemistry* **31**: 4959–4964.

Chen L, Durley RCE, Mathews FS, Davidson VL. (1994) Structure of an electron transfer complex: methylamine dehydrogenase, amicyanin, and cytochrome c-551$_i$. *Science* **264**: 86–89.

Christensen HEM, Conrad LS, Mikkelsen KV, Nielsen MK, Ulstrup J. (1990) Direct and superexchange electron tunneling at the adjacent and remote sites of higher plant plastocyanins. *Inorg. Chem.* **29**: 2808–2816.

Christensen HEM, Conrad LS, Hammerstad-Pedersen JM, Ulstrup J. (1992) Resonance effects in strongly exothermic long-range electron transfer and their possible implications for the behaviour of site-directed mutant proteins. *FEBS Lett.* **296**: 141–144.

Colman PM, Freeman HC, Guss JM, Murata M, Norris VA, Ramshaw JAM, Venkatappa MP. (1978) X-ray crystal analysis of plastocyanin at 2.7 Å resolution. *Nature* **272**: 319–324.

Cookson DJ, Hayes MT, Wright PE. (1980) NMR studies of the interaction of plastocyanin with chromium(III) analogues of inorganic electron transfer reagents. *Biochim. Biophys. Acta* **591**: 162–176.

Cotton FA, Wilkinson G. (1988) *Advanced Inorganic Chemistry.* J.Wiley and Sons, New York.

da Silva DGAH, Beoku-Betts D, Kyritsis P, Govindaraju K, Powls R, Tomkinson NP, Sykes AG. (1992) Protein-protein cross-reactions involving plastocyanin, cytochrome *f* and azurin: self-exchange rate constants and related studies with inorganic complexes. *J. Chem. Soc. Dalton Trans.* 2145–2151.

Dooley DM, McGuirl MA, Brown DE, Turowski PN, McIntire WS, Knowles PF. (1991) A Cu(I)-semiquinone state in substrate-reduced amine oxidases. *Nature* **349**: 262–264.

Duine JA. (1991) Quinoproteins: enzymes containing the quinoid cofactor pyrroloquinoline quinone, topaquinone or tryptophan-tryptophan quinone. *Eur. J. Biochem.* **200**: 271–284.

Farver O, Pecht I. (1981a) Affinity labeling of an electron transfer pathway in azurin by Cr(II) ions. *Isr. J. Chem.* **21**: 13–17.

Farver O, Pecht I. (1981b) Identification of an electron transfer locus in plastocyanin by chromium(II) affinity labeling. *Proc. Natl Acad. Sci. USA* **78**: 4190–4193.

Farver O, Pecht I. (1984) Reactivity of copper sites in the blue copper proteins. In: *Copper Proteins and Copper Enzymes,* Vol. 1 (ed. R Lontie). CRC, Boca Raton, FL, pp. 183–214.

Farver O, Pecht I. (1989a) Structure–reactivity studies of blue copper proteins. Affinity labeling of electron transfer proteins by transition metal coordination. *Coord. Chem. Rev.* **94**: 17–45.

Farver O , Pecht I. (1989b) Long-range intramolecular electron transfer in azurins. *Proc. Natl Acad. Sci. USA* **86**: 6968–6972.

Farver O, Pecht I. (1990) Energetics of the intramolecular electron transfer in ruthenium modified stellacyanin. *Inorg. Chem.* **29**: 4855–4858.

Farver O, Pecht I. (1991) Electron transfer in proteins; in search of preferential pathways. *FASEB J.* **5**: 2554–2559.

Farver O, Pecht I. (1992a) Long range intramolecular electron transfer in azurins. *J. Am. Chem. Soc.* **114**: 5764–5767.

Farver O, Pecht I. (1992b) Low activation barriers characterize the intramolecular electron transfer in ascorbate oxidase. *Proc. Natl Acad. Sci. USA* **89**: 8283–8287.

Farver O, Pecht I. (1994) Blue copper proteins as a model for investigating electron transfer processes within polypeptide matrices. *Biophys. Chem.* **50**: 203–216.

Farver O, Blatt Y, Pecht I. (1982a) Resolution of two distinct electron transfer sites on azurin. *Biochemistry* **21**: 3556–3561.

Farver O, Shahak Y, Pecht I. (1982b) Electron uptake and delivery sites on plastocyanin in its reactions with the photosynthetic electron transport system. *Biochemistry* **21**: 1885–1890.

Farver O, Licht A, Pecht I. (1987) Electron transfer pathways in stellacyanin: a possible homology with plastocyanin. *Biochemistry* **26**: 7317–7321.

Farver O, Skov LK, van de Kamp M, Canters GW, Pecht I. (1992) The effect of driving force on intramolecular electron transfer in proteins; studies on single-site mutated azurins. *Eur. J. Biochem.* **210:** 399–403.

Farver O, Skov LK, Pascher T, Karlsson BG, Nordling M, Lundberg LG, Vänngård T, Pecht I. (1993) Intramolecular electron transfer in single site mutated azurins. *Biochemistry* **32:** 7317–7322.

Farver O, Wherland S, Pecht I. (1994) Intramolecular electron transfer in ascorbate oxidase is enhanced in the presence of oxygen. *J. Biol. Chem.* **269:** 22933–22936.

Farver O, Skov LK, Gilardi G, van Pouderoyen G, Canters GW, Wherland S, Pecht I. (1996a) Structure–function correlation of intramolecular electron transfer in wild type and single-site mutated azurins. *Chem. Phys.* **204** (in press).

Farver O, Bonander N, Skov LK, Pecht I. (1996b) The pH dependence of intramolecular electron transfer in azurins. *Inorg. Chim. Acta* **243:** 1–7.

Fee JA, Malmström BG. (1968) The redox potential of fungal laccase. *Biochim. Biophys. Acta* **153:** 299–302.

Fridovich I. (1989) Superoxide dismutases. *J. Biol. Chem.* **264:** 7761–7764.

Fromme P, Schubert W-D, Krauss N. (1994) Structure of photosystem I: suggestions on the docking sites for plastocyanin, ferredoxin and the coordination of P700. *Biochim. Biophys. Acta* **1187:** 99–105.

Garrett TPJ, Clingeleffer DJ, Guss JM, Rogers SJ, Freeman HC. (1984) The crystal structure of poplar apoplastocyanin at 1.8 Å resolution. The geometry of the copper-binding site is created by the polypeptide. *J. Biol. Chem.* **259:** 2822–2825.

Goldberg M, Pecht I. (1976) Kinetics and equilibria of the electron transfer between azurin and the hexairon (II/III) couple. *Biochemistry* **15:** 4197–4208.

Gray HB. (1986) Long-range electron-transfer in blue copper proteins. *Chem. Soc. Rev.* **15:** 17–30.

Gray HB, Malmström BG. (1983) On the relationship between protein-forced ligand fields and the properties of blue copper centers. *Comments Inorg. Chem.* **2:** 203–209.

Gray HB, Solomon EI. (1981) Electronic structures of blue copper centers in proteins. In: *Copper Proteins* (ed. TG Spiro). J. Wiley and Sons, New York, pp. 1–39.

Groeneveld CM, Canters GW. (1988) NMR structure and electron transfer mechanism of *Pseudomonas aeruginosa* azurin. *J. Biol. Chem.* **263:** 167–173.

Groeneveld CM, Dahlin S, Reinhammar B, Canters GW. (1987) Determination of the self-exchange rate of azurin from *Pseudomonas aeruginosa* by a combination of fast-flow/rapid-freeze experiments and EPR. *J. Am. Chem. Soc.* **109:** 3247–3250.

Gross EL. (1993) Plastocyanin: structure and function. *Photosynth. Res.* **37:** 103–116.

Guss JM, Freeman HC. (1983) Structure of oxidized poplar plastocyanin at 1.6 Å resolution. *J. Mol. Biol.* **169:** 521–563.

Guss JM, Harrowell PR, Murata M, Norris VA, Freeman HC. (1986) Crystal structure analyses of reduced poplar plastocyanin at six pH values. *J. Mol. Biol.* **192:** 361–387.

Guss JM, Merritt EA, Phizackerley RP, Hedman B, Murata M, Hodgson KO, Freeman HC. (1988) Phase determination by multiple-wavelength X-ray diffraction: crystal structure of a basic 'blue' copper protein from cucumbers. *Science* **241:** 806–811.

Haehnel W, Jansen T, Gause K, Klösgen RB, Stahl B, Michl D, Huvermann B, Karas M, Herrmann RG. (1994) Electron transfer from plastocyanin to photosystem I. *EMBO J.* **13:** 1028–1038.

Hazzard JT, Marchesini A, Curir P, Tollin G. (1994) Direct measurement by laser flash photolysis of intramolecular electron transfer in the three-electron reduced form of ascorbate oxidase from zucchini. *Biochim. Biophys. Acta* **1208:** 166–170.

He S, Modi S, Bendall DA, Gray JC. (1991) The surface-exposed residue tyrosine Tyr83 of pea plastocyanin is involved in both binding and electron transfer reactions with cytochrome *f*. *EMBO J.* **10:** 4011–4016.

Ito N, Phillips SEV, Stevens C, Ogel ZB, McPherson MJ, Keen JN, Yadev KDS, Knowles PF. (1991) Three dimensional structure of galactose oxidase: an enzyme with a built-in secondary cofactor. *Nature* **350:** 87–90.

Jackman MP, McGinnis J, Powls R, Salmon GA, Sykes AG. (1988) Preparation and characterization of two His-59 ruthenium modified plastocyanins and an unusually slow rate constant for ruthenium(II)–copper(II) intramolecular electron transfer over ~12 Å. *J. Am. Chem. Soc.* **110:** 5880–5887.

Jortner J, Bixon M. (1993) Electron transfer in supermolecules. *Mol. Crystals Liquid. Crystals* **234**: 29–41.

Kalverda AP, Wymenga SS, Lommen A, van der Ven FJM, Hilbers CW, Canters GW. (1994) Solution structure of the type 1 blue copper protein amicyanin from *Thiobacillus versutus*. *J. Mol. Biol.* **240**: 358–371.

Karlin KD, Tyeklar Z, eds. (1993) *Bioinorganic Chemistry of Copper*. Chapman & Hall, New York/London.

Kostic NM, Margalit R, Che C-M, Gray HB. (1983) Kinetics of long-distance ruthenium-to-copper electron transfer in [pentaammineruthenium histidine-83]azurin. *J. Am. Chem. Soc.* **105**: 7765–7767.

Krauss N, Hinrichs W, Witt I, Fromme P, Pritzkow W, Dauter Z, Betzel C, Wilson KS, Witt HT, Saenger W. (1993) Three-dimensional structure of system I of photosynthesis at 6 Å resolution. *Nature* **361**: 326–331.

Kroneck PMH, Armstrong FA, Merkle H, Marchesini A. (1982) Ascorbate oxidase: molecular properties and catalytic activity. *Adv. Chem. Ser.* **220**: 223–248.

Kyritsis P, Messerschmidt A, Huber R, Salmon GA, Sykes AG. (1993) Pulse-radiolysis studies of the oxidised form of the multicopper enzyme ascorbate oxidase: evidence for two intermolecular electron-transfer steps. *J. Chem. Soc. Dalton Trans.* 731–735.

Lowery MD, Solomon EI. (1992) Axial ligand bonding in blue copper proteins. *Inorg. Chim. Acta* **198-200**: 233–243.

Lowery MD, Guckert JA, Gebhard MS, Solomon EI. (1993) Active-site structure contributions to electron-transfer pathways in rubredoxin and plastocyanin: direct versus superexchange. *J. Am. Chem. Soc.* **115**: 3012–3013.

Magnus KA, Ton-That H, Carpenter JE. (1993) Three-dimensional structure of the oxygenated form of the hemocyanin subunit II of *Limulus polyphemus* at atomic resolution. In: *Bioinorganic Chemistry of Copper* (eds KD Karlin, Z Tyeklar). Chapman & Hall, NewYork/London, pp. 143–150.

Malmström BG. (1994) Rack-induced bonding in blue-copper proteins. *Eur. J. Biochem.* **223**: 711–718.

Marchesini A, Kroneck PMH. (1979) Ascorbate oxidase from *Cucurbita pepo medullosa*. *Eur. J. Biochem.* **101**: 65–76.

Marcus RA, Sutin N. (1985) Electron transfer in chemistry and biology. *Biochim. Biophys. Acta* **811**: 265–322.

Margalit R, Kostic NM, Che C-M, Blair DF, Chiang H-J, Pecht I, Shelton JB, Shelton JR, Schroeder WA, Gray HB. (1984) Preparation and characterization of pentaammine-ruthenium-(histidine-83) azurin: thermodynamics of intramolecular electron transfer from ruthenium to copper. *Proc. Natl Acad. Sci. USA* **81**: 6554–6558.

Messerschmidt A. (1993) Ascorbate oxidase structure and chemistry. In: *Bioinorganic Chemistry of Copper* (eds KD Karlin, Z Tyeklar). Chapman & Hall, NewYork/London, pp. 471–484.

Messerschmidt A, Huber R. (1990) The blue oxidases, ascorbate oxidase, laccase and ceruloplasmin. *Eur. J. Biochem.* **187**: 341–352.

Messerschmidt A, Rossi A, Ladenstein R, Huber R, Bolognesi M, Gatti G, Marchesini A, Petruzzelli R, Finazzi-Agró A. (1989) X-ray crystal structure of the blue oxidase ascorbate oxidase from zucchini. *J. Mol. Biol.* **206**: 513–529.

Messerschmidt A, Ladenstein R, Huber R, Bolognesi M, Avigliano L, Marchesini A, Petruzzelli R, Rossi A, Finazzi-Agró A. (1992) Refined crystal structure of ascorbate oxidase at 1.9 Å resolution. *J. Mol. Biol.* **224**: 179–205.

Messerschmidt A, Luecke H, Huber R. (1993) X-ray structures and mechanistic implications of three functional derivatives of ascorbate oxidase from zucchini. *J. Mol. Biol.* **230**: 997–1014.

Meyer TE, Marchesini A, Cusanovich MA, Tollin G. (1991) Direct measurement of intramolecular electron transfer between type I and type III copper centers in the multi-copper enzyme ascorbate oxidase and its type II copper-depleted and cyanide-inhibited forms. *Biochemistry* **30**: 4619–4623.

Mikkelsen KV, Skov LK, Nar H, Farver O. (1993) Electron self-exchange in azurin; calculation of the superexchange electron tunneling rate. *Proc. Natl Acad. Sci. USA* **90**: 5443–5445.

Modi S, Nordling M, Lundberg LG, Hansson Ö, Bendall DS. (1992) Reactivity of cytochromes c and f with mutant forms of spinach plastocyanin. *Biochim. Biophys. Acta* **1102**: 85–90.

Morpurgo G, Pecht I. (1982) Affinity labeling of stellacyanin by Cr(II) aquo ions. *Biochem. Biophys. Res. Commun.* **104**: 1592–1596.

Moser CC, Keske JM, Warncke K, Farid RS, Dutton PL. (1992) Nature of biological electron transfer. *Nature* **355**: 796–802.

Mu D, Janes SM, Smith AJ, Brown DE, Dooley DM, Klinman JP. (1992) Tyrosine codon corresponds to topa quinone at the active site of copper amine oxidases. *J. Biol. Chem.* **267**: 7979–7982.

Murphy LM, Strange RW, Karlsson BG, Lundberg LG, Pascher T, Reinhammar B, Hasnain SS. (1993) Structural characterization of azurin from *Pseudomonas aeruginosa* and some of its methionine121 mutants. *Biochemistry* **32**: 1965–1975.

Nar H, Messerschmidt A, Huber R, van de Kamp M, Canters GW. (1991a) Crystal structure analysis of oxidized *Pseudomonas aeruginosa* azurin at pH 5.5 and pH 9.0 – a pH-induced conformational transition involves a peptide bond flip. *J. Mol. Biol.* **221**: 765–772.

Nar H, Messerschmidt A, Huber R, van de Kamp M, Canters GW. (1991b) X-ray crystal structure of the two site-specific mutants His35Gln and His35Leu of azurin from *Pseudomonas aeruginosa*. *J. Mol. Biol.* **218**: 427–447.

Nar H, Messerschmidt A, Huber R, van de Kamp M, Canters GW. (1992) Crystal structure of *Pseudomonas aeruginosa* apo-azurin at 1.85 Å resolution. *FEBS Lett.* **306**: 119–124.

Nocera DG, Winkler JR, Yocom KM, Bordignon E, Gray HB. (1984) Kinetics of intramolecular electron transfer from Ru(II) to Fe(III) in ruthenium-modified cytochrome c. *J. Am. Chem. Soc.* **206**: 5145–5150.

Nordling M, Sigfridsson K, Young S, Lundberg LG, Hansson Ö. (1991) Flash-photolysis studies of the electron transfer from genetically modified spinach plastocyanin to photosystem I. *FEBS Lett.* **291**: 327–330.

Pascher T, Bergström J, Malmström BG, Vänngård T, Lundberg LG. (1989) Modification of the electron-transfer sites of *Pseudomonas aeruginosa* azurin by site-directed mutagenesis. *FEBS Lett.* **258**: 266–268.

Pascher T, Karlsson BG, Nordling M, Malmström BG, Vänngård T. (1993) Reduction potentials and their pH dependence in site-directed-mutant forms of azurin from *Pseudomonas aeruginosa*. *Eur. J. Biochem.* **212**: 289–296.

Pelletier H, Kraut J. (1992) Crystal structure of a complex between electron transfer partners, cytochrome c peroxidase and cytochrome c. *Science* **258**: 1748–1755.

Plato M, Michel-Beyerle ME, Bixon M, Jortner J. (1989) On the role of tryptophan as a superexchange mediator for quinone reduction in photosynthetic reaction centers. *FEBS Lett.* **249**: 70–74.

Qin L, Kostic NM. (1992) Electron-transfer reactions of cytochrome f with flavin semiquinones and with plastocyanin: importance of protein–protein electrostatic interactions and donor–acceptor coupling. *Biochemistry* **31**: 5145–5150.

Regan JJ, Risser, SM, Beratan DN, Onuchic JN. (1993) Protein electron transport: single versus multiple pathways. *J. Phys. Chem.* **97**: 13083–13088.

Romero A, Hoitink CWG, Nar H, Huber R, Messerschmidt A, Canters GW. (1993) X-ray analysis and spectroscopic characterization of M121Q azurin. A copper site model for stellacyanin. *J. Mol. Biol.* **229**: 1007–1021.

Shephard WEB, Anderson BF, Lewandoski DA, Norris GE, Baker EN. (1990) Copper coordination geometry in azurin undergoes minimal changes on reduction of copper(II) to copper(I). *J. Am. Chem. Soc.* **112**: 7817–7819.

Siddarth P, Marcus RA. (1993) Correlation between theory and experiment in electron-transfer reactions in proteins: electronic couplings in modified cytochrome c and myoglobin derivatives. *J. Phys. Chem.* **97**: 13078–13082.

Sigfridsson K, Hansson Ö, Karlsson BG, Baltzer L, Nordling M, Lundberg LG. (1995) Spectroscopic and kinetic characterization of the spinach plastocyanin mutant Tyr83-His: a histidine residue with a high pK value. *Biochim. Biophys. Acta* **1228**: 28–36.

Skourtis SS, Regan JJ, Onuchic JN. (1994) Electron transfer in proteins: a novel approach for the description of donor–acceptor coupling. *J. Phys. Chem.* **98**: 3379–3388.

Skov LK, Christensen U, Olsen K, Farver O. (1993) Influence of chromium binding on the kinetics of aqua Cr(II) reduction of the blue copper proteins, azurin and stellacyanin. *Inorg. Chem.* **32:** 4762–4765.

StClair CS, Ellis WR, Gray HB. (1992) Spectroelectrochemistry of blue copper proteins: pH and temperature dependences of the reduction potential of five azurins. *Inorg. Chim. Acta* **191:** 149–155.

Tainer JA, Getzoff ED, Beem KM, Richardson JS, Richardson DC. (1982) Determination and analysis of the 2 Å structure of copper, zinc superoxide dismutase. *J. Mol. Biol.* **160:** 181–217.

Tainer JA, Getzoff ED, Richardson JS, Richardson DC. (1983) Structure and mechanism of copper, zinc superoxide dismutase. *Nature* **306:** 284–287.

Tollin G, Meyer TE, Cusanovich MA, Curir P, Marchesini A. (1993) Oxidative turnover increases the rate constant and extent of intramolecular electron transfer in the multicopper enzymes, ascorbate oxidase and laccase. *Biochim. Biophys. Acta* **1183:** 309–314.

Tsai L-C, Sjölin L, Langer V, Pascher T, Nar H. (1995) Structure of azurin mutant Phe114Ala from *Pseudomonas aeruginosa* at 2.6 Å resolution. *Acta Crystallogr.* **D51:** 168–176.

Valentine JS, Freitas DMD. (1985) Copper–zinc superoxide dismutase: a unique biological 'ligand' for bioinorganic studies. *J. Chem. Educat.* **62:** 990–997.

Valentine JS, Pantoliano MW. (1981) Protein–metal ion interactions in cuprozinc protein (superoxide dismutase). In: *Copper Proteins* (ed. TG Spiro). J. Wiley and Sons, New York, pp. 291–358.

van de Kamp M, Canters GW. (1992) Protein-mediated electron transfer. *Curr. Opin. Struct. Biol.* **2:** 859–869.

van de Kamp M, Silvestrini MC, Brunori M, Beeumen JV, Hali FC, Canters GW. (1990) Involvement of the hydrophobic patch of azurin in the electron transfer reactions with cytochrome *c*-551 and nitrite reductase. *Eur. J. Biochem.* **194:** 109–118.

van de Kamp M, Canters GW, Andrew CR, Sanders-Loehr J, Bender CJ, Peisach J. (1993) Effect of lysine ionisation on the structure and electrochemical behaviour of the Met44→Lys mutant of the blue-copper protein azurin from *Pseudomonas aeruginosa*. *Eur. J. Biochem.* **218:** 229–238.

Volbeda A, Hol WGS. (1989) Crystal structure of hexameric hemocyanin from *Panulirus interruptus* refined at 3.2 Å resolution. *J. Mol. Biol.* **209:** 249–279.

Winkler JR, Gray HB. (1992) Electron transfer in ruthenium-modified proteins. *Chem. Rev.* **92:** 369–379.

Wuttke DS, Gray HB. (1993) Protein engineering as a tool for understanding electron transfer. *Curr. Opin. Struct. Biol.* **3:** 555–563.

Wuttke DS, Bjerrum MJ, Winkler JR, Gray HB. (1992) Electron-tunneling pathways in cytochrome *c*. *Science* **256:** 1007–1009.

Haemoproteins

G.R. Moore

8.1 Introduction

Haemoproteins can be grouped into a variety of classes depending on their structures and functions. However, whatever their biological functions, all haemoproteins undergo redox reactions to yield Fe(II)/Fe(III) species provided they are supplied with redox partners operating at the necessary potentials. The cytochrome class of haemoprotein includes all those whose function is to act solely as electron carriers. However, this designation encompasses enzymatic haemoproteins also, such as cytochrome c peroxidase, cytochrome P450, cytochrome oxidase and cytochrome cd_1 (or nitrite reductase), since the name cytochrome refers to any haemoprotein whose normal biological function involves a reversible change in the haem iron oxidation state (Meyer and Kamen, 1982). In this chapter, we shall consider a range of topics concerning the rates at which haemoprotein-mediated electron transfers occur. They include haemoprotein structural dynamics, redox potentials, coupled electron–ion transfers, electron transfer within multiheaded haem-containing enzymes and the formation of reactive interprotein complexes involving cytochromes.

In recent years, the pace of developments in the haemoprotein electron transfer field has been extremely rapid. One of the principal causes of this has been the increase in the number of haemoproteins whose structures have been determined by X-ray crystallography, culminating in the recent determination of the structures of *Paracoccus denitrificans* and bovine heart cytochromes aa_3 (Iwata *et al.*, 1995; Tsukihara *et al.*, 1995). In addition, synthetic haemoproteins based on a symmetric all-parallel four-α-helical bundle structure have been described in which the four-α-helical bundle binds up to four b-type haems in bis-histidine ligation sites, with some of the haems exhibiting cooperativity in their redox potentials analogous to native haemoproteins, such as the dihaem component of the cytochrome bc_1 complex (Farid *et al.*, 1994). These new structures go hand in hand with theoretical descriptions of electron transmission in proteins, described elsewhere in this book (see Chapters 1 and 2), that appear to be consistent with the majority of reported mechanistic studies on haemoproteins.

Protein Electron Transfer, Edited by D.S. Bendall
© 1996 BIOS Scientific Publishers Ltd, Oxford

Interestingly, the biological functions of some of the best studied electron transfer haemoproteins have not been defined. The demonstration of *in vitro* reactivity with other redox proteins is not sufficient for this, because many electron transfer proteins exhibit low specificity profiles in their reactions. Perhaps the most compelling evidence for a physiological relationship is when two proteins that react with each other *in vitro* are encoded by the same operon and expressed under the same conditions. Redox proteins located in the periplasm of *Pseudomonas aeruginosa* provide a typical example. The blue copper protein azurin, monohaem cytochrome *c*-551, dihaem cytochrome *c* peroxidase and cytochrome cd_1 have all been isolated from cells grown under similar conditions, though maximum production of some requires different conditions, and cytochrome *c*-551 has been shown to react with all *in vitro* (Pettigrew and Moore, 1987, and references therein; Silvestrini *et al.*, 1990). Dihaem cytochrome *c* peroxidase was given its name as a consequence of its reaction with cytochrome *c*-551 (Rönnberg *et al.*, 1980), although it now appears that its gene is unrelated to that of the monohaem cytochrome (Rideout *et al.*, 1995). The gene for cytochrome *c*-551 is part of the nitrite reductase operon that also encodes cytochrome cd_1, giving support to the view that cytochrome *c*-551 is a physiological partner for cytochrome cd_1 (Nordling *et al.*, 1990). This is supported further by another biochemical approach involving construction of mutant strains by gene deletion. Vijgenboom *et al.* (1995) have applied this to *P. aeruginosa* and shown by comparison of wild-type, azu^-, c-551^- and azu^-/c-551^- cells that cytochrome *c*-551 has an essential role in NO_3^-/NO_2^- respiration but azurin does not. Thus cytochromes *c*-551 and cd_1 are established as natural reaction partners, and the extensive literature on the reaction between cytochrome *c* peroxidase and cytochrome *c*-551 may not be describing a physiological reaction. The multihaem class III cytochromes c_3 provide another example of where the application of molecular genetics has given functional information regarding a complex electron transfer system. The nucleotide sequence of the *hmc* operon from *Desulphovibrio vulgaris* subsp. *vulgaris* Hildenborough contains genes for a 65.5-kDa cytochrome containing 16 *c*-haems, 15 of which are probably bis-histidine coordinated in cytochrome c_3-like domains (Pollock *et al.*, 1991), as well as for several proteins that are apparently transmembrane, including one with four Fe–S clusters, and a cytoplasmic redox protein containing two Fe–S clusters (Rossi *et al.*, 1993). Rossi *et al.* proposed that these interact to form a transmembrane complex that links electron flow from periplasmic hydrogenases to cytoplasmic enzymes. The significance of this is that this is probably the best defined cytochrome c_3 as far as its functional role is concerned (see Moore and Pettigrew, 1990, and references therein). At this point, it is appropriate to note that in the subsequent sections of this chapter we shall consider various physico-chemical and structural aspects of haemoproteins with the aim of defining key factors that affect their rates of electron transfer. Rarely will this analysis be placed into a biological context, and this represents one of the major challenges in haemoprotein research.

As discussed in Chapter 1, the minimal reaction scheme for a bimolecular electron transfer reaction is the one shown in *Figure 8.1*. The reaction proceeds in three main stages: formation of a bimolecular precursor complex, followed

Association:	$A_o + B_r$	$\rightarrow$	$(A_o \text{---} B_r)$	(8.1.1)
Equalization of energy levels:	$(A_o \text{---} B_r)$	$\rightarrow$	$(A_o \text{---} B_r)^*$	(8.1.2)
Electron transfer:	$(A_o \text{---} B_r)^*$	$\rightarrow$	$(A_r \text{---} B_o)^*$	(8.1.3)
Relaxation:	$(A_r \text{---} B_o)^*$	$\rightarrow$	$(A_r \text{---} B_o)$	(8.1.4)
Dissociation:	$(A_r \text{---} B_o)$	$\rightarrow$	$A_r + B_o$	(8.1.5)

Figure 8.1. Individual steps of a simple outer sphere electron transfer reaction. o and r indicate oxidized and reduced species respectively. When A and B are the same, the reaction is self-exchange; where they are different it is a cross-reaction.

by electron transfer and dissociation of the bimolecular product complex. To allow the electron transfer to occur isoenergetically, the bimolecular collision complex needs to be activated and this gives rise to the activation and relaxation steps 8.1.2 and 8.1.4 shown in *Figure 8.1*. The nature of the activated complex is central to the overall rate of electron transfer but there are a number of factors that may be rate determining, including the following.

(i)   The protein association–dissociation steps. These are governed by the interaction work terms.
(ii) The reactant rearrangement energy (λ). This may be electronic or conformational in origin and is not restricted to events at the redox centres.
(iii) The redox driving energy (ΔG°_{AB}). This is usually taken to be the difference in redox potentials between the acceptor and donor.
(iv) Electron transmission within the activated complex. This depends upon the overlap between the donor and acceptor electronic wave functions, which is determined by the electronic properties of the redox centres, the medium between them and their separation.

The relationships between the terms listed above and the rate of electron transfer are generally considered within the framework of the theory pioneered by Marcus and others (see Marcus and Sutin, 1985; Moser *et al.*, 1992; see Chapter 1, and references therein). Haemoproteins are like other classes of electron transfer proteins: for rapid electron transfer the driving force should be optimized and redox-linked structural changes minimized; good overlap between the donor and acceptor orbitals is required, and, for bimolecular reactions, non-productive collisions need to be minimized. All of these topics have been investigated, but much of the recent mechanistic emphasis within the field has concerned the overlap between donor and acceptor orbitals and the problem of electron transmission within proteins. In this chapter, we shall start by considering structural features of haemoproteins and move on to look at experimental and computational investigations of various factors that influence electron transfer rates. The aim is to give a brief overview of a range of factors and systems rather than an in-depth analysis of any one.

8.2   Structures of haems

The four common types of haem are shown in *Figure 8.2*. Haems *b* and *c* are found in many types of haemoprotein, including enzymes, O_2 carrier proteins

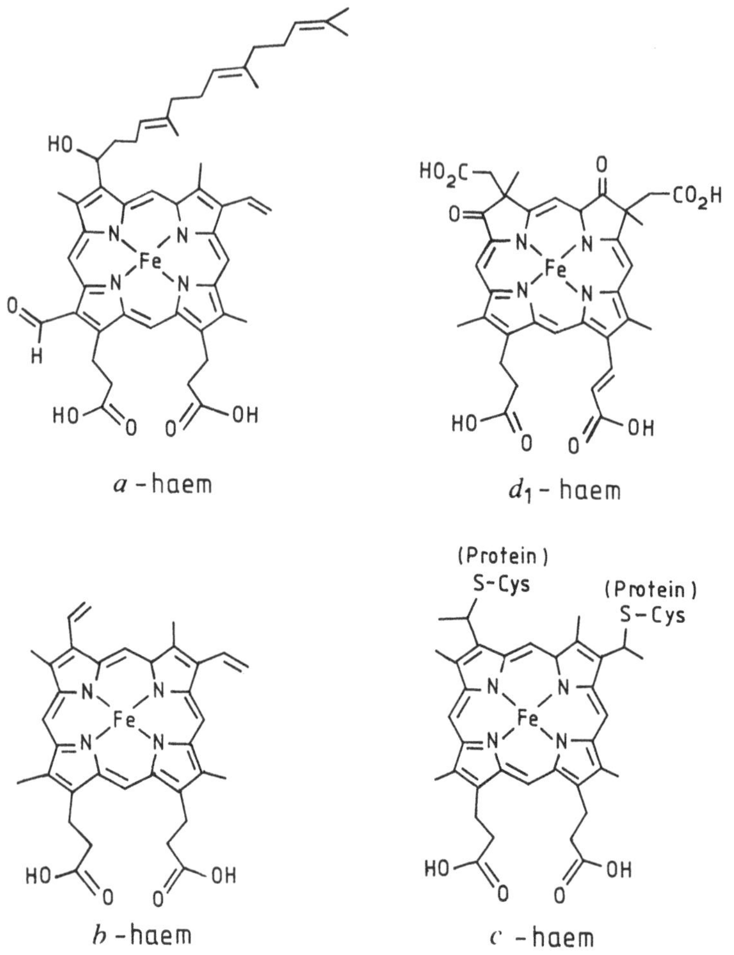

Figure 8.2. Haems a, b, c and d_1.

and electron transfer proteins but, to-date, haems a and d_1 have only been found in enzymes, although they may not always have a direct catalytic role. For example, in the dihaem a-containing cytochrome c oxidases, haem a_3, together with the dinuclear Cu_B, is the O_2 binding and reduction site, but haem a is just involved in electron transfer (Gennis and Ferguson-Miller, 1995). The key variability in the structural properties of the different haem groups resides with the nature of their substituents, the number and characteristics of their axial ligands and, in some cases, with the oxidation state of their central iron. The latter two subjects have been reviewed for cytochromes c by Moore and Pettigrew (1990). Briefly, for a six-coordinate low spin iron the haem plane is largely planar, but with a haem in which the iron sits outside of the central cavity substantial distortions to planarity can occur. Even for haems with low spin iron some distortion of the haem plane is occasionally seen and, if this is redox state dependent, then it might constitute a major contribution to reaction reorganization energies.

The replacement of the central iron of the haem by other metals has brought a range of spectroscopic techniques to bear on mechanistic questions pertaining to haemoprotein function (Moore and Pettigrew, 1990, and references therein) but perhaps of more significance is that they have allowed high driving force electron transfer reactions to be investigated by exploiting photochemically active metalloporphyrins which have a range of redox potentials and excited state lifetimes. This approach is typified by the replacement of the central iron of myoglobin by Zn(II), Cd(II), Pd(II) and Mg(II) (Winkler and Gray, 1992). The non-covalent interaction of the haem of b-type cytochromes and globins allows relatively facile metalloporphyrin replacement but, even with c-type haems, metal replacement, most commonly by Zn(II), is possible (Moore and Pettigrew, 1990).

It is not clear whether the spin state of the iron is an important rate-determining factor, provided there is not a high spin to low spin conversion accompanying the electron transfer, though electron transfer between the Fe(II) and Fe(III) states of a low spin haemoprotein requires less electronic rearrangement than the corresponding high spin reaction. Gray and his colleagues have studied extensively electron transfer reactions of myoglobins, in which not only are the ferric and ferrous states both high spin but also H_2O is coordinated to the six-coordinate ferric ion whilst the ferrous iron of the deoxy form is only five-coordinate. Thus it might be envisaged that the reorganization energy, λ, would be substantially higher in this system than for many other haemoproteins, though in the compilation of data given by Winkler and Gray (1992) the reorganization energy for ruthenated myoglobins in which the haem undergoes the normal Fe(III)-H_2O/Fe(II) reaction, (i.e. 1.48 eV), is not much higher than the reorganization energy of myoglobins containing metal-substituted porphyrins in which there is no change in coordination number (i.e. 1.26 eV). Nor do these values seem substantially different from the reorganization energies determined for reactions of low spin cytochromes (e.g. see *Figure 8.5*). However, it is notable that virtually all the simple electron transfer cytochromes are low spin (Moore and Pettigrew, 1990) and, conversely, most non-electron transfer haemoproteins are high spin. The haem transporter haemopexin is an exception; its tightly bound haem b is coordinated by two strong field ligands, probably histidines (Cox *et al.*, 1995).

Considerable attention has been paid to determining the electronic structures of haem groups to as high a resolution as possible, partly because this is essential for a full understanding of the spectroscopic properties of haemoproteins. However, it is also possible that there is a close relationship between the detailed electronic structure of a haem and its electron transfer properties. In assessing likely electron transfer pathways, the orbital(s) in which the electron 'hole' is located may indicate a low energy transfer route (Wüthrich, 1969); but, in the best characterized haemoprotein in this respect, mitochondrial cytochrome c (Palmer, 1983; Taylor, 1977), the difference in energy between the ground state electronic structure and the first excited state electronic structure is unlikely to represent a major reactivity determinant (Moore *et al.*, 1983).

Molecular orbital descriptions of porphyrins are well advanced (Gouterman, 1978; Loew, 1983) and, though the details of these are beyond the scope of this chapter, some general points are relevant. As would be expected, the HOMOs and LUMOs of interest are located at the periphery of the porphyrin with the π-orbital extension greatest above and below the plane of the haem. Thus, orbital overlap will be greater for two parallel haems compared with two co-planar haems at the same distance apart. The significance of this is that co-planarity has often been associated with optimal electron transfer (see, for example, Poulos and Kraut 1980; Salemme, 1976), though this may also be a consequence of steric constraints to putting parallel haems close together. In fact, in none of the reported X-ray structures of multihaem proteins (*Table 8.1*) are the haems co-planar, though they are generally relatively close to each other: interhaem distances are frequently as short as 5 Å, with Fe–Fe distances being 12–16 Å. Therefore, haems appear to represent regions of high electron tunnelling density with gradations in tunnelling density being a relatively minor perturbation.

There is an interesting difference between b- and c- haems and d_1-haems that has recently been discovered. In b- and c-haems, the ground state electronic configuration of the Fe(III) is $(d_{xy})^2(d_{xz},d_{yz})^3$ with, in proteins, rhombic distortions probably causing the $(d_{xz},d_{yz})^3$ pair to split to give $(d_{xz})^2(d_{yz})^1$, but in tetramesitylporphyrinates and d_1-haems, the ground state configuration appears to be $(d_{xz},d_{yz})^4(d_{xy})^1$ (M.R. Cheesman and F.A. Walker, submitted). This may account for some of the NO-binding properties of d_1-haems, but whether it has a significance for electron transfer has not been determined.

8.3   Tertiary structures of haemoproteins

Tertiary structures of numerous cytochromes b and c have been determined, but only two structures of haem a-containing proteins, *P. denitrificans* and bovine heart cytochromes aa_3 (Iwata *et al.*, 1995; Tsukihara *et al.*, 1995), and one structure for a haem d-containing protein, *Thiosphaera pantotropha* cytochrome cd_1 (nitrite reductase) (Fülöp *et al.*, 1995a), have been reported. The favoured method for structure determination has been X-ray diffraction, but nuclear magnetic resonance (NMR) structures of small cytochromes have also been reported. The NMR method was first tested on cytochromes whose X-ray structures were known, but it has since been applied to cytochromes for which X-ray structures are not available, for example, *D. vulgaris* cytochrome c-553 (Blackledge *et al.*, 1995). However, the current limitations of the NMR method restrict its application in structure determination to water-soluble cytochromes with a mass of less than 30 kDa containing haems with relatively uncomplicated magnetic properties. There is insufficient space in this chapter to describe adequately the many structures that are available, and we restrict ourselves to noting in *Table 8.1* those recently reported structures that have particularly important features.

Cytochrome f is important partly because of its unique haem axial ligation. Spectroscopic studies had indicated that its Fe was coordinated by a histidine and an amine, though lysine was the favoured amine (Moore and Pettigrew, 1990, and references therein). The N-terminal amine ligation satisfactorily accounts for the spectroscopic data and also provides a solution to the problem of how an amine could be a strong ligand to the Fe at pH 5: the unperturbed

Table 8.1. Recently reported X-ray structures of selected electron transfer haemoproteins

Cytochrome	Electron transfer haem	Other characteristic features	References
Turnip chloroplast ferricytochrome f	His-c-N-terminal NH_2		Martinez et al. (1994)
Chromatium vinosum flavocytochrome c sulphide dehydrogenase	The two c-type haems of the dihaem subunit each have His/Met coordination. Two propionic acid substituents, one from each haem, are hydrogen bonded to each other in a solvent-inaccessible region	The cytochrome subunit is tightly associated with an FAD-containing subunit that also contains a catalytically important disulphide bond	Chen et al. (1994)
T. pantotropha ferricytochrome cd_1	His-c-His	Tyr-d_1-His; site of O_2 and NO_2^- reduction	Fülöp et al. (1995a)
Bovine heart and P. denitrificans ferricytochrome c oxidases	His-a_3-His; the hydroxyfarnesylethyl substituent is almost fully extended	His-a_3/$Cu_B O_2$ binding and reduction site; the haem a_3 hydroxyfarnesylethyl substituent is twisted to form a U-shaped arm; an electron transfer dinuclear Cu_A centre and non-redox metal centres are also present. The bovine enzyme consists of 13 subunits and the Paracoccus enzyme consists of four subunits	Iwata et al. (1995); Tsukihara et al. (1995)
Escherichia coli 'non-haem-iron-containing' ferricytochrome b-557 (bacterioferritin)	Met-b-Met located in up to 12 intersubunit sites	Di-nuclear metal centres in each of the 24 subunits; non-haem-iron mineral deposit in central cavity	Cheesman et al. (1993); Ford et al. (1984); Frolow et al. (1994)
P. aeruginosa ferricytochrome c peroxidase	His-c-Met; Fe(III) spin state is in S=1/2,5/2 thermal equilibrium, probably because Met ligand is not always bound	His-c-His is peroxide reaction site. When electron transfer haem is reduced, the peroxidatic Fe(III) haem is high spin probably as a result of one of the His ligands dissociating	Fülöp et al. (1995b, 1993); Gilmour et al. (1994); Greenwood et al. (1988)

N-terminal amine pK_a is approximately 7.2 compared with that of lysine at about 10 (Martinez et al., 1994). The polypeptide folding pattern and domain structure of the 252-amino acid cytochrome f is also unique among the cytochromes, with a predominance of β-sheet over other secondary structure types and a bilobed structure with the haem located in the larger domain.

Martinez *et al.* (1994) have analysed the structure in mechanistic terms and suggested areas of potential interaction for its acidic reaction partners.

One of the bacterial analogues of ferritin, bacterioferritin or non-haem-iron-containing cytochrome b-557, contains haem as isolated (Stiefel and Watt, 1979), up to a maximum of 12 per molecule at intersubunit sites in which the haem iron is coordinated by two methionines (Cheesman *et al.*, 1990; Frolow *et al.*, 1994). This ligation state has not been detected in other cytochromes and its six-coordinate low spin character suggests that it has an electron transfer role, though this has yet to be demonstrated convincingly. Other features of the bacterioferritin structure (Frolow *et al.*, 1994) were anticipated from previous studies. Thus, ferritins consist of 24 subunits, of molecular mass 18–24 kDa each, that pack into a spherical structure to provide a 20–25 Å thick protein coat encompasing a central cavity of approximately 80 Å diameter, into which a non-haem-iron-containing mineral is deposited (Ford *et al.*, 1984). The bacterioferritin structure determined by Frolow *et al.*, shows a similar arrangement, with the haems exposed to solvent in the central cavity of the molecule. As with structures of ferritins, the non-haem-iron core was not defined, though two dinuclear metal centres close to each haem were identified. Glu51 in each of the 24 subunits is a ligand to the non-haem-iron at the dinuclear centres and Met52 from each subunit are the two axial ligands of the haem (Andrews *et al.*, 1995; Frolow *et al.*, 1994; Le Brun *et al.*, 1995). The involvement of these dinuclear centres in oxidative Fe(II) uptake is now firmly established (Le Brun *et al.*, 1995).

The sulphide dehydrogenase, cytochrome cd_1 and cytochrome c peroxidase structures are all important as the first members of complex multiheaded enzyme families to be structurally determined. Only the latter two, however, have been significantly characterized mechanistically. The 1.55 Å crystal structure of cytochrome cd_1 allows considerable atomic detail to be resolved and, based on this, Fülöp *et al.* (1995a) proposed a detailed mechanism for NO_2^- reduction. The two types of haem are located in different domains of the 567-amino acid polypeptide fold with the shortest distance between the two haems of one subunit at 11.0 Å. The intact protein is a dimer but the shortest interhaem distance between subunits is 38.5 Å, which is probably too long for intersubunit electron transfer to be significant. Unlike the cytochrome f and bacterioferritin structures, details of the haem sites predicted from spectroscopic studies of *P. aeruginosa* cytochrome cd_1 were not in agreement with the actual structure. Whether these differences arise from misinterpretations of spectroscopic data, from redox-dependent conformational differences or from species differences between the cytochromes is not clear yet, though Fülöp *et al.* (1995a) do note that sequence comparisons within the cytochrome cd_1 family indicate that key haem ligands are not conserved. Fülöp *et al.* (1995a) also draw attention to a likely redox-dependent conformational change involving concerted movement of the haem c and haem d_1 binding domains. Such a change is probably associated with release of the product NO.

Cytochrome c peroxidase has particularly complex solution behaviour. Gilmour *et al.* (1993, 1994) have shown that the *P. denitrificans* enzyme as isolated in the fully oxidized state is inactive and that the active enzyme is a dimer mixed-valence form stabilized by Ca(II) binding. The dimer dissociates at high dilution leading to a loss of activity. The *P. aeruginosa* cytochrome *c* peroxidase structure has not yet been reported in the same detail as the cytochrome cd_1 structure but, as with the latter cytochrome, it is clear from previous spectroscopic studies that there are redox-dependent conformational changes involving changes in haem spin states and axial ligation (Gilmour *et al.,* 1993, 1994; Greenwood *et al.,* 1988; Rönnberg *et al.,* 1980). The structure shows the haems to be approximately 16 Å apart at their closest, with an Fe–Fe distance of about 8 Å and a Ca(II) binding site located between the haems (Fülöp *et al.,* 1995b).

The two recent structures of the cytochrome *c* oxidases reveal atomic details of the various metal centres as well as suggest pathways for movement of protons (Iwata *et al.,* 1995; Tsukihara *et al.,* 1995). These latter proposals are considered elsewhere in this book (see Chapter 9), while the most significant features for the present chapter are the structural details relevant to the haem-mediated electron transfer processes. The internal electron transfer steps in mammalian cytochrome oxidase have been explored extensively (for a review, see Gennis and Ferguson-Miller, 1995) and though it is clear that there are different kinetic forms of cytochrome oxidase, it is also established that the main route for the passage of electrons is from cytochrome *c* to the dinuclear Cu_A centre and then on to the haem *a* and finally the haem a_3/Cu_B centre. Thus, the arrangement of the haems becomes critical for defining the electron transfer pathways. In both structures the two haems are close together, approximately 5 Å apart at their closest point and with an Fe–Fe distance of 13–14 Å. Therefore, facile interhaem electron transfer can be readily envisaged. The hydroxyfarnesylethyl chain of the haem *a* groups is almost fully extended, but it points in the opposite direction to the Cu_A centre. Thus, it is more likely to be a hydrophobic anchor for the haem than a through-bond electron transfer pathway. It will be interesting to see the electron 'flight-paths' for these enzymes predicted by the electron tunnelling density procedure of Moser *et al.* (1995).

Structural characterization of eukaryotic cytochrome *c* has continued apace (see Chapter 5), largely through the systematic crystallographic work of Brayer and his co-workers (Berghuis and Brayer, 1992; Berghuis *et al.,* 1994; Louie and Brayer, 1990). This work has served to give a structural basis to numerous mechanistic and mutagenesis studies of *Saccharomyces cerevisiae* iso-1-cytochrome *c* (see, for example, Cutler *et al.,* 1989; Davies *et al.,* 1993; Langen *et al.,* 1992; Northrup *et al.,* 1993) as well as to confirm the view of the redox-dependent conformational change of this class of protein described by Takano and Dickerson (1981) and Moore and Pettigrew (1990). Most importantly, the conformational differences are focused on the pyrrole ring A propionate group, an internal water molecule and the Met80 haem ligand. All are part of an extensive hydrogen bonding network involving residues Arg38, Tyr48, Trp59, Asn52, Thr78 and Tyr67, and the importance of these residues is indicated further by their invariance, or conservative substitution, within the cytochrome *c* family (see compilation in Moore and Pettigrew, 1990). Thus, it came as something of a surprise to discover that many of these residues could be replaced by site-directed mutagenesis without preventing the cytochrome from carrying out its physiological electron transfer (Section 8.4).

One of the common features of the structural studies referred to above is that redox-linked conformational changes, albeit small in the case of eukaryotic cytochrome c, appear to be the rule not the exception. This has major implications for electron transfer rates by influencing reorganization energies, λ, and for determining haemoprotein redox potentials, and hence electron transfer driving forces (Chapter 1). We shall return to these points later, but point out now that what is important is to obtain an energetic description of the structural changes in order to assess their mechanistic importance. Churg et al. (1983) and Langen et al. (1992) have carried out such calculations for eukaryotic cytochromes c, and indicate that the contribution redox-linked conformational changes in this protein make to reaction reorganization energies is relatively small. Reorganization energies for cytochrome c peroxidase and cytochrome cd_1 are likely to be much greater.

8.4   Control of redox potentials

Determining the factors that control the level of cytochrome redox potentials has been a popular topic of research for many years. In part this is because redox potentials are relatively straightforward to measure, and correlating these with protein structural features appears to be both simple and informative. It is also because determining the factors that control the driving force for electron transfer, that is ΔG_{AB}^{o}, is critical to a full understanding of protein electron transfer.

The Gibbs energy change for the reduction of a ferrihaemoprotein, ΔG_{redox}, is composed of four terms (Moore et al. 1986):

$$\Delta G_{redox} = \Delta G_{cen} + \Delta G_{el} + \Delta G_{lig} + \Delta G_{conf} \qquad (8.1)$$

ΔG_{cen} is the Gibbs energy difference between the two redox states resulting from bonding interactions at the redox centre; ΔG_{el} is the Gibbs energy difference resulting from electrostatic interactions between the redox centre charge and polar groups within both the protein and the solvent (*Figure 8.3*). Both of these terms are likely to be substantial. ΔG_{lig} arises from differential binding of ligands to the oxidized and reduced forms of the protein. This term is often negligible under physiological conditions, though for certain cytochromes where the ligand is H^+ it can contribute significantly to ΔG_{redox}. ΔG_{conf} arises from Gibbs energy changes due to oxidation state conformational differences. *Figure 8.4* illustrates the relationship between ΔG_{redox} and ΔG_{conf}. Electron binding to a given ferrihaem is assumed to have an intrinsic binding energy. If there are no changes in the protein or solvent structure on reduction that require a net change in the Gibbs energy, ΔG_{conf} is zero and the observed ΔG_{redox} equals the intrinsic ΔG_{redox}. Clearly, ΔG_{conf} is related to the reorganization energy of an electron transfer reaction, λ.

There are a number of factors that may contribute to ΔG_{el}, as indicated by *Figure 8.3* and Equation 8.2.

$$\Delta G_{el} = \Delta G_{ion} + \Delta G_{dip} + \Delta G_{int} + \Delta G_{surf} \qquad (8.2)$$

where ΔG_{ion} describes the effect of ions in solution on the redox energy, and arises mainly from the non-specific Debye–Hückel screening effect. Specific ion binding only affects ΔG_{redox} if the binding is dependent on the oxidation state

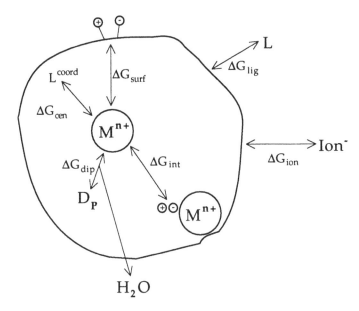

Figure 8.3. Factors affecting the stability of a metal ion (M^{n+}) buried within a protein. Gibbs free energy changes are shown which result from interaction of the metal ion with its ligands (L^{coord}, ΔG_{cen}), dipoles within and surrounding the protein (D_p, H_2O, ΔG_{dip}), buried charges (M^{n+}, +, −, ΔG_{int}), surface charges (+, −, ΔG_{surf}), ions in solution (Ion⁻, ΔG_{ion}) and ligands which bind to the protein non-covalently (L, ΔG_{lig}).

of the protein, and such effects are incorporated into ΔG_{lig}. ΔG_{dip} is the interaction energy between the redox centre and dipoles within both the protein and solvent. This term corresponds to the Born solvation energy of a charge, or its reaction field contribution, and was first recognized by Kassner (1973) to be an important component of ΔG_{redox}. ΔG_{int} is the Gibbs energy of interaction between the redox centre charge and internal charges that are not exposed to solvent. In cases where the redox centre is buried as well, ΔG_{int} may be large. ΔG_{surf} is the Gibbs energy of interaction between the redox centre charge and the charges on the surface exposed to solvent.

To analyse the separate Gibbs energy terms, the measured ΔG_{redox} must be separated into its various components. This is generally not possible, though ΔG_{lig} can usually be determined experimentally when the ligand is H^+. An empirical approach to the problem is to consider one class of related proteins in which the redox centre is the same and to assume that differences in ΔG_{redox} reflect differences in only one of the contributing factors. This was, for example, the basis for many of the early studies of class I cytochromes *c*; ΔG_{cen} was assumed to be constant and ΔG_{el} variable (Moore and Pettigrew, 1990, and references therein). The widespread use of site-directed mutagenesis allows this approach to be refined, as *Tables 8.2* and *8.3* indicate.

The electron donor–acceptor bonding properties of axial ligands are known to be influential in determining haem redox potentials, and the general view that comes from small molecule studies is that replacing a histidine ligand by methionine increases the redox potential by about 160 mV if all other factors

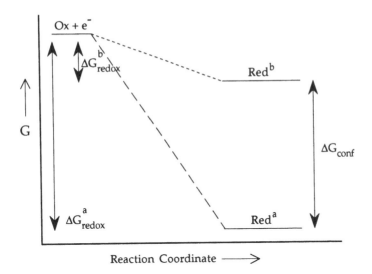

Figure 8.4. Gibbs free energy change for the reaction Ox + e⁻ → Red when there is no conformational change (large dashed line) and a substantial conformational change (small dashed line). The intrinsic redox energy is the same in both cases.

remain constant (Moore and Pettigrew, 1990). However, axial ligands do not appear to be the sole major determinants of cytochrome redox potentials, as illustrated by the class I cytochromes *c*, which have common haem ligation characteristics and redox potentials spanning from approximately 100 to 500 mV. The tetrahaem cytochrome of the *Rhodopseudomonas viridis* photosynthetic reaction centre discussed below provides another example. Perhaps the most striking case is that of the bis-methionine-ligated *b*-haems of *Azotobacter vinelandii* bacterioferritin which have redox potentials of –225 mV in the absence of a non-haem-iron core and –475 mV in its presence (Watt *et al.*, 1986). The mutagenesis studies summarized in *Table 8.2* confirm the effect of replacing a histidine by a methionine ligand in a protein, and indicate further that replacing methionine by lysine will lower the redox potential even further than replacing it with histidine. This has long been assumed to be so from studies of the alkaline isomerization of eukaryotic cytochromes *c*, but previously it has not been so clearly demonstrated. However, Barker and Mauk (1992) have shown that the high pH His/Lys-ligated form of mitochondrial cytochrome *c* (Ferrer *et al.*, 1993) has a redox potential of –205 mV. The effect of replacing a neutral ligand by an anionic one has long been known to cause a substantial reduction in the redox potential, since the additional charge on the ferrihaem is stabilized by the anion, and the tyrosine-ligated myoglobin and cysteine-ligated cytochrome are further confirmation of this relationship (*Table 8.2*).

The structural origins of the redox potential differences between class I cytochromes *c* have been investigated extensively, but clear, quantitative relationships between structural features and redox potential differences have not been established. The protein fold is clearly important, but whether it is the combined effects of all the protein dipoles and charges that control the level of the potential, or the effect of a few residues that dominate is unclear. Since it

Table 8.2. Effect of axial ligand mutations on redox potentials of haemoproteins

Protein	Axial ligands	E_m (mV)[a]	References
Horse cytochrome c	His/Met	260	
M80H	His/His	41	Raphael and Gray (1989)
M80L	His/H$_2$O[b]	−45	Raphael and Gray (1991)
M80C	His/Cys	−390	Raphael and Gray (1991)
Thiobacillus versutus			
cytochrome c-550	His/Met	252	Ubbink *et al.* (1994)
M100K	His/Lys	−77	Ubbink *et al.* (1994)
Desulphovibrio vulgaris			
cytochrome c_3	His/His	−280	Mus-Veteau *et al.* (1992)
H70M	His/Met	−80	Mus-Veteau *et al.* (1992)
Horse myoglobin	His/H$_2$O[b]	61 (pH 7)	Lloyd *et al.* (1995)
		45 (pH 8)	Hildebrand *et al.* (1995)
V68H	His/His	−110 (pH 7)	Lloyd *et al.* (1995)
H93Y	Tyr	−208 (pH 8)	Hildebrand *et al.* (1995)

[a] The conditions under which the redox potentials have been measured vary between the different studies. In the table all variant proteins are compared with their corresponding wild-type proteins under identical solution conditions.

[b] The haem iron of myoglobin is six-coordinate in the Fe(III) protein and five-coordinate in the deoxy-Fe(II) protein, in which there is an axial His ligand. It is not known what the axial ligation of the Fe(II) form of the M80L variant of cytochrome c is.

was observed that some of the evolutionarily unvaried residues of mitochondrial cytochrome c were haem propionate contact residues, and that in cytochromes c_2 lacking an arginine equivalent to Arg38 a haem propionate ionized with pK_a values in the range 6–8 with a consequent reduction in redox potential (Moore and Pettigrew, 1990), a mutagenesis study was undertaken to replace the contact residues Arg38, Tyr48 and Trp59. The results of this study are summarized in *Table 8.3*, which indicates that though the redox potentials varied, substantial perturbations were not seen. Furthermore, in none of the variants did a haem propionate ionize in the pH range 5–7, showing that local

Table 8.3. Redox potentials of *S. cerevisiae* iso-1-cytochrome c and variants of it produced by site-directed mutagenesis

Cytochrome	E_m at pH 5 (mV)	pK_{695}[a]
Control: R38,Y48,W59	285	8.8
H38	259	8.3
N38	248	8.2
Q38	245	8.4
A38	237	8.3
F48	263	8.6
F48,F59	215	7.6
A38,F48,F59	220	7.0

[a] pK_{695} is an indicator of the stability of the ferricytochrome and is the apparent pK_a for loss of the axial Met ligation to the Fe(III).
Data from Cutler *et al.* (1989) and Davies *et al.* (1993).

environment is not the dominant factor in determining haem propionate pK_a values. Since NMR studies showed that the amino acid substitutions did not produce large accompanying conformational changes, the redox potential decreases were ascribed to increased solvent accessibility of the haem and/or to variations in dipole interactions with the ferrihaem charge resulting from an altered polarity of haem contact residues (Cutler *et al.*, 1989; Davies *et al.*, 1993).

Many other variants of iso-1-cytochrome *c* have been described, including the Asn52Ile variant, which also involves a haem propionate contact residue. The substitution resulted in a small conformational perturbation that led to the displacement of an internal water molecule close to the haem, and caused the redox potential to drop by 55 mV at pH 7 (Langen *et al.*, 1992). This is a particularly significant result because the replacement of a polar group by a hydrophobic residue, and the displacement of an internal water molecule, might have been anticipated to lead to an increase in the relative stability of the ferrohaem and thus to an increase in the redox potential. That it did not do so is ascribed by Langen *et al.* (1992) to the loss of a repulsive electrostatic interaction between the Asn52 dipole and the ferrihaem positive charge.

A major significant advance has been the development of computational approaches for investigating haemoprotein redox potentials (Moore and Pettigrew, 1990, and references therein). Gunner and Honig (1991) have provided an excellent example of this approach applied to the tetrahaem cytochrome of the *R. viridis* photosynthetic reaction centre. Though there are a number of approximations that need to be made to enable the calculations to be carried out, the results do provide a guide to the important redox potential contributions (*Table 8.4*). The largest individual components are the axial ligand contributions, but the effect of the reaction fields of the haems and the influence of ionized groups are also significant. The effect of polar, but uncharged, backbone and side chain groups and of the ferrihaem charges are mixed, but it is an interplay of all these terms that gives rise to the redox potential variation. Unfortunately, the structure of the reduced form of the photosynthetic reaction centre cytochrome is not known, so it is not possible to assess whether redox state conformational changes will affect the redox potential (*Figure 8.3*). However, it is likely that, in the reaction centre itself, conformational changes are minimized to allow rapid charge separation and thus similar considerations might apply to the cytochrome. Certainly there is reasonably good agreement between the calculated and experimental redox potentials ignoring redox-linked conformational changes (*Table 8.4*).

8.5   Intraprotein electron transfer

As indicated earlier, intraprotein electron transfer has been a major area of study in recent years in order to determine values of β (see Chapter 1, Equation 1.1). For such analyses to be carried out, it is important that the electron transfer rate be limited by the distance the electron has to travel and the medium through which it travels. This means that interaction work terms between reactants must be negligible, and that driving force effects and reorganization energies be corrected for. Generally, measurements have been

Table 8.4. Experimental and calculated midpoint redox potentials of the *R. viridis* reaction centre tetrahaem cytochrome *c*

	Haem 1	Haem 2	Haem 3	Haem 4
Axial ligation	His–Met	His–His	His–Met	His–Met
Experimental potentials (mV)	370–400	10–50	295–320	−80 to −50
Calculated potentials (mV)	319	16	320	41
Scaling factor (mV)	623	623	623	623
Axial ligand (meV)	−254	−453	−224	−254
Back-bone and polar groups (meV)	71	24	34	35
Reaction field (meV)	−104	−136	−119	−172
Ionized groups (meV)	−113	−52	−108	−191
Oxidized haems (meV)	96	10	114	0

The haem group nomenclature is taken from the X-ray structure of Deisenhofer *et al.* (1984). The calculated contributions to the potential were obtained from calculations of electrostatic free energies of oxidation. The overall calculated potentials are the sums of the individual components plus the scaling factor, which converts the free energies to potentials relative to the standard hydrogen electrode. The contributions from ionized groups were estimated from calculated pK_a values for all 289 ionizable groups.
Calculated values from Gunner and Honig (1991); experimental values from compilation given by Gunner and Honig (1991) (see references therein and also Nitschke and Rutherford, 1994).

made over a range of driving forces and the maximum rate for a system determined from a Marcus analysis (Chapter 1, Equation 1.4) (see *Figure 8.5*). The ΔG°_{AB} variation at relatively high driving forces has been achieved using a range of quinones in photosynthetic reaction centres (Gunner and Dutton, 1989), labelling proteins with inorganic redox reagents (Durham and Millett, 1994; Winkler and Gray 1992; Winkler *et al.*, 1982, and references therein) and replacing the haem groups in some proteins with porphyrin or zinc-containing porphyrin, both of which have useful photochemical properties (see Cheung *et al.*, 1986; Durham and Millett, 1994; Ho *et al.*, 1985; McLendon *et al.*, 1987; Moore and Pettigrew, 1990; Winkler and Gray, 1992, and references therein). The latter two procedures have been particularly extensively explored with haemoproteins: *Table 8.5* gives a selection of the relevant studies with mitochondrial cytochrome *c*.

A major advantage to the use of proteins derivatized with redox-active groups is that intermolecular interaction work terms can be eliminated from the kinetic analysis. This allows first order rates of electron transfer to be measured, though in some cases analysis of the reaction kinetics requires considerable experimental ingenuity before these can be obtained (e.g. Chang *et al.*, 1991; Willie *et al.*, 1992). Ru(II)-containing species have been the redox reagents of choice for derivatization with a variety of compounds investigated, partly because, as with Zn(II)-substituted porphyrins, these provide photochemically active redox centres with a range of driving forces and excited state lifetimes. Initially the modifications were made to wild-type proteins but, with the development of site-directed mutagenesis, proteins engineered to have suitable Ru(II)-binding groups, usually histidines, at specific locations have been used increasingly (e.g. Willie *et al.*, 1992; Wuttke *et al.*, 1992). More recently, semi-synthetic methods have been adopted to engineer Ru(II)-binding groups into proteins that are not found in native proteins. An excellent example of this

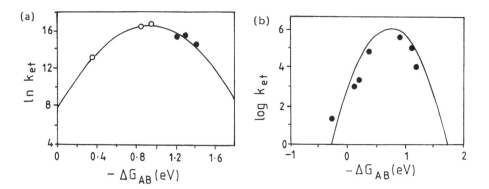

Figure 8.5. Plots of the electron transfer rate constants against reaction driving forces for two redox systems. (a) Intramolecular electron transfer in Ru(II)-65-cytochrome b_5. The Thr65Cys variant of cytochrome b_5 was labelled with ruthenium-containing compounds having different redox potentials. The unimolecular rates of electron transfer between the photoexcited species, *Ru(II), and the ferrihaem (open circles), and the thermal back reaction to generate the ground state Ru(II)-Fe(III) species (closed circles) are shown. The solid line is calculated for a theoretical dependence with $\lambda = 0.94$ eV. Data and theoretical line taken from Scott et al. (1993).
(b) Intramolecular electron transfer in complexes formed between cytochromes c and b_5. The unimolecular rates of electron transfer in the cytochrome b_5 reactions with native, Zn(II)-substituted and iron-free porphyrin cytochromes c were determined. The different cytochrome c derivatives covered a range of redox potentials. The solid line is calculated for a theoretical dependence with $\lambda = 0.8$ eV. Data and theoretical line taken from McLendon and Hake (1992).

approach is the introduction of bipyridyl-derivatized amino acids into cytochrome c (Wuttke et al., 1993). An initially separate approach to studying haemoprotein electron transfer (Ho et al., 1985; McLendon and Miller, 1985; Moore and Pettigrew, 1990), but one that is now routinely employed in conjunction with ruthenation (Axup et al., 1988; Winkler and Gray, 1992), involves replacement of the haem iron by another metal, typically Zn(II).

Figure 8.5a shows a typical example of the use of ruthenated proteins to investigate the driving force dependence of a series of related electron transfer reactions. The Thr65Cys variant of rat liver cytochrome b_5 labelled with various Ru(II)-polypyridine complexes has the two internal electron transfer processes described by Equation 8.5 (*Table 8.5*). By studying a range of derivatives with different Ru(II) ligands, Scott et al. (1993) were able to obtain six unimolecular rate constants for essentially similar reactions at different driving forces. The plot for these is characteristic of the Marcus 'inverted region' (Marcus and Sutin, 1985) in which the maximum rate of electron transfer occurs when $\Delta G^\circ_{AB} = \lambda$. Thus this plot allows the reorganization energy of the electron transfer reactions to be obtained, in this case 0.94 eV, and directly gives a measurement of the donor–acceptor electronic coupling terms (Scott et al., 1993; Winkler and Gray 1992; see Chapter 1). From such rate data, the distance dependence of the electron transfer can be determined, and hence the value β.

Results from the studies described above have been discussed widely and the consensus view appears to be that long-range electron transfer within a protein matrix occurs with a β of approximately 1.4 Å^{-1} and with only a minimal medium effect (Durham and Millett, 1994; Moser et al., 1994; Winkler and Gray 1992, see Chapter 1). However, the number of cases where long-range electron transfer occurs at rates determined by the $\exp(-\beta R)$ decay of the donor–acceptor coupling interaction may not be that great. Leaving aside the photosynthetic reaction centre (Moser et al., 1992), none of the other biologically significant multiheaded electron transfer haemoproteins that have been characterized show evidence of this. Table 8.6 summarizes relevant data for flavocytochrome b_2 and cytochrome cd_1.

The data in Table 8.6 for flavocytochrome b_2 led Chapman et al. (1994) and Chapman and Mount (1995) to conclude that the rate of intramolecular electron transfer between the flavin and haem of flavocytochrome b_2 is governed by the structural mobility of the domains of flavocytochrome b_2. This was particularly evident from the effect of altering the interdomain hinge region of flavocytochrome b_2 by protein engineering (Sharp et al., 1994). The haem and flavin domains are linked by a flexible 15-amino acid hinge, and removing three of the amino acids from this reduced the flavin to haem electron transfer rate from 445 to 91 s^{-1}. Similarly, McLendon et al. (1987) concluded that conformational changes within flavocytochrome b_2 determined electron transfer rates from a study of the reactions of flavocytochrome b_2 with native cytochrome c and its metal-free and Zn(II)-substituted porphyrin derivative. McLendon et al. (1987) found that the rates of reaction did not depend on their driving forces.

The data in Table 8.6 for cytochrome cd_1 are also striking. Clearly the redox-linked conformational changes, that may involve haem ligation differences as well as domain movements (Fülöp et al., 1995a), are a greater influence than the distance the electron has to travel. The significance of ligand changes with respect to electron transfer rates has been known for many years, with the alkaline isomerization of cytochrome c providing an early example (Moore and Pettigrew, 1990), and with the characterization of enzymes such as cytochrome cd_1 and cytochrome c peroxidase now under way this aspect of haemoprotein electron transfer may come to the fore again.

8.6   Interprotein electron transfer

All haemoprotein electron transfer processes, even those involving transmembrane complexes, require a bimolecular step to either inject electrons into a centre or remove electrons from it. The minimal scheme for such reactions is the one shown in Figure 8.1 and the overall rate for this, k_{obs}, is given by:

$$k_{obs} = K_a \cdot k_{et} \cdot [A_o][B_r] \qquad (8.7)$$

where K_a is the association constant for the formation of the precursor complex in Figure 8.1, step 8.1.1, and k_{et} is the rate constant for the electron transfer in step 8.1.2. It is k_{et} that has been the subject of much of the recent work in the haemoprotein electron transfer field, though it is K_a that often determines the overall bimolecular rate of electron transfer.

In general, physiological reaction partners are oppositely charged so that there are attractive electrostatic interactions to guide the proteins together and

Table 8.5. Electron transfer rate constants for reactions of mitochondrial cytochrome c[a] covalently labelled with ruthenium-containing redox-active species

Ru-containing species	Residue modified	Electron transfer rates (s^{-1})	Electron transfer distance (Å)	References
$[Ru(NH_3)_5]-$	His 33	$k_1 = 30; 55$[b,c]	11	Bechtold et al. (1986a); Winkler and Gray (1992); Winkler et al. (1982)
$[Ru(NH_3)_5]-$	His33 and Zn(II)-haem	$k_1 = 1.6 \times 10^6$[d]	11	Elias et al. (1988)
$[Ru(NH_3)_4(isn)]-$[e]	His33	$k_2 < 10^{-4}$[b]	11	Bechtold et al. (1986b)
$[Ru(NH_3)_4(isn)]-$[e]	His33 and Zn(II)-haem	$k_1 = 2 \times 10^5$[d]	11	Winkler and Gray (1992)
$[Ru(bipy)_2(im)]-$[e]	His33	$k_1 = 2 \times 10^5$, $k_2 = 2.6 \times 10^6$[f]	11	Chang et al. (1991); Wuttke et al. (1992)
	His39	$k_1 = 1.4 \times 10^6$, $k_2 = 3.2 \times 10^6$[f]	12	Wuttke et al. (1992)
	residue 62[g]	$k_1 = 1.1 \times 10^5$, $k_2 = 1 \times 10^4$[f]	15	Wuttke et al. (1992)
	residue 72[g]	$k_1 = 3.4 \times 10^5$, $k_2 = 9 \times 10^5$[f]	8	Wuttke et al. (1992)
$[Ru(bipy)_2(4bpa)]-$[e]	residue 72[g]	$k_1 = 6 \times 10^6$, $k_2 = 6.5 \times 10^6$[f]	8	Wuttke et al. (1993)
$[Ru(bipy)_2(dcbipy)]-$[e]	Lys7	$k_1 = 3 \times 10^5$, $k_2 = 6 \times 10^5$, $k_3 = 2 \times 10^6$[h]	6–16	Durham et al. (1989)
	Lys13	$k_1 = 16 \times 10^6$, $k_2 = 26 \times 10^6$, $k_3 = 8 \times 10^6$[h]	6–10	Durham et al. (1989)
	Lys25	$k_1 = 1 \times 10^6$, $k_2 = 1.5 \times 10^6$, $k_3 = 6 \times 10^6$[h]	6–16	Durham et al. (1989)
	Lys27	$k_1 = 20 \times 10^6$, $k_2 = 30 \times 10^6$, $k_3 = 20 \times 10^6$[h]	6–12	Durham et al. (1989)
	Lys72	$k_1 = 14 \times 10^6$, $k_2 = 24 \times 10^6$, $k_3 = 6 \times 10^6$[h]	8–16	Durham et al. (1989)

[a] Except for the [Ru(bipy)$_2$(Im)]- modifications of His39 (*Candida krusei*) and residue 62 (*S. cerevisiae*), horse cytochrome *c* was used. The cytochromes do not differ substantially in conformation (Moore and Pettigrew, 1990).
[b] The reactions are:

$$\text{Ru(II)-Fe(III)} \underset{k_2}{\overset{k_1}{\rightleftharpoons}} \text{Ru(III)-Fe(II)} \tag{8.3}$$

[c] The different k_1 values come from different studies. The driving force for the forward reaction (k_1) is 0.18 eV.
[d] As well as the His33 ruthenation, the haem iron has been replaced by Zn(II). The reaction is:

$$\text{Ru(II)-}^+\text{Zn(II)} \overset{k_1}{\rightarrow} \text{Ru(III)-Zn(II)} \tag{8.4}$$

$^+$Zn(II) is the triplet-excited state of Zn(II)-cytochrome *c* generated photochemically. The driving force is 1.01 eV for the [Ru(NH$_3$)$_5$]- derivative and 0.66 eV for the [Ru(NH$_3$)$_4$(isn)]- derivative.
[e] isn = isonicotinamide; bipy = 2,2'-bipyridyl; im = imidazole; 4bpa = (S)-2-amino-3-(2,2'-bipyridyl)propionic acid; dcbipy = 4,4'-dicarboxybipyridine.
[f] The reactions are:

$$^*\text{Ru(II)-Fe(III)} \overset{k_1}{\rightarrow} \text{Ru(III)-Fe(II)} \overset{k_2}{\rightarrow} \text{Ru(II)-Fe(III)} \tag{8.5}$$

*Ru(II)-Fe(III) are photochemically generated charge-transfer excited states containing bipy radical anions. Their driving forces are 1.17 and 1.13 eV respectively for the [Ru(bipy)$_2$(im)]-His and [Ru(bipy)$_2$(4bpa)] reactions. The driving forces for the k_2 reactions are 0.74 and 1.0 eV respectively.
[g] The normal residues 62 and 72 were replaced by histidine using site-directed mutagenesis or semi-synthesis.
[h] The reactions are:

$$\text{Ru(II)-Fe(III)} \underset{k_3}{\overset{h\nu}{\rightleftharpoons}} {}^*\text{Ru(II)-Fe(III)} \tag{8.6}$$

with k_2 and k_1 leading to Ru(III)-Fe(II).

The driving forces for the reactions are 0.98 and 1.05 eV for reactions k_1 and k_2, respectively.

---

stabilize the complex long enough for the electron to be transferred. Usually the attractive forces are not so great that there is a high binding affinity, presumably because this would lead to dead-end complex formation in which product complexes did not dissociate. However, this presents a general problem for those attempting to characterize the structures of reactive complexes in that, with few exceptions, crystals suitable for X-ray diffraction studies have not been reported. Even when crystals have been obtained and structures determined it is not always clear how relevant the structure is to the solution state electron transfer event. The yeast cytochrome *c* peroxidase–cytochrome *c* complex provides an excellent example of this: the X-ray structures reported by Pelletier and Kraut (1992) are not consistent with solution NMR data (Jeng *et al.*, 1994; Moore *et al.*, 1995), although the differences may result from the NMR methods employed monitoring a range of related structures that are minor variants of the crystallographically defined structures.

The most widely used approach to map interaction sites on proteins is computer graphics molecular modelling based on factors such as optimizing interprotein electrostatic interactions. However, these methods have had limited

Table 8.6. Electron transfer distances and rates for multiheaded haem enzymes

Cytochrome	Distances between redox centres	Intramolecular electron transfer rates	References
S. cerevisiae flavocytochrome b_2	9.7 Å (shortest flavin to haem distance)	F: $b^{III} \rightarrow$ F. b^{II} $k = 1500 \pm 500$ s^{-1} F. $b^{II} \rightarrow$ F: b^{III} $k = 270 \pm 90$ s^{-1}	Chapman *et al.* (1994) and references therein
T. pantotropha cytochrome cd_1	20.6 Å (Fe–Fe) and 11.0 Å (haem edge to edge)	$c^{II}d_1{}^{III}.NO \rightarrow c^{III}d_1{}^{II}.NO$, $k = 1$ s^{-1}	Silvestrini *et al.* (1990); Fülöp *et al.* (1995a)

F: and F. are reduced FMN and FMN semiquinone, respectively.

success in defining precise regions. A typical case is the model proposed by Tegoni *et al.* (1993) for the cytochrome *c*–flavocytochrome b_2 complex. This model suggested several interprotein electrostatic interactions to stabilize the complex with a *c*–*b* edge-to-edge distance of about 14 Å. However, mutagenesis studies to test this model showed that proposed key residues could be substituted with no apparent effect on the reaction (Daff *et al.*, 1995). An alternative model of the complex proposed by Moser *et al.* (1995) from considerations of surface regions of high electron tunnelling density in both partners is, however, supported by the mutagenesis data (Daff *et al.*, 1995).

Numerous studies of interprotein complexes of electron transfer proteins point to some degree of motion at the interprotein interface, consistent with a 'rolling or sliding ball' model of the complex. This kind of model was first suggested for the reaction of small inorganic redox reagents with proteins (reviewed by Moore *et al.*, 1983) and later extended to the complex formed between cytochrome *c* and cytochrome b_5 (Eley and Moore, 1983). A central feature of this model is that the binding surfaces of immobilized proteins would have to fit together precisely to avoid formation of unproductive complexes, though if the proteins retained some independent motion within the complex the interacting surfaces could move against one another and so increase the probability of a productive interaction. This has been investigated by Brownian dynamics modelling which specifically takes account of the dynamic behaviour of protein complexes (see Chapter 4). Northrup *et al.* (1993) calculated the first order rate constant for electron transfer, k_{et}, for a variety of cytochrome *c* variants reacting with ferrocytochrome b_5 under second order conditions and then used Brownian dynamics to compute the second order rate constants. The excellent agreement between the observed and calculated rate constants (*Table 8.7*) indicates that the actual electron transfer event is not perturbed by the mutations of cytochrome *c* lysines proposed to be involved in stabilizing the complex. Perhaps just as striking as the invariance of the rate of electron transfer in the different cytochrome *c*–b_5 complexes is that the overall bimolecular rate of reaction is relatively little affected by the replacement of supposedly key lysines of cytochrome *c* responsible for the formation of the complex in the first place (Salemme, 1976). This is a further indication that there are more than four interprotein contacts critical for complex formation.

Table 8.7. Theoretical and experimental rate constants for the oxidation of bovine ferrocytochrome b_5 by variants of yeast iso-1-ferricytochrome c at $I = 0.19$ M

Ferricytochrome c	Experimental ($\times 10^6$ M^{-1} s^{-1})	Calculated ($\times 10^6$ M^{-1} s^{-1})	k_{et} (ns^{-1})
Wild-type	136 ± 3	136 ± 17	130
TML72A	37 ± 3	59 ± 6	121
K79A	68 ± 3	112 ± 9	133
TML72A/K79A	30 ± 2	46 ± 6	127

Data from Northrup *et al.* (1993) and Mauk *et al.* (1995).

The recent description of a novel procedure to map interaction sites on proteins, based on likely electron ingress and egress points deduced from calculated electron tunnelling densities (Moser *et al.*, 1995), holds out the prospect of interprotein reaction sites, as opposed to interprotein binding sites, being readily identified. From those already suggested, Moser *et al.* note that they are relatively broad areas defining multiple binding geometries, consistent with the dynamic picture emerging from NMR studies and Brownian dynamics calculations.

The investigation of interprotein electron transfer reactions has led to the development of numerous non-physiological experimental systems. As with the study of intraprotein electron transfer described above, these include the introduction of Zn(II)-containing porphyrin in place of haem in some proteins and the use of ruthenated proteins in complex with unmodified proteins. The work of Willie *et al.* (1992, 1993) and Liu *et al.* (1995) exemplifies the latter approach. Willie *et al.* (1992) replaced Thr65 of rat liver cytochrome b_5 with cysteine and attached a 4-methylene-4′-methyl-2,2′-bipyridine(bipy)$_2$Ru(II) to the thiol. This placed the Ru(II) about 12 Å from the cytochrome b_5 haem iron in a region of the protein surface where it would not interfere with interprotein complex formation involving cytochromes c according to the Salemme (1976) proposal for the complex geometry. Laser excitation of the Ru(II) group in a bimolecular Ru(II)-65-cytochrome b_5–cytochrome c complex then gave the following electron transfer reactions:

$$\overset{k_1}{*Ru(II) \rightarrow b_5\text{-Fe(III)}} \quad \overset{k_f, k_s}{\rightarrow} \quad c\text{-Fe(III)} \tag{8.8}$$

k_1 was greater than 10^6 s^{-1} and the biphasic reduction of ferricytochrome c had rate constants of $k_f = 4 \times 10^5$ s^{-1} and $k_s = 3 \times 10^4$ s^{-1}. This system was then extended to variants of cytochrome c produced by site-directed mutagenesis to explore whether the nature of individual side chains close to the haem was important for influencing the rate of electron transfer. The chief conclusion from this work was that an aromatic group at position 82 of cytochrome c was not essential for rapid electron transfer (Willie *et al.*, 1993), in contrast to expectations from previous structural and kinetic studies (Liang *et al.*, 1987; Poulos and Kraut 1980). The biphasic kinetics of cytochrome c reduction in this system was attributed to different conformational states of the Ru(II)-65-cytochrome b_5–cytochrome c complex (Willie *et al.*, 1992), in keeping with the 'rolling or sliding ball' model of the complex described above.

Liu *et al.* (1995) have extended this approach to cytochrome *c* peroxidase (CCP) complexed to cytochromes *c* in which individual lysines had been attached to dicarboxy-2,2'-bipyridine(bipy)$_2$Ru(II). The reaction cycle was initiated by photoexcitation of the Ru(II) to generate the charge transfer state *Ru(II) which reduced the cytochrome *c* haem iron, leaving a strongly oxidizing Ru(III) group. This oxidized the indole ring of the cytochrome *c* peroxidase Trp191 to generate its distinctive free radical form, and this in turn was reduced by the ferrohaem of cytochrome *c*. Thus, the reaction scheme was:

$$
\begin{array}{c}
\xrightarrow{ k_1 } \\
\text{*Ru(II)-}c\text{-Fe(III):CCP-W191} \rightarrow \text{Ru(III)-}c\text{-Fe(II)-W191} \\
\downarrow k_2 \\
\text{Ru(II)-}c\text{-Fe(III):CCP-W191} \leftarrow \text{Ru(II)-}c\text{-Fe(II)-W191}^+ \\
k_3
\end{array}
\qquad (8.9)
$$

With $k_1 = 2.3 \times 10^7$ s^{-1}, $k_2 = 7 \times 10^6$ s^{-1} and $k_3 = 6.1 \times 10^4$ s^{-1} for the reaction in which Lys27 of cytochrome *c* bears the Ru(II) label. This reaction system allowed Liu *et al.* (1995) to study the effect of substituting several cytochrome *c* peroxidase residues by site-directed mutagenesis and allows the formation of the peroxidase free radical to be carried out in the absence of peroxides.

Just as with individual ruthenated proteins (Section 8.5), Marcus-type analyses of interprotein complexes allows the maximum electron transfer rates, reorganization energies and electronic coupling factors to be determined (McLendon and Hake, 1992; Winkler and Gray 1992). An example of this is given in *Figure 8.5b* for the reaction of cytochrome *c* with cytochrome *b$_5$* (McLendon and Hake, 1992; McLendon and Miller, 1985). In this case, the Marcus analysis gave a reorganization energy of about 0.8 eV and a maximum rate of about 10^6 s^{-1}, well above the experimentally determined rate for the native proteins of approximately 10^3 s^{-1} (McLendon and Miller, 1985; Willie *et al.*, 1993). This emphasizes that physiologically significant electron transfer does not have to occur at the maximum rate allowed by high driving forces.

8.7 Coupled electron–ion transfers

Coupled electron–proton transfer reactions are widespread in haemoproteins, though not all may be biologically important. They are important in transmembrane complexes coupling electron transfer with proton pumping (Wikström, 1989; see Chapter 9), and in redox enzymes involved in transforming organic and inorganic substrates, but in most simple electron transfer proteins carrying out scalar electron–proton transfer reactions such coupling is probably not biologically important.

Coupled electron–metal ion transfers occur in the iron storage proteins, the ferritins, and may be important in other biomineralization processes and metal transport and storage systems involving redox-active metals. Ferritins contain up to 4500 iron ions per molecule in their central cavity, usually as Fe(III) under physiological conditions; however, in order to get the iron to enter or leave the protein, at least at relatively rapid rates, the iron has to be Fe(II) (Fatemi *et al.*, 1991, and references therein; Ford *et al.*, 1984). The mechanistic consequences of this are not clear yet, but it is clear that long-range electron transfer is important in some reactions. Animal ferritins do not contain haem as isolated but up to 16 haem groups per molecule can be bound *in vitro* in some preparations, and the bound haem can be shown to be effective in promoting iron release (Kadir *et al.*, 1992; Moore *et al.*, 1994; Prècigoux *et al.*, 1994) (*Table 8.8*).

The basis of the iron release assay is that incubation of ferritin with a reductant produces Fe(II) in the core of the protein which can be detected after its complexation with coloured ligands outside the protein. The general reaction scheme is:

$$\text{Core - Fe}_n^{3+} + e^- \xrightarrow{\text{slow}} \text{Core - Fe}_{n-1}^{3+} + \text{Fe}^{2+} \tag{8.10}$$

$$\text{Fe}^{2+} + x\text{ligand} \xrightarrow{\text{fast}} \text{Fe}^{2+}(\text{ligand})_x \tag{8.11}$$

Provided the reaction in Equation 8.11 is much faster than that in Equation 8.10, the overall rate measured corresponds to the rate of the reaction in Equation 8.10. For the ligands used by Kadir *et al.* (1992), that was the case. Two points are striking about the data in *Table 8.8*. Firstly, the rate of iron release from haem-free ferritin depends upon the reductant, increasing with the driving force for the redox stage of the reaction, whereas the rate of release from the haem-bound ferritin does not; and secondly, the rate of release from the haem-bound ferritin is faster than that from the haem-free ferritin. The proposal put forward by Kadir *et al.* (1992) for these data is that, in the absence of haem, electron transfer across the protein coat from reductant located outside the protein was the rate-limiting step for release of Fe(II), and in the presence of haem the rate of electron transfer across the coat was increased to such an extent that it was no longer the limiting step. Similar behaviour has been observed in reductive iron release assays from haem-containing and haem-free non-haem-iron-containing cytochrome *b*-557 (bacterioferritin) and this, together with the observation that haem is not required for oxidative uptake of Fe(II) by bacterioferritin (Andrews *et al.*, 1995; Le Brun *et al.*, 1995), suggests that the role of the intrinsic haem of bacterioferritin is in iron release. Alternative roles could be envisaged, especially if bacterioferritin turned out to be an electron store (Stiefel and Watt, 1979).

Table 8.8. Initial rates[a] of iron release from horse spleen ferritin

Reductant	E_o (mV)	No. of haems	
		0	16
Caffeic acid	570	0.48	0.65
Chlorogenic acid	610	0.41	0.65
3,4-Dihydroxyphenyl acetic acid	700	0.36	0.65
3,4-Dihydroxybenzoic acid	740	0.34	0.65
Ferulic acid	790	0.31	0.62

[a] The rates are reported as: nM Fe(II) released min^{-1} [mM initial Fe(III)]$^{-1}$ [µM added electron]$^{-1}$.
Data from Kadir *et al.* (1992).

Acknowledgements

I thank Dr S.K. Chapman (Edinburgh) and Professor A.G. Mauk (Vancouver) for many stimulating and helpful discussions concerning cytochrome electron transfer; NATO for providing travel funds to assist collaboration between UEA and UBC (Vancouver); the Wellcome Trust for providing me with a Research Leave Fellowship and generous experimental support; and the EPSRC and BBSRC for supporting the UEA Centre for Metalloprotein Spectroscopy and Biology via their Biomolecular Sciences Panel.

References

Andrews SC, Le Brun NE, Barynin V, Thomson AJ, Moore GR, Guest JR, Harrison PM. (1995) Site directed replacement of the coaxial-heme ligands of bacterioferritin generates heme-free variants. *J. Biol. Chem.* **270**: 23268–23274.

Axup AW, Albin M, Mayo SL, Crutchley RJ, Gray HB. (1988) Distance dependence of photoinduced long-range electron transfer in zinc/ruthenium-modified myoglobins. *J. Am. Chem. Soc.* **110**: 435–439.

Barker PD, Mauk AG. (1992) pH-linked conformational regulation of a metalloprotein oxidation–reduction equilibrium: electrochemical analysis of the alkaline form of cytochrome *c*. *J. Am. Chem. Soc.* **114**: 3619–3624.

Bechtold R, Kuehn C, Lepre C, Isied SS. (1986a) Directional electron transfer in ruthenium-modified horse heart cytochrome *c*. *Nature* **322**: 286–288.

Bechtold R, Gardineer MB, Kazmi A, van Hemelryck B, Isied SS. (1986b) Ruthenium-modified horse heart cytochrome *c*: effect of pH and ligation on the rate of intramolecular electron transfer between Ru(II) and heme(III). *J. Phys. Chem.* **90**: 3800–3804.

Berghuis AM, Brayer GD. (1992) Oxidation state-dependent conformational changes in cytochrome *c*. *J. Mol. Biol.* **223**: 959–976.

Berghuis AM, Guillemette JG, McLendon G, Sherman F, Smith M, Brayer GD. (1994) The role of a conserved internal water molecule and its associated hydrogen bond network in cytochrome *c*. *J. Mol. Biol.* **236**: 786–799.

Blackledge MJ, Medvedeva S, Poncin M, Guerlesquin G, Bruschi M, Marion D. (1995) Structure and dynamics of ferrocytochrome *c*-553 from *Desulfovibrio vulgaris* studied by NMR spectroscopy and restrained molecular dynamics. *J. Mol. Biol.* **245**: 661–681.

Chang I-J, Gray HB, Winkler JR. (1991). High-driving-force electron transfer in metalloproteins. *J. Am. Chem. Soc.* **113**: 7056–7057.

Chapman SK, Mount AR. (1995) Electron transfer in proteins. *Nat. Prod. Res.* **12**: 93–100.

Chapman SK, Reid GA, Daff S, Sharp RE, White P, Manson FDC, Lederer F. (1994) Flavin to haem electron transfer in flavocytochrome b_2. *Biochem. Soc. Trans.* **22**: 711–715.

Cheesman MR, Thomson AJ, Greenwood C, Moore GR, Kadir FHA. (1990) Bis-methionine axial ligation of haem in bacterioferritin from *Pseudomonas aeruginosa*. *Nature* **346**: 771–773.

Cheesman MR, Le Brun NE, Kadir FHA , Thomson AJ, Moore GR, Andrews SC, Guest JR, Harrison PM, Smith JMA, Yewdall SJ. (1993) Haem and non-haem iron sites in *E. coli* bacterioferritin. *Biochem. J.* **292**: 47–56.

Chen Z-W, Koh M, van Driessche G, van Beeumen JJ, Bartsch RG, Meyer TE, Cusanovich MA, Mathews FS. (1994) The structure of flavocytochrome *c* sulfide dehydrogenase from a purple phototrophic bacterium. *Science* **266**: 430–432.

Cheung E, Taylor K, Kornblatt JA, English AM, McLendon G, Miller JR. (1986) Direct measurements of intramolecular electron transfer rates between cytochrome *c* and cytochrome *c* peroxidase: effects of exothermicity and primary sequence on rate. *Proc. Natl Acad. Sci. USA* **83**: 1330–1333.

Churg AK, Weiss RM, Warshel A, Takano T. (1983) On the action of cytochrome *c*: correlating geometry changes upon oxidation with activation energies of electron transfer. *J. Phys. Chem.* **87**: 1683–1694.

Cox MC, Le Brun N, Thomson AJ, Smith A, Morgan WT, Moore GR. (1995) MCD, EPR and NMR spectroscopic studies of rabbit hemopexin and its heme binding domain. *Biochim Biophys. Acta* 253: 215–223.

Cutler RL, Davies AM, Creighton S, Warshel A, Moore GR, Smith M, Mauk AG. (1989) Role of arginine-38 in regulation of the cytochrome *c* oxidation–reduction equilibrium. *Biochemistry* 28: 3188–3197.

Daff S, Sharp RE, Short DM, Bell C, White P, Manson FDC, Reid GA, Chapman SK. (1995) The interaction of cytochrome *c* with flavocytochrome b_2. *Biochem. J.* (in press).

Davies AM, Guillemette JG, Smith M, Greenwood C, Thurgood AGP, Mauk AG, Moore GR. (1993) Redesign of the interior hydrophilic region of mitochondrial cytochrome *c* by site-directed mutagenesis. *Biochemistry* 32: 5431–5435.

Deisenhofer J, Epp O, Miki K, Huber R, Michel H. (1984) X-ray structure analysis of a membrane protein complex: electron density map at 3Å resolution and a model of the chromophores of the photosynthetic reaction center from *Rhodopseudomonas viridis. J. Mol. Biol.* 180: 385–398.

Durham B, Millet FS. (1994) Iron: heme proteins and electron transfer. In: *Encyclopedia of Inorganic Chemistry*, Vol. 4 (ed. RA Scott). John Wiley & Sons, Chichester, pp. 1642–1661.

Durham B, Pan LP, Long J, Millet FS. (1989) Photoinduced electron-transfer kinetics of singly labeled ruthenium bis(pyridine)dicarboxybipyridine cytochrome *c* derivatives. *Biochemistry* 28: 8659–8665.

Eley CGS, Moore GR. (1983) ^{1}H-n.m.r. investigation of the interaction between cytochrome *c* and cytochrome b_5. *Biochem. J.* 215: 11–21.

Elias H, Chou MH, Winkler JR. (1988) Electron-transfer kinetics of Zn-substituted cytochrome *c* and its $Ru(NH_3)_5$(histidine-33) derivative. *J. Am. Chem. Soc.* 110: 429–434.

Farid RS, Robertson DE, Moser CC, Pilloud D, DeGrado WF, Dutton PL. (1994) Design and synthesis of simplified energy-converting proteins. *Biochem. Soc. Trans.* 22: 687–691.

Fatemi SJA, Fadir FHA, Williamson DJ, More GR. (1991) The uptake, storage, and mobilization of iron and aluminium in biology. *Adv. Inorg. Chem.* 36: 409–448.

Ferrer JC, Guillemette GJ, Bogumil R, Inglis SC, Smith M, Mauk AG. (1993) Identification of Lys79 as a iron ligand in one form of alkaline yeast iso-1-ferricytochrome *c. J. Am. Chem. Soc.* 115: 7507–7508.

Ford GC, Harrison PM, Rice DW, Smith JMA, Treffry A, White JL, Yariv J. (1984) Ferritin: design and formation of an iron-storage molecule. *Philos. Trans. R. Soc. Lond.* 304: 551–565.

Frolow F, Kalb AJ, Yariv J. (1994) Structure of a unique twofold symmetric haem-binding site. *Struct. Biol.* 1: 453–460.

Fülöp V, Little R, Thompson A, Greenwood C, Hajdu J. (1993) Crystallization and preliminary analysis of the dihaem cytochrome *c* peroxidase from *Pseudomonas aeruginosa. J. Mol. Biol.* 232: 1208–1210.

Fülöp V, Moir JWB, Ferguson SJ, Hajdu J. (1995a) The anatomy of a bifunctional enzyme: structural basis for reduction of oxygen to water and synthesis of nitric oxide by cytochrome cd_1. *Cell* 81: 369–377.

Fülöp V, Rideout CJ, Greenwood C, Hajdu J. (1995) Crystal structure of the dihaem cytochrome *c* peroxidase from *Pseudomonas aeruginosa. Structure* 3: 1225–1233.

Gennis R, Ferguson-Miller S. (1995) Structure of cytochrome *c* oxidase, energy generator of aerobic life. *Science* 269: 1063–1064.

Gilmour R, Goodhew CF, Pettigrew GW, Prazeres S, Moura I, Moura JJG. (1993) Spectroscopic characterization of cytochrome *c* peroxidase from *Paracoccus denitrificans. Biochem. J.* 294: 745–752.

Gilmour R, Goodhew CF, Pettigrew GW, Prazeres S, Mours JJG, Moura I. (1994) The kinetics of the oxidation of cytochrome *c* by *Paracoccus* cytochrome *c* peroxidase. *Biochem. J.* 300: 907–914.

Gouterman M. (1978) Optical spectra and electronic structure of porphyrins and related rings. In: *The Porphyrins*, Vol. IIIA (ed. Dolphin D). Academic Press, New York, pp. 1–165.

Greenwood C, Foote N, Gadsby PMA, Thomson AJ. (1988) A di-haem cytochrome *c* peroxidase (*P. aeruginosa*). *Chem. Scripta* 28A: 79–84.

Gunner MR, Dutton PL. (1989) Temperature and $-\Delta G^o$ dependence of the electron transfer from BPh^{-} to Qa$_A$ in reaction centre protein from *R. sphaeroides* with different quinones as Qa$_A$. *J. Am. Chem. Soc.* **111**: 3400–3412.

Gunner MR, Honig B. (1991) Electrostatic control of midpoint potentials in the cytochrome subunit of the *Rhodopseudomonas viridis* reaction center. *Proc. Natl Acad. Sci. USA* **88**: 9151–9155.

Hildebrand DP, Burk DL, Maurus R, Ferrer JC, Brayer GD, Mauk AG. (1995) The proximal ligand variant His93Tyr of horse heart myoglobin. *Biochemistry* **34**: 1997–2005.

Ho PS, Sutoris C, Liang N, Margoliash E, Hoffman BM. (1985) Species specificity of long-range electron transfer within the complex between zinc-substituted cytochrome *c* peroxidase and cytochrome *c*. *J. Am. Chem. Soc.* **107**: 1070–1071.

Iwata S, Ostermeier C, Ludwig B, Michel H. (1995) Structure at 2.8Å resolution of cytochrome *c* oxidase from *Paracoccus denitrificans*. *Nature* **376**: 660–669.

Jeng M-F, Englander SW, Pardue K, Rogalskyj JS, McLendon G. (1994) Structural dynamics in an electron-transfer complex. *Struct. Biol.* **1**: 234–238.

Kadir FHA, Al-Massad FK, Moore GR. (1992) Haem binding to horse spleen ferritin and its effect on the rate of iron release. *Biochem. J.* **282**: 867–870.

Kassner RJ. (1973) A theoretical model for the effects of local nonpolar heme environments on the redox potentials in cytochromes. *J. Am. Chem. Soc.* **95**: 2674–2677.

Langen R, Brayer GD, Berghuis AM, McLendon G, Sherman F, Warshel A. (1992) Effect of the Asn52Ile mutation on the redox potential of yeast cytochrome *c*. *J. Mol. Biol.* **224**: 589–600.

Le Brun NE, Andrews SC, Guest JR, Harrison PM, Moore GR, Thomson AJ. (1995) Identification of the ferroxidase centre of *E.coli* bacterioferritin *Biochem. J.* **312**: 385–392.

Liang N, Pielak GJ, Mauk AG, Smith M, Hoffman BM. (1987) Yeast cytochrome *c* with phenylalanine or tyrosine at position 87 transfers electrons to (zinc cytochrome *c* peroxidase)$^+$ at a rate ten thousand times that of the serine-87 or glycine-87 variants. *Proc. Natl Acad. Sci. USA* **84**: 1249–1252.

Liu R-Q, Hahm S, Miller M, Durham B, Millet F. (1995) Photooxidation of Trp-191 in cytochrome *c* peroxidase by ruthenium-cytochrome *c* derivatives. *Biochemistry* **34**: 973–983.

Lloyd E, Hildebrand DP, Tu KM, Mauk AG. (1995) Conversion of myoglobin into a reversible electron transfer protein that maintains bishistidine axial ligation. *J. Am. Chem. Soc.* **117**: 6434–6438.

Loew GH. (1983) Theoretical investigations of iron porphyrins. In: *Iron Porphyrins,* Vol. 1 (eds ABP Lever, HB Gray). Addison-Wesley, Reading, MA, pp. 1–87.

Louie GV, Brayer GD. (1990) High-resolution refinement of yeast iso-1 cytochrome *c* and comparisons with other eukaryotic cytochromes *c*. *J. Mol. Biol.* **214**: 527–555.

Marcus RA, Sutin N. (1985) Electron transfers in chemistry and biology. *Biochim. Biophys. Acta* **811**: 265–322.

Martinez SE, Huang D, Szczepaniak A, Cramer WA, Smith JL. (1994) Crystal structure of chloroplast cytochrome *f* reveals a novel cytochrome fold and unexpected heme ligation. *Structure* **2**: 95–105.

Mauk AG, Mauk MR, Moore GR, Northrup SH. (1995) Experimental and theoretical analysis of the interaction between cytochrome *c* and cytochrome *b₅*. *J. Bioenerg. Biomembr.* **27**: 311–330.

McLendon G, Hake R. (1992) Interprotein electron transfer. *Chem. Rev.* **92**: 481–490.

McLendon G, Miller JR. (1985) The dependence of biological electron transfer rates on exothermicity: the cytochrome *c*/cytochrome *b₅* couple. *J. Am. Chem. Soc.* **107**: 7811–7816.

McLendon G, Pardu K, Bak P. (1987) Electron transfer in the cytochrome *c*/cytochrome *b₂* complex: evidence for conformational gating. *J. Am. Chem. Soc.* **109**: 7540–7541.

Meyer TE, Kamen MD. (1982) New perspectives on *c*-type cytochromes. *Adv. Prot. Chem.* **35**: 105–212.

Moore GR, Pettigrew GW. (1990) *Cytochromes* c: *Evolutionary, Structural and Physicochemical Aspects.* Springer-Verlag, Berlin.

Moore GR, Eley CGS, Williams G. (1983) Electron transfer reactions of class I cytochromes *c*. *Adv Inorg. Bioinorg. Mech.* **3**: 1–96.

Moore GR, Pettigrew GW, Rogers NK. (1986) Factors influencing redox potentials of electron transfer proteins. *Proc. Natl Acad. Sci. USA* **83**: 4998–4999.

Moore GR, Kadir FHA, Al-Massad FK, Le Brun NE, Thomson AJ, Greenwood C, Keen, JN, Findlay JBC. (1994) On the structural heterogeneity of *Pseudomonas aeruginosa* bacterioferritin. *Biochem. J.* 304: 493–497.

Moore GR, Cox MC, Crowe D, Osborne MJ, Mauk AG, Wilson MT. (1995) NMR studies of paramagnetic systems to characterise small molecule:protein and protein:protein interactions. In: *Nuclear Magnetic Resonance of Paramagnetic Macromolecules* (ed. GN La Mar). NATO ASI Series C, Vol. 457; pp. 95–122.

Moser CC, Keske JM, Warncke K, Farid RS, Dutton PL. (1992) Nature of biological electron transfer. *Nature* 355: 796–802.

Moser CC, Page CC, Farid RS, Dutton PL. (1995) Biological electron transfer. *J. Bioenerg. Biomembr.* 27: 263–274.

Mus-Veteau I, Dolla A, Guerlesquin F, Payan F, Czjzek M, Haser R, Bianco P, Haladjian J, Rapp-Giles B, Wall JD, Voordouw G, Bruschi M. (1992) Site-directed mutagenesis of tetraheme cytochrome c_3. *J. Biol. Chem.* 267: 16851–16858.

Nitschke W, Rutherford AW. (1994) The tetrahaem cytochromes associated with photosynthetic reaction centres: a model system for intraprotein redox centre interactions. *Biochem. Soc. Trans.* 22: 692–697.

Nordling M, Young S, Karlsson BG, Lundberg LG. (1990) The structural gene for cytochrome *c*-551 from *Pseudomonas aeruginosa*. *FEBS Lett.* 259: 230–232.

Northrup SH, Thomasson KA, Miller CM, Barker PD, Eltis LD, Guillemette JG, Inglis SC, Mauk AG. (1993) Effects of charged amino acid mutations on the bimolecular kinetics of reduction of yeast iso-1-ferricytochrome *c* by bovine ferrocytochrome b_5. *Biochemistry* 32: 6613–6623.

Palmer G. (1983) Electron paramagnetic resonance of hemoproteins. In: *Iron Porphyrins*, part two (eds ABP Lever, HB Gray) Addison-Wesley, London/Amsterdam, pp. 43–88.

Pelletier H, Kraut J. (1992) Crystal structure of a complex between electron transfer partners, cytochrome *c* peroxidase and cytochrome *c*. *Science* 258: 1748–1755.

Pettigrew GW, Moore GR. (1987) *Cytochromes* c: *Biological Aspects*. Springer-Verlag.

Pollock WBR, Loutfi M, Bruschi M, Rapp-Giles BJ, Wall JD, Voordouw G. (1991) Cloning, sequencing and expression of the gene encoding the high-molecular-weight cytochrome *c* from *Desulfovibrio vulgaris* Hildenborough. *J. Bacteriol* 173: 220–228.

Poulos TL, Kraut J. (1980) A hypothetical model of the cytochrome *c* peroxidase: cytochrome *c* electron transfer complex. *J. Biol. Chem.* 255: 10322–10330.

Prècigoux G, Yariv J, Gallois B, Dautant A, Courseille C, Langlois D'Estaintot B. (1994) A crystallographic study of haem binding to ferritin. *Acta Crystallogr.* D50: 739–743.

Raphael AL, Gray HB. (1989) Axial ligand replacement in horse heart cytochrome *c* by semisynthesis. *Proteins: Struct. Function Genet.* 6: 338–340.

Raphael AL, Gray HB. (1991) Semisynthesis of axial-ligand (Position 80) mutants of cytochrome *c*. *J. Am. Chem. Soc.* 113: 1038–1040.

Rideout CJ, James R, Greenwood C. (1995) Nucleotide sequence encoding the di-haem cytochrome *c*-551 peroxidase from *Pseudomonas aeruginosa*. *FEBS Lett.* 365: 152–154.

Rönnberg M, Osterland K, Ellfolk N. (1980) Resonance Raman spectra of *Pseudomonas* cytochrome *c* peroxidase. *Biochim. Biophys. Acta* 626: 23–30.

Rossi M, Pollock WBR, Reij MW, Keon RG, Fu R, Voordouw G. (1993) The *hmc* operon of *Desulfovibrio vulgaris* subsp. *vulgaris* Hildenborough encodes a potential transmembrane redox protein complex. *J. Bacteriol.* 175: 4699–4711.

Salemme FR. (1976) An hypothetical structure for an intermolecular electron transfer complex of cytochrome *c* and cytochrome b_5. *J. Mol. Biol.* 102: 563–568.

Scott JR, Willie A, McLean M, Stayton PS, Sligar SG, Durham B, Millet F. (1993) Intramolecular electron transfer in cytochrome b_5 labeled with ruthenium(II) polypyridine complexes. *J. Am. Chem. Soc.* 115: 6820–6824.

Sharp RE, White P, Chapman SK, Reid GA. (1994) Role of the interdomain hinge of flavocytochrome b_2 in intra- and inter-protein electron transfer. *Biochemistry* 33: 5115–5120.

Silvestrini MC, Tordi MG, Musci G, Brunori M. (1990) The reaction of *Pseudomonas* nitrite reductase and nitrite. *J. Biol. Chem.* 265: 11783–11787.

Stiefel EI, Watt GD. (1979) *Azotobacter* cytochrome *b*-557.5 is a bacterioferritin. *Nature* 279: 81–83.

Takano T, Dickerson RE. (1981) Conformation change of cytochrome *c*: ferricytochrome *c* refinement at 1.8Å and comparison with the ferrocytochrome structure. *J. Mol. Biol.* **153:** 95–115.

Taylor CPS. (1977) The EPR of low spin heme complexes. *Biochim. Biophys. Acta* **491:** 137–149.

Tegoni M, White SC, Roussel A, Mathews FS, Cambillau C. (1993) A hypothetical complex between crystalline flavocytochrome b_2 and cytochrome *c. Proteins: Struct. Function Genet.* **16:** 408–422.

Tsukihara T, Aoyama H, Yamashita E, Tomizaka T, Yamaguchi H, Shinzawa-Itoh K, Nakashima R, Yaono R, Yoshikawa S. (1995) Structures of metal sites of oxidized bovine heart cytochrome *c* oxidase at 2.8Å. *Science* **269:** 1069–1074.

Ubbink M, Campos AP, Teixeira M, Hunt NI, Hill HAO, Canters GW. (1994) Characterization of mutant Met100Lys of cytochrome *c*-550 from *Thiobacillus versutus* with lysine–histidine heme ligation. *Biochemistry* **33:** 10051–10059.

Vijgenboom E, Busch JE, Canters GW. (1995) Physiological role and expression of the blue copper protein azurin in *Pseudomonas aeruginosa. J. Inorg. Chem.* **59:** N65, 720.

Watt GD, Frankel RB, Papaefthymiou GC, Spartalian K, Stiefel EI. (1986) Redox properties and Mössbauer spectroscopy of *Azotobacter vinelandii* bacterioferritin. *Biochemistry* **25:** 4330–4336.

Wikström M. (1989) Identification of the electron transfers in cytochrome oxidase that are coupled to proton-pumping. *Nature* **338:** 776–778.

Willie A, Stayton PS, Sligar SG, Durham B, Millet F. (1992) Genetic engineering of redox donor sites: measurement of intracomplex electron transfer between ruthenium-65-cytochrome b_5 and cytochrome *c. Biochemistry* **31:** 7237–7242.

Willie A, McLean M, Liu R-Q, Hilgen-Willis S, Saunders AJ, Pielak GJ, Sligar SG, Durham B, Millet F. (1993) Intracomplex electron transfer between ruthenium-65-cytochrome b_5 and position 82 variants of cytochrome *c. Biochemistry* **32:** 7519–7525.

Winkler JR, Gray HB. (1992) Electron transfer in ruthenium-modified proteins. *Chem. Rev.* **92:** 369–379.

Winkler JR, Nocera DG, Yocom KM, Bordignon E, Gray HB. (1982) Electron-transfer kinetics of pentaammineruthenium(III)(His-33) ferricytochrome *c. J. Am. Chem. Soc.* **104:** 5798–5800.

Wüthrich K. (1969) High-resolution proton nuclear magnetic resonance spectroscopy of cytochrome *c. Proc. Natl Acad. Sci. USA* **63:** 1071–1078.

Wuttke DS, Bjerrum MJ, Winkler JR, Gray HB. (1992) Electron-tunneling pathways in cytochrome *c. Science* **256:** 1007–1009.

Wuttke DS, Gray HB, Fisher SL, Imperali B. (1993) Semisynthesis of bipyridyl-alanine cytochrome *c. J. Am. Chem. Soc.* **115:** 8455–8456.

Electron transfer complexes coupled to ion translocation

Peter R. Rich

9.1 Introduction

The essential process of biological ATP synthesis is highly endergonic. Some ATP synthesis occurs by classical enzymological mechanisms in which an energy-releasing chemical transformation is strictly coupled to ADP phosphorylation within a single enzyme complex. Perhaps best known of these 'substrate level phosphorylations' are those that are part of the glycolytic pathway. In most cases, however, the majority of ATP synthesis is provided by respiratory and photosynthetic pathways. In all organisms these are composed of a series of multiprotein electron transfer enzymes embedded in an ion-impermeable lipid membrane. They catalyse a sequence of electron transfer reactions between an electron source and an electron sink, driven by exergonic substrate transformations in the case of respiration or by light absorption in the case of photosynthesis. The electron transfer steps themselves are exergonic, but the energy is not lost as heat but is instead coupled to the generation of an ionmotive force across the membrane in which they are located. The ionmotive force is used subsequently to drive the synthesis of ATP via enzymes plugged into the same membrane system or to power a wide variety of other energy-requiring reactions from metabolite transfer to mechanical movements.

The purpose of this chapter is to review the current knowledge of structure of some of the enzymes that catalyse these ionmotive reactions and to explore the types of mechanisms by which the coupling of electron transfer to ion translocation across a membrane can be achieved.

9.2 Principles of coupling

9.2.1 *The chemiosmotic theory*

The central features and underlying principles of the chemiosmotic mechanism by which energy-generating electron transfer processes can be coupled to those

Protein Electron Transfer, Edited by D.S. Bendall
© 1996 BIOS Scientific Publishers Ltd, Oxford

which require energy were described by Mitchell in the 1960s (Mitchell, 1961, 1966, 1968). Mitchell realized that vectorial ion transport across natural membranes could occur only if the metabolic reactions were themselves arranged vectorially in the membrane. In this way, the classical scalar free energy change of the reaction could produce a real force field across the membrane with energy conserved in the form of the electrochemical potential gradients of the chemical groups along their pathways of translocation. The term chemiosmotic was introduced to describe this type of process.

The chemiosmotic hypothesis was applied particularly to the generation of an electrochemical gradient of protons (the protonmotive force) during respiratory and photosynthetic electron flow, and to its subsequent use in ATP generation. It had four basic proposals in its original form (Mitchell, 1961), namely:

(i)   the electron transfer chain is vectorially arranged across a membrane in a manner such that electron transfer is linked to proton translocation across the membrane;
(ii) the ATP synthase is arranged anisotropically across the same membrane in a manner such that the proton gradient can drive ATP synthesis;
(iii) the membrane is impermeable to protons and is osmotically closed;
(iv) there are proton-linked porter systems for osmotic stabilization and metabolite transport.

9.2.2 *The vectorial redox loop*

Importantly, Mitchell also developed a realistic and testable model of the chemistry that would be required in a series of membrane-bound electron transfer components so that electron transfer would necessarily result in coupled proton translocation across the membrane. This elegant and simple proposal was termed the vectorial redox loop (*Figure 9.1a*). It was postulated that the components of the redox chain alternated between electron carriers and hydrogen atom carriers and that these were spatially arranged in a definite way in the membrane. The hydrogen carriers have redox-linked pK values such that reduction causes protonation and oxidation causes deprotonation. It was postulated that the hydrogen atom carriers are effectively mobile in the membrane, and diffuse from a site of reduction (and proton uptake) on one side of the membrane to a site of oxidation (and proton release) close to the opposite membrane surface. Since a proton and an electron are transferred across the membrane together, the process is electroneutral. The loop is completed, however, by the transfer of the electron back across the membrane via appropriately arranged electron carriers. In this case, an uncompensated charge is moved across the membrane, and so generates an electric field across the insulating membrane, that is the reaction is electrogenic.

The net result is the translocation of a proton across the membrane, and the conversion of redox energy into an electrochemical potential gradient of protons across the membrane, composed of a chemical potential difference because of the proton concentration difference (the ΔpH) and an electric field component ($\Delta\psi$). It is this protonmotive force which is used subsequently to drive energy-requiring processes such as ATP synthesis.

(a)

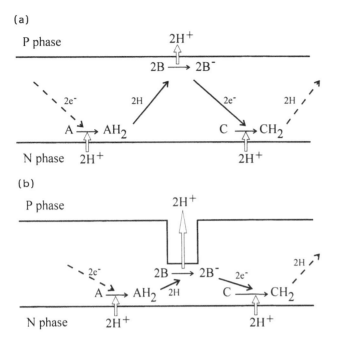

(b)

Figure 9.1. The vectorial redox loop. (a) Redox loop with mobile H carrier; (b) redox loop with proton well.

9.2.3 *Protein structure and the proton well*

Understanding of the mechanism in terms of the redox loop can be complicated by the role of the protein structures in the generation of the electrical and chemical potential components of the protonmotive force. Many of the key chemical and redox transformations may take place in a relatively small part of the protein structure, with channels in the protein structure allowing access of ions to this core. If the ion movement is uncompensated by charge from a counterion or from local protein structure, then the ion movement itself can be electrogenic, and the channel may be termed an 'ion well' (*Figure 9.1b*). The notion of a proton well is inherent in the early chemiosmotic proposals and its properties have been discussed further in Mitchell (1968, 1969), particularly its ability to interconvert a proton concentration gradient and an electric field. In the example in *Figure 9.2b*, the highly localized redox loop results in an electron transfer arm which does not generate an electric field. Instead, protons are released into proton channels leading to the aqueous P phase. Movement of protons along these channels is driven by the redox reaction, which essentially creates a pH gradient along the channel. The protons diffuse along the pH gradient and, in so doing, generate an electric field. Hence, in this case, it is proton movement, rather than electron transfer, which is electrogenic.

In practice, it is likely that the behaviour in many systems lies between the two extreme possibilities shown in *Figure 9.1*. In any case, the precise properties of charged ions moving through such channels may be influenced by possible dynamic aspects. It is quite likely, for example, that the dielectric properties of

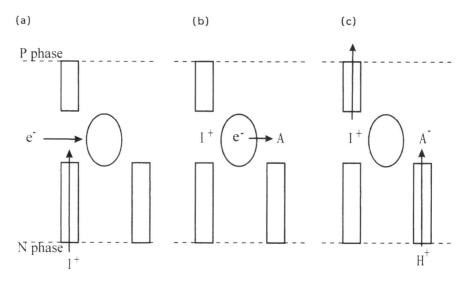

Figure 9.2. A general mechanism for coupling of ion translocation to chemical transformation in an ionmotive enzyme. The basic elements involve (a) the neutralization of charge in a region of low dielectric strength by uptake of a counterion, I^+, from the N phase; (b) reduction of an acceptor, A, at a locus separated from I^+; (c) protonation of A^- and repulsion of I^+ into the P phase. Further details and variants are given in the text.

the channel may alter between initiation of turnover and attainment of a steady-state as fixed and movable charges redistribute. Further questions concerning the net movement of charged species through an ion well have been discussed elsewhere (Rich, 1991).

9.2.4 *The importance of local electroneutrality within the reaction core*

I have emphasized recently a further important consideration when exploring the detailed mechanism of coupling of ion and electron transfer within the reaction cycle of the protein. This concerns the need for local charge neutralization of those catalytic intermediates which are produced in a region of low dielectric strength, a condition likely to be the case in a number of important ionmotive enzymes. In the absence of appropriate solvent or protein charge rearrangements, electron transfer into such systems can provide a strong driving force for uptake of a charge-compensating proton or, possibly, another cation.

The notion of a need for local electroneutrality within regions of proteins of low dielectric strength arose from our empirical measurements of protonic changes associated with electronic or ligand state changes in cytochrome oxidase and cytochrome *bo* (Mitchell and Rich, 1994; Mitchell *et al.*, 1992; Moody and Rich, 1994). However, in addition to the oxidases and other ionmotive systems (see below), the consideration can also be relevant to soluble enzymes with occluded reaction cores, and recently we have discussed this notion in relation to the ligand-associated proton chemistry of horseradish peroxidase (Meunier *et al.*, 1995, 1996).

From a physical point of view, the requirement for charge compensation originates in free energy considerations and the factors that control redox potentials. Several factors affect the redox potentials of groups in proteins (see Chapter 8 and Churg and Warshel, 1986; Gunner and Honig, 1991; Moore and Pettigrew, 1990) and, in general, no single dominating factor can be identified. Moore *et al.* (Moore and Pettigrew, 1990; Moore *et al.*, 1986) have divided these factors into energetic changes associated with electrostatic interactions, bonding interactions and protein conformational changes. The last factor is often negligible, with large contributions confined to a few specific proteins, and will not be considered further here.

The electrostatic interactions can involve several components: the ions in solution (non-specific screening and any specific ion–protein complex formation at the protein surface); the surface charges on the protein; the solvent itself; and the charges in the protein around the redox centre. Moore (Chapter 8) has discussed these components in greater detail. For proteins in which the catalytic centre is buried in a region of relatively low dielectric strength, as is likely to be the case for many of the enzymes involved in ion translocation, it is this latter factor which can become particularly important.

In terms of the Born model, in which the ion is treated as a rigid sphere and the medium as a structureless continuum, the potential, ψ, at the surface of a spherical charge, q, of radius r_i, in a medium of dielectric strength ε is given by:

$$\psi \propto q/\varepsilon r_i = \text{Constant}.q/\varepsilon r_i \tag{9.1}$$

and the work of charging the sphere, W, is therefore

$$W = \text{Constant}.q^2/2\varepsilon r_i \tag{9.2}$$

Hence, the energetic cost of the movement of a single electrical charge is dependent on the effective radius of the charge, r_i, and on the dielectric constant of the surroundings. The energy increases as the charge becomes more localized and as the dielectric constant of the surrounding medium decreases.

A quantitative estimate may be made by substitution of a value for the constant in the above equation, which in S.I. units gives

$$W = (1/4\pi\varepsilon_0).q^2/\varepsilon r_i \tag{9.3}$$

where ε_0 = permittivity of free space (8.8542×10^{-12} C^2 N^{-1} m^{-2}) so that force is in Newtons. Hence, one can arrive at the type of equations for the smeared charge model described in Moore and Pettigrew (1990), where the difference in electrostatic energy between a reduced (with one net charge) and oxidized (with zero net charge) species can be expressed by the equation:

$$\Delta G = N_A e^2.(1/4\pi\varepsilon_0).q^2/\varepsilon r_i \quad \text{J mol}^{-1} \tag{9.4}$$

where N_A is Avogadro's constant and e = electronic charge of a single electron (1.6×10^{-19} C).

For a redox centre of radius 3 Å, initially with a zero net charge, and embedded in a protein of dielectric strength 3.5 [an assumed average value for the interior of proteins (Moore and Pettigrew, 1990)], this indicates that a single electron transfer into it would contribute more than 130 kJ mol^{-1} (31 kcal mol^{-1} or >1350 mV) in electrostatic energy! In practice, the cost would be less in a microscopic situation where the macroscopic concept of dielectric constant is not properly applicable. Nevertheless, it illustrates the point that the introduction of a charged species into a protein of low dielectric strength would, in the absence of other factors, be energetically costly, and will instead be accompanied by any of a number of energy-minimizing processes, whose balance is dependent on the protein structure.

(i) If the metal centre is exposed to the aqueous medium, which has a high dielectric strength, a non-specific ion redistribution will occur in the medium in the vicinity of the charge.

(ii) If the metal centre is buried inside the protein, then local charges and hydrogen bonds within the protein can redistribute so as to minimize the charge imbalance. The greater the dielectric strength of the protein structure surrounding the charge, the more effective the charge compensation.

These two factors can be related to the inner and outer sphere reorganization energies of classical Marcus theory (see Chapter 1). Both of the above factors can be important in systems which exhibit pH-independent midpoint potentials, for example in the redox behaviour of cytochrome c under certain conditions of pH and ionic composition.

(iii) If the protein structure surrounding the introduced charge has a low dielectric strength, then charge redistribution is limited and the energetic cost of introduction of the charge on its own becomes high. Delocalization of the charge itself can help to decrease the energy cost, and such delocalization is quite feasible within larger cofactor structures such as the porphyrin ring macrocycle or in a multimetal redox centre such as an iron–sulphur cluster. Indeed, it seems possible that some biological redox centres may have become large, multicentre entities simply in order to delocalize charge for this reason. However, if there is an appropriate site in the protein structure, a specific counterion may bind in response to the introduction of the charge, that is the affinity of the site for the counter ion is increased in response to the charge. In the special case of a protonation, a redox-linked pK shift is the result. Furthermore, a covalent bond is formed, and this could provide further energy minimization because of the electronic rearrangements and interactions that comprise the bond formation.

Some of the best known examples of the latter behaviour are the redox-linked protonation phenomena that occur extensively in chemistry (Clark, 1960) and in biology (Dutton and Wilson, 1974) and lead to the characteristic pH dependencies of redox potentials, as in the aqueous solution behaviour of p-benzoquinones, where reduction with two electrons results in extensive orbital rearrangement in conjunction with the formation of two covalent oxygen–hydrogen bonds. The requirement for an associated protonation means that the protonation reaction can control the electron transfer process itself, and is therefore a factor additional to those considered in electron transfer theory. Such control might be particularly important if the proton has to move through the protein structure to arrive at its binding site. More generally, in protein–ligand reactions, there are a number of good examples of anion-linked protonation phenomena which are reflected in the pH dependency of the anion dissociation constants (Ellis and Dunford, 1968; Erman, 1974; Yamazaki *et al.*, 1978). Furthermore, some biological systems have a counterion selectivity for species other than protons, for example sodium ions, and in such cases the energy minimization is presumably provided mostly by the electrostatic, rather than the bond-forming, factors.

In the discussions of coupling mechanisms below, I will describe how such local charge neutralization is likely to form a central feature of the coupling mechanism of several of the more complex ionmotive enzymes. In such mechanisms (*Figure 9.2*) electron transfer into a region of locally low dielectric strength is electrically counterbalanced by the uptake of a proton (or other cation). Subsequent electron transfer to an acceptor, A (in the specific examples below, the acceptor is ubiquinone or oxygen), at later stages in the reaction cycle provides a strong driving force for protonation of the reduced electron acceptor. However, the charge-counterbalancing cations are spatially separated from the chemically transformed acceptor species (the system may be said to be 'electrostatically locked' in a charge-balanced state incapable of net production of protonated, reduced acceptor) so that protons must be taken up. As they are taken up from the N phase, these 'substrate protons' electrostatically repulse the counterbalancing cations into the aqueous P phase. In the general example shown in *Figure 9.2*, the cation, I^+, and protons are shown to remain separate as they enter by separate channels, which seems quite likely if I^+ is not itself a proton. However, when I^+ is a proton, mechanisms may also be considered in which I^+ and protons enter through a single channel, and with separation of I^+ from substrate by ligand rearrangements in the catalytic site (see Section 9.3.4).

Critical to the mechanism outlined in *Figure 9.2* is the nature of the 'gating' process by which the charge-counterbalancing cations are taken from the N phase in the initial reduction step, but are repulsed into the P phase during the final chemical transformation. Uptake from the N phase during reduction is likely to be ensured by a large permanent activation barrier for passage of protons through the P phase channel. Repulsion at step c through the N phase channel may be prevented by a number of possible mechanisms. Electrostatic interaction with the incoming substrate protons may in itself be enough. Other possibilities include the physical movement of cations away from the N phase channel or physical channel closing by ligand rearrangements, or perhaps a strict dependency in step a of the movement of a proton through the N phase channel on the passage of an electron from donor to redox centre. Some further details are given in the specific examples below.

9.3   Structure and mechanism of complexes

In order to illustrate the roles of the factors discussed above, several enzymes have been chosen for detailed discussion. I have begun with a short note on some of those enzymes which simply catalyse a quinone redox reaction on one side of the membrane or catalyse the movement of an electron across the membrane to a terminal acceptor. Appropriate arrangement of such components in series provides a simple and widespread means of coupling electron transfer to the generation of a protonmotive force across the membrane, conforming to the vectorial redox loop and often with some degree of proton well contributions. I have gone on to review the coupling mechanism of the *bc* complexes where several of these features are present in a single enzyme to yield a particularly elegant and still classical coupling mechanism. Finally, I discuss the protonmotive oxidases and the ion-coupled NADH dehydrogenases, enzymes in which the principle of electroneutrality by counterion uptake may provide a basis for understanding the more intricate coupling mechanisms that may be operative.

9.3.1 *Enzymes catalysing components of larger chemiosmotic systems*

Many of the enzymes that form part of photosynthetic and respiratory chains catalyse a reaction which acts as one component of a larger vectorial redox loop. In fact, such systems are common and account for a large part of the protonmotive processes that are known in biology. For example, a wide range of dehydrogenases act simply to ensure that ubiquinone (or menaquinone) is reduced at a location such that its redox-linked protons are taken up from one specific side of the membrane. Examples include succinate dehydrogenase and the simple bacterial NADH-ubiquinone oxidoreductase, NQR-2. The reduced quinone is then able to dissociate from the protein and physically diffuse to oxidation sites on other proteins such that the redox-linked protons are released at the opposite membrane surface.

In the case of photosystems (PS) II and I and the bacterial reaction centre, the photoactivated electron transfer chain is spatially arranged so that initial charge separation results in a downhill series of electron transfers across the membrane dielectric and part of the energy is conserved by charging of the membrane potential. In addition to this electrogenic function, other processes can result in proton changes in the aqueous media. For example, in oxygenic photosynthesis the photo-oxidized chlorophyll, $P680^+$, of PSII is reduced with electrons derived ultimately from water, resulting in the formation of oxygen and the release of protons into the P phase thylakoid lumen. In PSII (and also in many bacterial reaction centres), the ultimate acceptor of the electron which has been transferred across the membrane from P680 is a quinone. Full reduction of the quinone will result in proton uptake, in this case from the aqueous N phase in which it is in contact. In the case of the bacterial reaction centre, whose structure is known to atomic resolution (Deisenhofer and Michel, 1989; Feher *et al.*, 1989), the proton uptake occurs through part of the protein structure (Paddock *et al.*, 1990; Takahashi and Wraight, 1992) and is associated with generation of a small electric field (Shinkarev *et al.*, 1993), indicative of a proton well effect. The other thylakoid reaction centre, PSI, reduces NADP which will also consume protons from the N phase as it is reduced.

A further example of enzymes providing a relatively simple means of protonmotive force generation are several oxidases and reductases that act as the terminal steps of many respiratory chains. For example, cytochrome *bd* is one of two major bacterial quinol oxidases. It oxidizes ubiquinol close to the P phase side of the membrane into which phase the protons are released. The electrons are transferred to oxygen at a reaction site which is in a position such that the protons required for water production are derived from the opposite N phase side of the membrane. Hence, both a pH gradient due to proton changes, and an electric field, due to transfer of charge across the membrane, are formed. A related system is nitrate reductase, which again carries charge across the membrane from ubiquinol to nitrate, in this case depositing protons from quinol oxidation into the P phase and generating an electric field due to the transmembrane electron transfer to nitrate.

9.3.2 The bc complexes

The mitochondrial and bacterial bc_1 *complex.*
(a) *Core structure.* The structure of a *bc* complex at atomic resolution has not yet appeared. Although the mitochondrial enzyme is composed of 11 subunits (Schägger *et al.*, 1986), only three of these (iron–sulphur protein, cytochrome c_1 and cytochrome *b*) have redox centres. The bacterial forms of the enzyme are much simpler, being composed of homologues of these three redox-active subunits, together with a fourth non-redox polypeptide (Robertson *et al.*, 1993). As far as can be ascertained, they display all of the electron and proton transfer reactions of the larger mitochondrial counterpart, and so point to a more minimal catalytic core structure of this superfamily. The functions of the additional mitochondrial subunits are unknown, except for a processing protease activity associated with one or more of the large 'core subunits' (Gencic *et al.*, 1991; Glaser *et al.*, 1994; Hosokawa *et al.*, 1990).

Despite a lack of direct structural information, the very large database of sequence information has enabled prediction of convincing likely structures of the redox-carrying polypeptides. Many reviews have appeared recently on these possible structures (Colson, 1993; Crofts *et al.*, 1992; Degli Esposti *et al.*, 1993; Frank and Trebst, 1995; Gennis *et al.*, 1993; Knaff, 1993; Link *et al.*, 1993; Malkin, 1992; Trumpower and Gennis, 1994), and there is no need to reiterate the details here. The iron–sulphur subunit is likely to have one or two membrane-spanning α-helices at its N terminus, followed by a more globular region containing the iron–sulphur centre. The iron–sulphur centre has a high midpoint potential ($E_{m7} \sim 300$ mV) and is a [2Fe–2S] structure which is ligated by two histidines and two cysteines (Gurbiel *et al.*, 1991; Kuila *et al.*, 1993). The two histidines have been assigned definitively in the sequence and, of the four conserved cysteines, the two most likely to act as ligands have been identified (Davidson *et al.*, 1992). Cytochrome c_1 is also predicted to have a membrane-spanning α-helix, although in this case near the C terminus. The hydrophilic domain preceding this anchor has the CxxCH motif for covalent binding of haem *c*. Magnetic circular dichroism (MCD) spectroscopy, and the high midpoint potential ($E_{m7} \sim 300$ mV), suggest histidine–methionine ligation of the haem iron (Gray *et al.*, 1992; Simpkin *et al.*, 1989). In the case of the corresponding cytochrome *f* subunit of the homologous chloroplast cytochrome *bf* complex, the structure of the hydrophilic domain has been solved to 2.3 Å resolution and the haem is ligated by a histidine and, unexpectedly, by the amino group of the N-terminal tyrosine residue (Martinez *et al.*, 1994).

Because of its central role in proton–electron coupling in these enzymes, the possible structure of cytochrome *b* has received particular attention. A variety of recent reviews present a similar consensus view of the important structural features (Crofts *et al.*, 1992; Degli Esposti *et al.*, 1993; Gennis *et al.*, 1993; Knaff, 1993; Link *et al.*, 1993; Malkin, 1992). Of particular note are the eight predicted membrane-spanning helices, the likelihood of an amphipathic helix between the third and fourth of these, and the ligation of two protohaems by two pairs of histidines in the second and fourth membrane-spanning helices. Especially fruitful has been the generation and analyses of a now very wide range of point mutations which confer inhibitor resistance on one or other of the two quinone-reactive sites, the Q_o site and the Q_i site. This has led to the confirmation of the vectorial location of these two sites which had been predicted by the Q cycle mechanism (Mitchell, 1976) and has allowed further refinement of the likely three-dimensional structure (*Figure 9.3*). This in turn has led to the generation of specific site-directed mutations to probe further the proposed structure. Overall, the data have provided a compelling 5–10 Å predicted structure for cytochrome *b*. In this model, a quinol oxidation site, Q_o, is formed between the iron–sulphur centre subunit and fairly well defined regions of cytochrome *b* involving the P phase ends of helices III and VI, and a face of the amphiphilic helix between helices III and IV. The two haems *b* electronically connect this site to the second Q site, Q_i, which is located close to the other side of the membrane and formed primarily from residues in the N phase ends of helices I, IV and V and the loop region connecting helices IV and V. Unfortunately, despite the considerable data now accumulated, it has not proved possible to define more exactly the residues which actually bind to the quinones whilst in these sites. Although the general features of such Q sites can

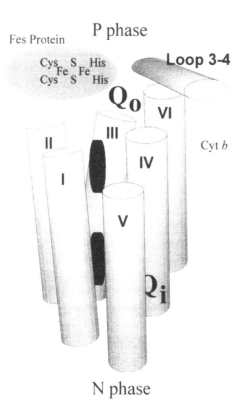

Figure 9.3. Predicted three-dimensional structure of parts of cytochrome *b* (Cyt *b*) of the bc_1 complexes. Haems *b* are ligated to helices II and IV. The Q_o site is likely to be formed from the tops of helices III and VI and interhelical loop 3–4 of cytochrome *b*, together with a region of the iron–sulphur protein (Fes Protein) close to the [2Fe–2S] centre. The Q_i site is formed from the lower ends of helices I, IV and V of cytochrome *b*.

be predicted with some confidence (Rich, 1996), no Q site motif can yet be recognized, and further resolution must await new information, especially from new crystal structures.

(b) *Internal electron transfer.* The general features of electron transfer in the *bc* complexes have been elucidated in some detail. A quinol binds to the Q_o site, in close proximity to the iron–sulphur centre and to the lower potential haem *b*, b_L, of the cytochrome *b*. Hydrogen bonding of one quinol hydroxyl to a histidine ligand of the iron–sulphur centre and of the other quinol hydroxyl to a residue in cytochrome *b* which is close to the iron or the porphyrin macrocycle is an attractive working model for this intermediate. The environment is such that the quinol undergoes a concerted or sequential two-step reaction in which the first electron is transferred to the iron–sulphur centre, and the second electron is donated to the haem b_L. It seems clear that this is an all-or-nothing process: when an inhibitor is in the other Q_i site to prevent haem reoxidation, the haem b_L becomes reduced after a few turnovers. This in turn prevents electron donation from a quinol in Q_o to the iron–sulphur centre and the enzyme ceases to turn over. Hence, only the first electron from the quinol can be transferred to the iron–sulphur centre. Since this first redox couple $(QH_2/Q \cdot^-)$ has a high redox potential whereas the second $(Q \cdot^-/Q)$ has a low one (Rich, 1984), this raises mechanistic questions on how this bifurcation of the electron transfer pathway is ensured. Consideration has been given to possible protein conformational changes after the first electron transfer which prevent a second electron transfer to the iron–sulphur centre (Brandt and von Jagow, 1991) and to the possibility that the two detected Q-binding sites within the Q_o site may provide a charge separation favouring the bifurcation (Ding *et al.*, 1992). However, it seems possible that the detailed energetics of the reaction steps might themselves be sufficient to ensure bifurcation even with the type of structure envisaged above. In this mechanism, the second reaction (haem b_L reduction) serves to remove rapidly an energetically unfavourable intermediate (an unstable semiquinone in the Q_o site) which would otherwise prevent any significant forward net rate of the first electron transfer. Only by having both steps acting in series can a substantial forward rate be established. It remains to be established whether anything else is operative in order to favour further the bifurcated process.

After bifurcation at the Q_o site, the electrons proceed independently. The electron transferred to the iron–sulphur centre equilibrates with the cytochrome c_1 haem. These centres are both likely to lie close to the P phase (Ohnishi *et al.*, 1989). A lower limit of this rate constant of around 5000 s^{-1} can be deduced, based on the 200–400 μs equilibration rate between the iron–sulphur centre and the haem *c* of cytochrome c_1 (Bowyer *et al.*, 1980) or of cytochrome *f* (Whitmarsh *et al.*, 1982) after flash-induced oxidation. From cytochrome c_1 the electron can be transferred onto a transiently bound cytochrome *c* and hence, ultimately, to cytochrome oxidase. The electron on haem b_L (E_{m7} around −50 to −100 mV) is donated across the membrane to haem b_H (E_{m7} 50–100 mV). Experimentally, haem b_L is not normally seen to be transiently reduced during turnover unless unusual conditions are imposed on the experimental system (Meinhardt and Crofts, 1983). This can be attributed to a haem–haem electron transfer process, which significantly exceeds the haem b_L reduction rate (a $k_{observed}$ for reduction of haem $b_L > 10^3$ s^{-1} can be measured under conditions where the other haem, b_H, is artificially pre-reduced). Certainly, the edge-to-edge separation of the haem groups of less than 10 Å, together with a driving force of perhaps 100 mV, is consistent with such rapid transfer (Moser et al., 1992). A conserved aromatic residue placed between the haems of the bc complexes was suggested to play a possible role in the haem–haem transfer (Furbacher et al., 1989; Widger and Cramer, 1991), but site-directed mutagenesis of this residue has ruled this out (Yun et al., 1992). Although this electron transfer pathway through haem b_L seems most reasonable, it does remain to be shown definitively to operate as the mechanism of transferring the second electron from the Q_o quinol to haem b_H.

Electrons on haem b_H are then transferred onto a quinone bound at the other Q site, Q_i. The structure of this site, and therefore the detailed mechanistics, remains unclear. The general regions of cytochrome b which form it have been defined by analyses of mutants and model building, and it seems clear that it is located towards the N phase side of the membrane. Furthermore, the thermodynamics of the quinone when bound in the site are well established by electron paramagnetic resonance and optical spectroscopy, and show that the semiquinone is stabilized to such an extent that the two electrons required for reduction are at roughly similar potentials (Rich et al., 1990; Salerno and Ohnishi, 1980). However, questions on whether the site is formed by two enzymes in a dimeric structure (Hope, 1993; Marres and de Vries, 1991), how buried the site is in the protein (Konstantinov, 1990) and the identity of the important amino acids, remain unresolved.

(c) *Coupled proton transfer.* The redox-linked protonation chemistry of the quinone substrate, and the vectorial nature of the redox centre and Q sites combine to provide a mechanism of proton–electron coupling of particular elegance and simplicity in this class of enzymes. This mechanism falls very much within the framework of ideas provided by the vectorial redox loop formulation, and indeed it was Mitchell who first outlined its 'Q cycle' mechanism of proton–electron coupling (Mitchell, 1976). Because of the vectorial location of the Q sites, together with the bifurcated electron transfer reaction at the Q_o site, the oxidation of two quinols results in four protons deposited in the P phase and two being taken up from the N phase when a quinol is regenerated at the Q_i site. Hence, it is still a mobile, protonated quinol which essentially moves the protons across the membrane. The internal reactions of the bc complex act to transfer one of the charges released by quinol oxidation back across the membrane, so increasing the coupling efficiency of the reaction above that which would result if both charges were passed through the iron–sulphur centre to cytochrome c.

In this simple formulation, protons are deposited and taken up in the scalar reactions of the Q sites and all electrogenic reactions are associated with the electron transfer mediated by the haems b between the two Q sites. However, some evidence has been presented to suggest that the Q sites may themselves be partly buried in the protein, so that the protonation reactions themselves may be electrogenic as protons are lost from the Q_o site to the P phase

(Konstantinov, 1990), or as they are taken up from the N phase into the Q_i site (Hope and Rich, 1989; Robertson and Dutton, 1988). Hence, proton wells, as in *Figure 9.2*, may play a role in the overall processes of charge separation. The quantitative extent of this remains to be established. In any case, I have argued previously that the extent to which a channel acts as a simple access route through which movement of charged ions is electrically counterbalanced, as opposed to a field-generating proton well, may not be a fixed characteristic of such structures.

Thylakoid bf *complex.* In thylakoids of higher plants and algae, a homologue of the *bc* complexes operates between the photosystems and catalyses the transfer of reducing equivalents from plastoquinol to plastocyanin, analogous to the ubiquinol-cytochrome *c* oxidoreductase activity of the *bc* complexes. The homology to the cytochrome *b* polypeptide is clear, although several significant differences are evident (Bendall, 1982; Frank and Trebst, 1995; Hope, 1993; Malkin, 1992; Widger and Cramer, 1991). Most prominent is the fact that the chloroplast enzyme has a smaller cytochrome *b*, homologous to the region containing only the first four membrane-spanning helices. However, another subunit is present with clear homology to a segment covering at least three of the other four membrane-spanning helices. A further difference is that an additional amino acid separates the two haem-ligating histidines in helix IV of the cytochrome *bf* complex, so that they are separated by 14 residues rather than the 13 residues found in helix II. This will tend to strain the haem environment and probably contributes to some of the different spectroscopic properties of the haem pair in comparison with their properties in the *bc* complexes. Also notable are the quite different sensitivities of the Q sites of the *bf* complex to inhibitors of Q sites in the *bc* complexes, the different operative potentials of the redox centres and the very low sequence homology between the primary sequences of cytochrome *f* and cytochrome c_1.

These differences have led some to consider that these enzymes might operate in a fundamentally different way to the *bc* complexes, particularly that they might be able to function without being coupled in the same way as *bc* complexes to proton translocation, as described above. If this is the case, then the easiest way to achieve this would be if the obligatory bifurcated reaction at the Q_o site could be substituted by a reaction in which both electrons are transferred to the iron–sulphur centre. This would result in a simple process which resulted only in the deposition of protons into the P phase as electrons were passed to plastocyanin. Whether this can occur at all has remained a matter of debate, although I have argued previously in detail against the possible evidence for such a reaction (Rich, 1988). In any case, it now appears accepted that, under many conditions, the chloroplast enzyme is likely to have the same coupling properties as the *bc* complexes and, therefore, can operate in essentially the same manner in terms of internal electron transfer processes.

9.3.3 *Ionmotive NADH dehydrogenases*

Several types of enzyme (NQR) are known to catalyse the respiratory reaction of oxidation of NADH by membrane-bound ubiquinone. Only one of these, termed complex I or NDH-1, is coupled to proton translocation. All homologues of complex I are presumed to be coupled to proton translocation and have been generically termed NDH-1 types (Yagi, 1991, 1993). Two other types of NADH dehydrogenase can be found in mitochondria from fungal and plant sources (Møller *et al.*, 1993). One of these oxidizes internally generated NADH and is likely to be homologous to a non-protonmotive NQR in bacteria, termed NDH-2 (Yagi, 1991). The enzyme may be a single polypeptide, whose sequence has been established (de Vries *et al.*, 1992; Xu *et al.*, 1991; Young *et al.*, 1981), and with FAD as the only redox cofactor (Yagi, 1991). A third NADH dehydrogenase in the inner membrane of plant and fungal mitochondria, which I have termed NDH-3, has an externally facing NADH site and can directly oxidize cytosolic NADH (Moore and Rich, 1985). A bacterial homologue has not been identified, and its detailed structure and composition remain unknown. Neither the NDH-2 (Yagi, 1993) nor the NDH-3 (Moore *et al.*, 1978) types of NQR are coupled to proton translocation. However, a fourth type of NQR, Na-NDH, is present in some bacteria (Unemoto and Hayashi, 1993). This enzyme is unrelated to other types of NQR and is coupled to the translocation of sodium ions across the membrane. I explore below the structure and coupling mechanisms of the NDH-1 and Na-NDH enzymes.

Complex I.

(a) *Structure.* A number of useful reviews on polypeptides, topography and redox centres of the bovine and bacterial homologues have appeared recently (Fearnley and Walker, 1993; Finel, 1993; Friedrich *et al.*, 1995; Sled *et al.*, 1993; Vinogradov, 1993; Weiss *et al.*, 1991; Yagi, 1993; Yagi *et al.*, 1993). Bovine complex I is remarkably complicated, with seven mitochondrially encoded and at least 34 further nuclear-encoded subunits (Fearnley and Walker, 1993). The enzyme from the fungus *Neurospora crassa* is composed of at least 30 subunits (Friedrich *et al.*, 1995). However, gene clusters coding for proton-translocating NADH-ubiquinone oxidoreductases have also been identified in the prokaryotes *Paracoccus denitrificans, Rhodobacter capsulatus* and *Escherichia coli* (Dupuis, 1992; Weidner *et al.*, 1993; Yagi *et al.*, 1993). These comprise homologues of the seven mitochondrially encoded and only seven of the nuclear-encoded mammalian subunits (Yagi, 1993), together with a number of other open reading frames of unknown significance. The 14 homologous subunits presumably represent a more minimal core catalytic structure.

To date, only a low resolution overall structure of the complex is available, from electron microscopy images of single particles and two-dimensional arrays, in both cases of the enzyme isolated from mitochondria of *N. crassa* (Hofhaus *et al.*, 1991). The enzyme appears to be L-shaped with a long membrane-embedded arm and a shorter arm which extends into the aqueous phase. These two domains appear to be assembled separately and can be produced independently in appropriately treated *N. crassa* cells. Sub-fractionation of purified enzymes from bovine and bacterial sources has given further insight into the roles of these domains (Finel *et al.*, 1992, 1994; Friedrich *et al.*, 1993). For example, the purified bovine enzyme can be subfractionated into three domains, namely: a fragment (FP) of three subunits which contains the NADH site, FMN and iron–sulphur centres; a more amphipathic fragment (IP) of seven or eight subunits also containing iron–sulphur centres; and a very hydrophobic fragment (HP) containing all seven mitochondrially encoded subunits together with the remainder of the nuclear-encoded polypeptides. Comparison with the homologous subunits of the *Paracoccus* enzyme (Yagi, 1993) reveals that only two subunits of FP, four of IP and eight (including all seven counterparts of the mitochondrially encoded subunits) of HP are present in bacteria. It has been suggested that the mitochondrial enzyme is a heterodimer, and different properties (van Belzen *et al.*, 1990) and functional (Vinogradov, 1993) aspects of each unit have been described.

(b) *Internal electron transfer*. Besides FMN and the NADH-binding site, the enzyme contains five well-defined iron–sulphur centres, and probably several less well-defined ones (Ohnishi, 1975). Analyses for motifs of the polypeptide sequences, together with analysis of subfragments, has allowed some assignments to specific locations (Fearnley and Walker, 1993). Perhaps most surprising is the finding that none of the seven homologues of mitochondrially encoded subunits contain conserved motifs for redox cofactors, in stark contrast to the mitochondrially encoded subunits of the bc_1 complex and cytochrome oxidase. Instead, most, if not all, of the iron–sulphur centres are contained in the hydrophilic HP and IP fragments, which also contains the subunit (51-kDa FP subunit; NQO1) with the FMN and associated NADH-binding site. This subunit is also likely to contain a [4Fe–4S] centre, centre N3. The other FP subunit (24 kDa or NQO2) is likely to contain the [2Fe–2S] centre N1b. Of the remaining FeS, the [4Fe–4S] centre N4 and the [2Fe–2S] are located in the 75-kDa (NQO3) subunit of the IP fragment (Fearnley and Walker, 1993; Yagi, 1993). Other less well-defined iron–sulphur centres may also be located on two further subunits (19- and 22-kDa subunits) of the bovine enzyme (Fearnley and Walker, 1993). The large hydrophobic arm is likely to contain only one polypeptide with an iron–sulphur cluster. This subunit (23 kDa; NQO9) is the only subunit in the bacterial HP fragment which is not a homologue of the mitochondrially encoded eukaryotic subunits. The iron–sulphur centre is equated with the tetranuclear centre N2, although two [4Fe–4S] binding motifs can be identified. One further subunit (ND1; NQO8) in the hydrophobic arm has been suggested to provide a site for binding of ubiquinone, based on labelling with radioactive rotenone and dicyclohexylcarbodiimide (DCCD) (Fearnley and Walker, 1993) and on a very weak homology with a glucose-ubiquinone oxidoreductase (Friedrich *et al.*, 1990). Functions in electron transfer for other subunits are unknown, although some are likely to house quite unrelated functions (Fearnley and Walker, 1993).

On the basis of this information, it may be concluded that the internal redox reactions involve electron transfer from NADH, via FMN, to centre N3, all within the 51-kDa subunit. Electron transfer is then connected through the IP fragment and its iron–sulphur centres to centre N2, the iron–sulphur centre most closely associated with the membrane domain, and which is the immediate donor to ubiquinone.

(c) *Coupled proton transfer*. Partly because of the complexity of redox components, uncertainty of topology and positions of redox centres, and the great difficulty in elucidating electron transfer pathways, models for coupled proton translocation remain speculative. Even the $H^+/2e^-$ ratio is not certain, although a ratio of four is most likely (Hinkle *et al.*, 1991; Ragan, 1990). This number has led to proposals of two separate sites for proton translocation, one centred around the redox cycle of the flavin, and the other around the reaction of centre N2 with ubiquinone.

Ragan (1987) postulated that oxidation of the reduced flavin, $FMNH_2$, occurs in a bifurcating manner, rather analogous to quinol oxidation at the Q_o site of the *bc* complexes. Protons released by $FMNH_2$ oxidation enter the P phase, whilst one of the two electrons is transferred back across the membrane via iron–sulphur centre(s). On a second cycle of the enzyme, $FMNH_2$ is again reduced by NADH and oxidized at the bifurcating site. However, the oxidation occurs only to the semiquinone stage (releasing one or two protons). The electron stored in the first cycle reduces it back to $FMNH_2$ with uptake of one or two protons from the N phase. Another scheme involving FMN has been outlined recently by Vinogradov (1993), where transmembrane conduction of an electron again occurs via iron–sulphur centre(s), but the flavin is envisaged to have a more classical 'mobile hydrogen atom carrier' function. In yet another model (Krishnamoorthy and Hinkle, 1988) the flavin is proposed to cycle only between the semiquinone (FMN^-) and reduced ($FMNH_2$) forms, with one electron reduction resulting in uptake of two N phase protons and reoxidation involving liberation of the two protons into the P phase.

Some variants of the models of Ragan and Vinogradov require an additional site to provide an $H^+/2e^-$ ratio of four. This second coupling site has been proposed to be located around the reactions of centre N2 with ubiquinone. Ragan has suggested another bifurcated oxidation site, this time for oxidation of ubiquinol (Ragan, 1987), where one electron is transferred back across the membrane to act again as a reductant for quinone. Vinogradov (1993) has suggested that a proton translocation process could result at this site from dismutation of a pair of appropriately configured protonated semiquinones which release protons to the P phase whilst generating a quinol which can enter the Q pool when it picks up N phase protons.

Most details of these possible schemes have yet to be tested experimentally and so all of them remain speculative at present. Particularly unclear is how the necessary strict sidedness of redox-linked protonations might be achieved, especially when protonation and deprotonation of the same, apparently firmly bound, cofactor must occur into different aqueous phases.

Considerations of a role for local electroneutrality in some of the reaction steps of complex I may be useful in relation to ideas on its protonmotive mechanism, as has been the case recently for formulation of a model for redox-linked sodium ion translocation by the Na-NDH (Rich *et al.*, 1995); and see below). Specifically for complex I, the pH dependency of the redox potential of centre N2 is indicative of a redox-linked protonation when the centre is reduced (Ohnishi, 1975). It is likely that N2 donates electrons to ubiquinone. Hence, a scheme illustrated in *Figure 9.4* may be envisaged. In this, the charge-compensating proton required for reduction of centre N2 is taken up from the

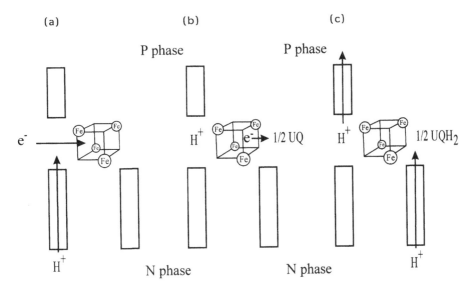

Figure 9.4. A possible basis for proton translocation in complex I. The model involves the redox reaction of the [4Fe–4S] centre, N2, with pool ubiquinone, and is based on the general scheme in *Figure 9.2*. (a) Redox-linked proton uptake; (b) ubiquinone reduction; (c) proton–proton repulsion. Details are given in the text.

N phase, but is located at a site which is spatially separated from the ubiquinone binding site. The extended structure of the [4Fe–4S] cluster may lend itself to such a spatial separation of sites. When ubiquinone is reduced to ubiquinol, further protons must be taken up from the N phase on to the oxygen atoms of the quinol. This in turn can lead to the electrostatic repulsion of the charge-compensating protons from their site close to centre N2 into the P phase. Such a scheme may also be imagined at the flavin site. In this case, the reduction of one component (flavin or centre N1a, both of which have a pH-dependent midpoint potential) binds protons which are subsequently repelled into the P phase as a subsequent acceptor (N1a of flavin) becomes protonated.

Na-NDH.

(a) *Structure.* A fourth type of NQR was first found in the marine bacterium *Vibrio alginolyticus*, an organism which produces an electrochemical gradient of sodium ions in aerobic respiration. The enzyme is induced at alkaline pH and is coupled to sodium ion translocation (Skulachev, 1992; Unemoto and Hayashi, 1993). Substantial evidence has been accumulated that this is the result of a primary coupling of electron transfer and sodium ion translocation, rather than to a secondary sodium/proton antiport system (Hayashi and Unemoto, 1984; Tokuda and Unemoto, 1984). This Na-NQR appears to be distinct from NDH-1, NDH-2 and NDH-3 on the basis of its sodium ion dependency, inhibitor sensitivity, NADH analogue specificity and polypeptide and cofactor composition (Asano *et al.*, 1985; Bourne and Rich, 1992; Hayashi and Unemoto, 1987). Early biochemical studies had indicated that the purified enzyme was composed of only three subunits, and contained only FAD and FMN as redox cofactors. A possible mechanism of ion coupling was proposed on this basis (Hayashi and Unemoto, 1987). However, the operon encoding these subunits suggests that six subunits may be involved (Beattie *et al.*, 1994; Hayashi *et al.*, 1995; Tan *et al.*, 1996). Three are rather hydrophobic, perhaps forming a membrane arm, whereas the other three are more hydrophilic.

(b) *Internal electron transfer.* Only one of these hydrophilic polypeptides has identifiable cofactor motifs, and appears to be homologous to the ferredoxin-NADP reductase superfamily (Karplus and Bruns, 1994). In common with other members of the superfamily, it has a binding site for NADH and one FAD in the hydrophilic C-terminal portion. In addition, however, there is an iron–sulphur motif between this region and an N-terminal domain which is predicted to have two membrane-spanning α-helices (Rich *et al.*, 1995). No other redox cofactor motifs have yet been identified. Hence, it seems reasonable to suppose that electron transfer occurs from NADH, via a single FAD, to the iron–sulphur centre, and that this iron–sulphur centre is able, directly or via the other hydrophobic parts, to reduce ubiquinone in the membrane.

(c) *Coupled ion transfer.* The above structural considerations and our previous finding of a lack of sodium ion dependency of the midpoint potential of the FAD (Bourne and Rich, 1992; and unpublished data) have been combined with considerations of electroneutrality of buried redox centres to produce a proposal for the essential features of coupling of electron transfer and sodium ion translocation (Rich *et al.*, 1995), analogous to the scheme for complex I shown in *Figure 9.4*.

In this model, the iron–sulphur centre is proposed to be in a relatively low dielectric environment such that electron transfer to it requires counter-ion uptake. It is proposed that the other subunits of this enzyme provide ion selectivity so that only sodium (or lithium) ions can provide such charge counterbalance. Binding and reduction of a quinone produces a quinol which must, inevitably, become protonated (whether the quinol is formed by a dismutation of two semiquinones, or whether there are two sequential electron transfers to the same bound quinone, remains a open point). Indeed, there will be a very strong driving force for such protonation. The spatial orientation of the reactants results in an electrostatic repulsion by the proton of the sodium ion into the positive aqueous phase, resulting in a net sodium ion translocation across the membrane. This is analogous to the type of process that might operate around centre N2 of complex I (*Figure 9.4*), substituting redox-linked sodium binding for the redox-linked protonation reaction of centre N2.

9.3.4 *Protonmotive oxidases*

Core structure of the oxidase superfamily. Mitochondrial cytochrome *c* oxidase and bacterial cytochrome *bo* are members of a diverse superfamily of homologous enzymes (Calhoun *et al.*, 1994; Saraste, 1990) which reduce oxygen to water. They catalyse a charge transfer across the membrane as in the simpler oxidases, such as cytochrome *bd* described above. However, there is an additional linked function by which further protons are translocated across the membrane so that the overall efficiency of energy coupling is significantly enhanced in comparison with their simpler counterparts.

Despite the fact that the substrate of some members is cytochrome c but ubiquinol in others, and the haem type may be a, b or o, it is now evident that all contain a copper–haem binuclear centre of similar structure. This binuclear centre is the site of oxygen binding and reduction to water. It has become widely accepted that local ligand movements within the binuclear centre (Mitchell, 1988), powered by the primary chemistry of oxygen reduction (Rich, 1991), also drives the associated proton translocation. Such models also incorporate a number of proton channels in the protein structure to link the binuclear centre to the aqueous surfaces on either side of the membrane.

As for other mitochondrial respiratory enzymes, bacterial homologues are much smaller than the beef heart form of cytochrome oxidase, which is composed of 13 subunits (Capaldi, 1990). A minimal unit of perhaps only three subunits, equivalent to the only three subunits that are mitochondrially encoded in eukaryotes, can carry out all known electron and proton transfer functions. One of these, subunit III, has no redox components and must provide a structural or stability factor and, possibly, forms part of the binding site for cytochrome c. A second essential subunit, subunit II, also provides part of the binding site for cytochrome c in the cytochrome oxidases, and has a binuclear copper centre, Cu_A, which functions as a pathway for electrons from cytochrome c into the catalytic core. Subunit II is also present in the quinol oxidases, although the Cu_A centre is absent, and it may instead provide the quinol oxidation site, although this has yet to be established. All members of the superfamily have a homologous subunit I which contains the binuclear haem–copper centre together with an additional low spin haem.

The ligand structure around the metal centres was deduced from a large range of studies of site-directed mutants in bacterial systems, in combination with a variety of spectroscopic techniques, reviewed in Hosler et al. (1993). This information could be combined with predictions of secondary structure by conventional methods and information from biophysical measurements to produce a model of the overall subunit I structure, a model which has been reviewed extensively (Babcock and Wikström, 1992; Brown et al., 1993; Hosler et al., 1993). Most recently, the structures of the metal sites in the four-subunit cytochrome oxidase from P. denitrificans (Iwata et al., 1995) and in the entire 13-subunit enzyme from beef heart mitochondria (Tsukihara et al., 1995) have been solved by X-ray crystallography to 2.8 Å resolution. The polypeptide backbone structure and some other details of the Paracoccus enzyme have also been reported. It is likely that a full atomic structure of both enzymes at this resolution should appear soon. These structures show that the consensus outline model of predicted structure in terms of polypeptide folding, ligand structure around the metals and the possibility of proton channels is, in essence, correct. However, considerable atomic detail has now been added to the regions of mechanistic importance around the metal centres and to the possible proton channels between the reaction core and the aqueous surfaces.

Important features that have emerged from these crystal structures to date in relation to energy coupling considerations are: the confirmation of the binuclear nature of the Cu_A centre and the ligands to the metal centres in general; accurate definition of the positions and orientations of the metal centres; high resolution detail around the binuclear centre; and specific evidence for three 'pores' (Iwata et al., 1995) in subunit I, two of which provide possible proton channels from the aqueous N phase. One of these ('pore B') leads from the N phase at least up to a conserved glutamate residue on helix VI, and a second ('pore A') leads from the N phase to the binuclear centre itself. Interestingly, an obvious route for protons from the binuclear centre to the P phase has not emerged, although a possible route between subunits I and II has been recognized (Iwata *et al.*, 1995).

The reaction cycle of cytochrome c *oxidase.* In comparison with the enzymes discussed above, the oxidases have a particularly complex reaction cycle. Oxygen reduction to water requires four electrons and four protons and must occur in a stepwise manner. In each of these steps, the thermodynamics and chemistry of the intermediates are quite different. In the case of cytochrome *c* oxidase, there has been considerable progress in identification by spectroscopic techniques of the stable catalytic intermediates which occur in the binuclear centre of the oxidases during the reaction cycle. The major stable intermediates and some of their other properties have been reviewed recently (Babcock and Wikström, 1992).

The first intermediate to be formed from the oxidized (O state) enzyme is the one-electron-reduced or E state, in which the electron is shared between the haem *a* and the binuclear centre (Moody *et al.*, 1991), in a manner consistent with the relative midpoint potentials of the four metal centres. At this stage, oxygen probably does not bind and the electron transfer from cytochrome *c* is reversible. A subsequent electron transfer produces the two-electron-reduced enzyme (R state) in which chemical reaction with oxygen can occur. It has been suggested that oxygen first binds to Cu_B before being transferred onto the ferrohaem (Oliveberg and Malmström, 1992). However, the first well-defined product is an oxygen-ferrous species similar to oxyhaemoglobin, and termed compound A by Chance who first observed it at low temperatures (Chance *et al.*, 1975). Compound A is very unstable at room temperature where it can only be observed as a transient intermediate in the microsecond time range (Han *et al.*, 1990; Ogura *et al.*, 1993; Orii, 1984), after which time an irreversible change occurs to produce a stable species, termed the peroxy or P state. This species has no direct counterparts in model chemical compounds, and its structure remains uncertain. It may contain a peroxide moiety, either bridged between the iron and copper metals of the binuclear centre or as an end-on peroxide bound to the haem (Babcock and Wikström, 1992; Varotsis *et al.*, 1995) or, as now seems more likely, the O–O bond may already be broken (Hirota *et al.*, 1994), perhaps to form an oxyferryl species with an accompanying radical in a manner analogous to the possible structures of compound I of the peroxidases. A third electron transfer from substrate to P leads to a stable oxyferryl or F intermediate (Varotsis and Babcock, 1990) in which the O–O bond has certainly been broken. A similar type of haem compound has been observed in compound II of peroxidases, myoglobin and catalase (see Yonetani, 1970; and references in Watmough *et al.*, 1994). The reaction is completed by a fourth electron transfer to F, which results in the generation of a second water molecule and reversion of the metal centres to their fully oxidized O state.

The homology between subunits I of the cytochrome *c* oxidases and the cytochrome *bo*-type quinol oxidases makes it very likely that the same basic mechanism of oxygen reduction operates in both. The major difference between them is the presence of an extra copper centre, Cu_A, in the cytochrome c oxidases, and the two-electron quinol donor in the quinol oxidases, which results in differences in detail of rate-limiting steps and the expected mixture of intermediates during turnover. Indeed, it has already been shown that only a single two-electron-reduced oxygen intermediate is observed during turnover with quinol (Moody and Rich, 1994; Moody et al., 1993), and the structure of this compound may be unusual, having an MCD spectrum similar to other oxyferryl species, and prompting the suggestion that a third electron may be removed from a radical-forming species, as is the case in compound I of horseradish peroxidase (Watmough et al., 1994).

Internal electron transfer. In the cytochrome oxidases, it must now be clear from the structure that Cu_A is indeed the entry pathway for electrons from cytochrome c. Interestingly, Cu_A is positioned over the metal centres in subunit I so that the distances of the nearest copper atom to the haem a iron (19.5 Å) and to the haem a_3 iron (22.1 Å) are similar. Hence, although some other factors (Iwata et al., 1995) may favour the generally supposed pathway of the electron firstly to haem a before it reaches the binuclear centre, this point may require further experimental testing. Certainly, the very close contact of the edges of the haems (4.7 Å) ensures very fast haem–haem electron transfer. Once on the binuclear centre, the close proximity of the haem a_3 and Cu_B (5.2 Å) ensures that the electronic rearrangements are unlikely to limit the chemical transformation of oxygen into water.

Coupled proton transfer. Although many diverse models have been proposed for the mechanism of coupling of electron and proton transfer in these enzymes, it was not until the proposal of Mitchell that attention became focused on the local binuclear centre chemistry. He proposed a model involving the rotation of water/hydroxide/oxide ligands around Cu_B (Mitchell, 1987), and attempts to fit such a scheme with the likely structures of the oxygen intermediates were made (Rich, 1991). Several other variants of binuclear centre ligand chemistry have been suggested, involving ligand exchanges and protonations on Cu_B (Mitchell, 1991; Wikström et al., 1994), on haem a_3 (Rousseau et al., 1993) or between both metals (Woodruff, 1993). All proposals share a common theme that a relatively small movement of ligands, linked to the oxygen reduction chemistry, can result in the movement of protons between the N phase and the P phase proton channels.

   In order to characterize further the oxygen intermediates in cytochrome c oxidase, and as a prelude to further defining the protonmotive chemistry, we tested empirically the degree to which the binuclear centre neutralizes charge changes by protonation changes. To date, we have found no exception to the general rule in this system that all stable electronation or ligand binding changes in the binuclear centre are electroneutralized by protonation changes (Mitchell and Rich, 1994; Mitchell et al., 1992). The pH dependency of midpoint potentials of the metals during classical redox potentiometry is also consistent with this view (reviewed in Wikström et al., 1981). Originally, we had assumed a simple model in which the net proton changes represented protonations of the oxygen intermediates themselves (Mitchell et al., 1992). Subsequently, however, the notion of spatial separation of countercharges was introduced, as outlined in *Figure 9.2*, to provide a basis for the coupling mechanism in this class of enzymes. The detailed way in which this can be combined with the known and assumed chemistry of the oxygen intermediates has been described in detail in Rich (1995). The major features of this model are shown in *Figure 9.5* and are given below:

(i)   only the last two steps of oxygen reduction, P → F and F → O, are assumed to be coupled, each to the translocation of two protons, as proposed by Wikström (1989);

(ii) a number of protonatable groups are located within a 'proton trap' which is physically separated from the oxygen intermediates but nevertheless provides the means of overall charge compensation;

(iii) reversible proton uptake from the N phase into this trap occurs as charge is accumulated in the binuclear centre;

(iv) chemical transformation of oxygen intermediates ultimately leads to the formation of oxide products which are physically separate form the trap protons. Protonation of the oxides with N phase protons to form water provides the driving force which electrostatically repulses the trap protons into the P phase.

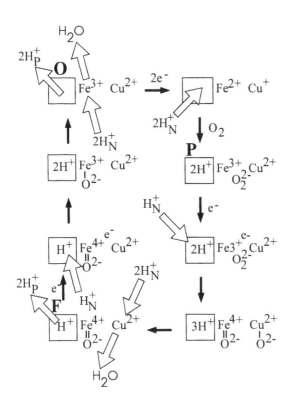

Figure 9.5. Oxygen reduction and proton translocation in cytochrome oxidase. The scheme is based on that in Rich (1995). Starting top left, the oxidized O state is reduced with two electrons and with associated protonation of the trap. Oxygen reacts to give the P state, shown here as a peroxide (but see text). P receives a further electron and trap proton. Chemical transformation into the F state is coupled with oxide protonation and repulsion of trap protons into the P phase (bottom left). Finally, a further electron and trap proton are added, chemical transformation occurs, and oxide protonation/trap deprotonation returns the system to the oxidized O state.

Some further details of some of the important steps in *Figure 9.5* are highlighted in *Figure 9.6*. *Figure 9.6a* illustrates an electrostatically locked P state in which charge-compensating protons are separate from the oxygen compound [here shown as a peroxide dianion, although recent work suggests that the O–O bond is already broken at this stage (Hirota *et al.,* 1994)]. *Figure 9.6b* illustrates the proton translocating step as P is converted to F. An electron has been donated to the binuclear centre of P, with an associated proton uptake into the trap to form an unstable intermediate, and chemical transformation has occurred to an oxyferryl species and an oxide. Protonation of this oxide results in proton repulsion from the trap into the P phase. *Figure 9.6c* shows a similar final step in the conversion of F into O.

Possible identity and chemistry of the proton trap. The recent structural information limits the possible identities of entities that might provide the protonatable sites of the proton trap. Two general possibilities seem evident (*Figure 9.6*), depending on whether the translocated protons enter the structure by a proton channel which is different from that for the substrate protons used to form water. Two separate channels (*Figure 9.7a*) had been indicated by site-directed mutants in which proton translocation could apparently be prevented whilst still allowing oxygen reduction (Fetter *et al.,* 1995; Thomas *et al.,* 1993). Although it is difficult to understand why prevention of one process would not also very much hinder the other, unless some rather major structural perturbations had been caused by the mutation, it has been noted that the putative 'pore A' of the crystal structure is consistent with the positions of mutations which cause the apparent 'uncoupling' effects (Iwata *et al.,* 1995). In this case, the trap site(s) might be equated with protonation on and around the conserved glutamate on helix VI (Glu278 in the *Paracoccus* sequence). Certainly, the necessary

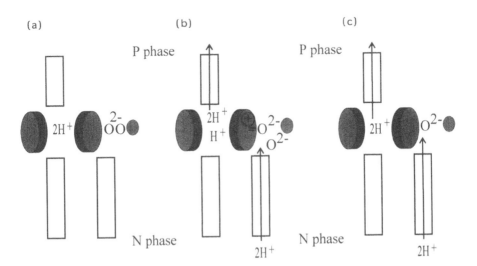

Figure 9.6. Oxygen reduction and proton translocation in cytochrome oxidase. Three steps of the scheme of *Figure 9.5* are shown in greater detail. (a) Electrostatically locked P state; (b) F formation; (c) O state formation after F reduction. See text for details.

(a)   (b)

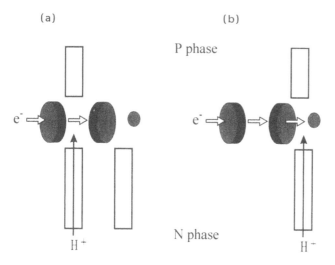

Figure 9.7. Oxygen reduction and proton translocation in cytochrome oxidase. Two general organizations of reactants may be possible in order to separate charge-balancing protons from oxygen intermediates. In the first of these (a) the protons enter the system by a physically separate proton channel [perhaps equivalent to 'pore A' in Iwata *et al.* (1995)]. In the second (b), only a single proton entry route is present, and local ligand rearrangements are required in order to provide a physical barrier between oxygen intermediates and translocated protons. See text for details.

spatial separation of these protons from the oxygen intermediates would seem to be satisfied by the intervening structure which separates it from the oxygen-binding site. However, whether enough sites could be supplied by surrounding residues (perhaps by water molecules or even by the haem propionate groups), and the identity of a route by which the protons would be repelled into the P phase, remains uncertain.

If, instead, only 'pore B' is operative in transferring protons from the N phase (*Figure 9.7b*), then trap sites more directly associated with the binuclear centre must be considered. This immediately raises the critical and chemically very difficult question of the nature of the barrier between the trap protons and the oxygen intermediates. Examination of the available crystal structural data shows that possible trap sites are limited. Only the direct ligands to haem a_3 and Cu_B, protonatable groups in 'pore B' itself, and possible protonatable sites in loop 9–10 of subunit I or perhaps in parts of subunit II close to the binuclear centre, appear to be reasonable candidates. However, an oxide–water ligand on Cu_B, in a manner similar to that proposed in the Mitchell model (Mitchell, 1987), cannot be excluded at this stage. With only one channel for proton entry, access to the proton trap would have to be gated by the electronation state of the metal centres: electronation would allow trap protonation, but subsequent electronic rearrangement during oxygen reduction would then have to preclude their physical availability to the final oxide products so that they are instead repelled into the P phase. A mechanism for the proton movement into a site which becomes physically separated only on subsequent electronic changes might involve a redox state-dependent ligand-switching process. Possibilities for ligand movement include the oxide–water ligand rotation model of Mitchell (1987), a tyrosine–histidine ligand exchange on the haem (Rousseau *et al.*, 1993) or on the Cu_B (Wikström *et al.*, 1994), or a possible ligand exchange between the haem and Cu_B (Rousseau *et al.*, 1993). In all cases, ligand mobility is facilitated by the presence of an electron in the metal system but is prevented when the electronic structure changes. The possibility of multiple orientations of one of the histidine ligands to Cu_B, His325, has led Iwata *et al.* (1995) to favour a model based on the 'histidine cycle' of Wikström (Wikström *et al.*, 1994) in which this flexible ligand cycles between the imidazolate/imidazolium states in response to local electroneutrality requirements (Rich, 1995).

In any of the models above, the subsequent route of the trap protons into the P phase, which has yet to be identified, could have a large activation barrier and may therefore be difficult to recognize, even perhaps only occurring transiently in the thermal protein motions. Most important for efficient coupling of proton translocation and electron transfer is the strict separation of the charge-compensating protons from the oxide products, a reaction which would otherwise effectively short-circuit the coupling process (Capitanio *et al.*, 1991). This undesired reaction can be prevented in one of two fundamental ways: by making the reaction kinetically slow or by making the reaction thermodynamically non-viable. In kinetic control, the reaction equilibrium can favour the undesired reaction products, but a large activation barrier ensures that the reaction proceeds too slowly to compete with the reactions of the coupled reaction cycle. Such a barrier has been invoked previously in the control of internal electron transfer in cytochrome oxidase (Brunori *et al.*, 1994; Gray and Malmström, 1989). In this case, slowing down of the coupled reaction cycle would allow an increasing percentage of non-coupled turnover. In thermodynamic control, the equilibrium position is such that essentially no undesirable products are able to form. If two N phase channels are operative (*Figure 9.7a*), it seems likely that the large spatial separation of the charge-compensating protons from the oxygen intermediates would provide a sufficient kinetic barrier. However, if a single N phase channel is used (*Figure 9.7b*), then their physical separation may be much smaller and, in this case, prevention of non-coupled turnover may require an element of thermodynamic control.

Clearly, many of the above questions may be resolved as structures of the intermediates are also solved, and as biophysical probes are used to test further the large range of mutant forms already available.

9.4 Applications to other systems and future prospects

In the above, I have chosen to illustrate the discussion of coupling mechanisms with several enzymes that have quite different chemistry and associated mechanisms of energy conservation. The principles outlined at the beginning are an attempt to define the basic elements that might be considered when trying to define such mechanisms. These may, perhaps, usefully be applied to other ion-coupled enzymes even when the reaction catalysed is unrelated to the above examples. In transhydrogenase (Jackson, 1991; Olausson *et al.*, 1995), for example, it seems likely that hydride transfer from NADPH to NAD^+ will occur in an occluded environment in which charge compensation by protonic rearrangements may be required. Hence, protonation of the site around NADH from the N phase might be coupled with proton repulsion into the P phase though a proton channel (Holmberg *et al.*, 1994; Olausson *et al.*, 1995) from the site around NADP$^+$.

One further quite unusual example of a coupled enzyme is provided by the Na$^+$-linked oxaloacetate decarboxylase described by Dimroth and colleagues in certain bacteria (Dimroth, 1987; Dimroth and Thomer, 1993). Again, one might consider a site which can accommodate a sodium ion which becomes filled in response to the electrostatic requirements of a reaction intermediate. Subsequent chemical transformation provides a force for the repulsion of this sodium ion into the P phase, perhaps in conjunction with a protonation process.

At this time, when new crystal structures are at last providing the framework to understand the large body of data that can now be obtained with the powerful combination of genetic and biophysical approaches, it seems likely that the chemical detail of many of the ideas above may soon be provided.

Acknowledgements

I am grateful to my colleagues at The Glynn Research Foundation Ltd for useful comments and criticisms during the preparation of this manuscript and to EPSRC (GR/J28148), BBSRC (CO2632), HFSP (RG-464/95 M) and The European Community (BIO2-CT93-0076) for financial support of experimental investigations of aspects of these ideas.

References

Asano M, Hayashi M, Unemoto T, Tokuda H. (1985) Ag$^+$-sensitive NADH dehydrogenase in the Na$^+$-motive respiratory chain of the marine bacterium *Vibrio alginolyticus*. *Agric. Biol. Chem.* **49**: 2813–2817.

Babcock GT, Wikström M. (1992) Oxygen activation and the conservation of energy in cell respiration. *Nature* **356**: 301–309.

Beattie P, Tan K, Bourne RM, Leach D, Rich PR, Ward FB. (1994) Cloning and sequencing of four structural genes for the Na$^+$-translocating NADH-ubiquinone oxidoreductase of *Vibrio alginolyticus*. *FEBS Lett.* **356**: 333–338.

Bendall DS. (1982) Photosynthetic cytochromes of oxygenic organisms. *Biochim. Biophys. Acta* **683**: 119–151.

Bourne RM, Rich PR. (1992) Characterization of a sodiummotive NADH: ubiquinone oxidoreductase. *Biochem. Soc. Trans.* **20**: 577–582.

Bowyer JR, Dutton PL, Prince RC, Crofts AR. (1980) The role of the Rieske iron–sulfur center as the electron donor to ferricytochrome c_2 in *Rhodopseudomonas sphaeroides*. *Biochim. Biophys. Acta* **592**: 445–460.

Brandt U, von Jagow G. (1991) Analysis of inhibitor binding to the mitochondrial cytochrome *c* reductase by fluorescence quench titration – evidence for a 'catalytic switch' at the Q$_o$ center. *Eur. J. Biochem.* **195**: 163–170.

Brown S, Moody AJ, Mitchell R, Rich PR. (1993) Binuclear centre structure of terminal protonmotive oxidases. *FEBS Lett.* **316**: 216–223.

Brunori M, Antonini G, Giuffre A, Malatesta F, Nicoletti F, Sarti P, Wilson MT. (1994) Electron transfer and ligand binding in terminal oxidases: impact of recent structural information. *FEBS Lett.* **350**: 164–168.

Calhoun MW, Thomas JW, Gennis RB. (1994) The cytochrome oxidase superfamily of redox-driven proton pumps. *Trends Biochem. Sci.* **19**: 325–330.

Capaldi RA. (1990) Structure and function of cytochrome *c* oxidase. *Annu. Rev. Biochem.* **59**: 569–596.

Capitanio N, Capitanio G, De Nitto E, Villani G, Papa S. (1991) H$^+$/e$^-$ stoichiometry of mitochondrial cytochrome complexes reconstituted in liposomes. Rate-dependent changes of the stoichiometry in the cytochrome c oxidase vesicles. *FEBS Lett.* **288:** 179–182.

Chance B, Saronio C, Leigh JS. (1975) Functional intermediates in the reaction of membrane-bound cytochrome oxidase with oxygen. *J. Biol. Chem.* **250:** 9226–9237.

Churg AK, Warshel A. (1986) Control of the redox potential of cytochrome c and microscopic dielectric effects in proteins. *Biochemistry* **25:** 1675–1681.

Clark WM. (1960) *Oxidation–Reduction Potentials of Organic Systems.* Bailliere, Tindall & Cox, Ltd, London.

Colson A-M. (1993) Random mutant generation and its utility in uncovering structural and functional features of cytochrome b in *Saccharomyces cerevisiae.* *J. Bioenerg. Biomembr.* **25:** 211–220.

Crofts A, Hacker B, Barquera B, Yun C-H, Gennis R. (1992) Structure and function of the bc-complex of *Rhodobacter sphaeroides.* *Biochim. Biophys. Acta* **1101:** 162–165.

Davidson E, Ohnishi T, Atta-Asafo-Adjei E, Daldal F. (1992) Potential ligands to the [2Fe–2S] Rieske cluster of the cytochrome bc_1 complex of *Rhodobacter capsulatus* probed by site-directed mutagenesis. *Biochemistry* **31:** 3342–3351.

Degli Esposti M, de Vries S, Crimi M, Ghelli A, Patarnello T, Meyer A. (1993) Mitochondrial cytochrome b: evolution and structure of the protein. *Biochim. Biophys. Acta* **1143:** 243–271.

Deisenhofer J, Michel H. (1989) The photosynthetic reaction center from the purple bacterium *Rhodopseudomonas viridis.* *Science* **245:** 1463–1471.

de Vries S, Van Witzenburg R, Grivell LA, Marres CAM. (1992) Primary structure and import pathway of the rotenone-insensitive NADH-ubiquinone oxidoreductase of mitochondria from *Saccharomyces cerevisiae.* *Eur. J. Biochem.* **203:** 587–592.

Dimroth P. (1987) Sodium ion transport decarboxylases and other aspects of sodium ion cycling in bacteria. *Microbiol. Rev.* **51:** 320–340.

Dimroth P, Thomer A. (1993) On the mechanism of sodium ion translocation by oxaloacetate decarboxylase of *Klebsiella pneumoniae.* *Biochemistry* **32:** 1734–1739.

Ding H, Robertson DE, Daldal F, Dutton PL. (1992) Cytochrome bc_1 complex [2Fe–2S] cluster and its interaction with ubiquinone and ubihydroquinone at the Q_o site: a double-occupancy Q_o site model. *Biochemistry* **31:** 3144–3158.

Dupuis A. (1992) Identification of two genes of *Rhodobacter capsulatus* coding for proteins homologous to the ND1 and 23-kDa subunits of the mitochondrial complex I. *FEBS Lett.* **301:** 215–218.

Dutton PL, Wilson DF. (1974) Redox potentiometry in mitochondrial and photosynthetic bioenergetics. *Biochim. Biophys. Acta* **346:** 165–212.

Ellis WD, Dunford HB. (1968) The kinetics of cyanide and fluoride binding by ferric horseradish peroxidase. *Biochemistry* **7:** 2054–2062.

Erman JE. (1974) Kinetic and equilibrium studies of cyanide binding by cytochrome c peroxidase. *Biochemistry* **13:** 39–44.

Fearnley IM, Walker JE. (1993) Conservation of sequences of subunits of mitochondrial complex I and their relationships with other proteins. *Biochim. Biophys. Acta* **1140:** 105–134.

Feher G, Allen JP, Okamura MY, Rees DC. (1989) Structure and function of bacterial photosynthetic reaction centres. *Nature* **339:** 111–116.

Fetter JR, Qian J, Shapleigh J, Thomas JW, Garcia-Horsman A, Schmidt E, Hosler J, Babcock GT, Gennis RB, Ferguson-Miller S. (1995) Possible proton relay pathways in cytochrome c oxidase. *Proc. Natl Acad. Sci. USA* **92:** 1604–1608.

Finel M. (1993) The proton-translocating NADH: ubiquinone oxidoreductase: a discussion of selected topics. *J. Bioenerg. Biomembr.* **25:** 357–366.

Finel M, Skehel JM, Albracht SPJ, Fearnley IM, Walker JE. (1992) Resolution of NADH: ubiquinone oxidoreductase from bovine heart mitochondria into two subcomplexes, one of which contains the redox centers of the enzyme. *Biochemistry* **31:** 11425–11434.

Finel M, Majander AS, Tyynelä J, De Jong AMP, Albracht SPJ, Wikström M. (1994) Isolation and characterisation of subcomplexes of the mitochondrial NADH: ubiquinone oxidoreductase (complex I). *Eur. J. Biochem.* **226:** 237–242.

Frank K, Trebst A. (1995) Quinone binding sites on cytochrome b/c complexes. *Photochem. Photobiol.* **61:** 2–9.

Friedrich T, Strohdeicher M, Hofhaus G, Preis D, Sahm H, Weiss H. (1990) The same domain motif for ubiquinone reduction in mitochondrial or chloroplast NADH dehydrogenase and bacterial glucose dehydrogenase. *FEBS Lett.* **265**: 37–40.

Friedrich T, Weidner U, Nehls U, Fecke W, Schneider R, Weiss H. (1993) Attempts to define distinct parts of NADH: ubiquinone oxidoreductase (Complex I). *J. Bioenerg. Biomembr.* **25**: 331–337.

Friedrich T, Steinmüller K, Weiss H. (1995) The proton-pumping respiratory complex I of bacteria and mitochondria and its homologue in chloroplasts. *FEBS Lett.* **367**: 107–111.

Furbacher PN, Girvin ME, Cramer WA. (1989) On the question of interheme electron transfer in the chloroplast cytochrome b_6 *in situ. Biochemistry* **28**: 8990–8998.

Gencic S, Schägger H, von Jagow G. (1991) Core I protein of bovine ubiquinol-cytochrome-*c* reductase; an additional member of the mitochondrial-protein-processing family – cloning of bovine core I and core II cDNAs and primary structure of the proteins. *Eur. J. Biochem.* **199**: 123–131.

Gennis RB, Barquera B, Hacker B, Van Doren SR, Arnaud S, Crofts AR, Davidson E, Gray KA, Daldal F. (1993) The bc_1 complexes of *Rhodobacter sphaeroides* and *Rhodobacter capsulatus. J. Bioenerg. Biomembr.* **25**: 195–209.

Glaser E, Eriksson A, Sjöling S. (1994) Bifunctional role of the bc_1 complex in plants: mitochondrial bc_1 complex catalyses both electron transport and protein processing. *FEBS Lett.* **346**: 83–87.

Gray HB, Malmström BG. (1989) Long-range electron transfer in multisite metalloproteins. *Biochemistry* **28**: 7499–7505.

Gray KA, Davidson E, Daldal F. (1992) Mutagenesis of Methionine-183 drastically affects the physicochemical properties of cytochrome c_1 of the bc_1 complex of *Rhodobacter capsulatus. Biochemistry* **31**: 11864–11873.

Gunner MR, Honig B. (1991) Electrostatic control of midpoint potentials in the cytochrome subunit of the *Rhodopseudomonas viridis* reaction center. *Proc. Natl Acad. Sci. USA* **88**: 9151–9155.

Gurbiel RJ, Ohnishi T, Robertson DE, Daldal F, Hoffman BM. (1991) Q-band ENDOR spectra of the Rieske protein from *Rhodobactor capsulatus* ubiquinol-cytochrome *c* oxidoreductase show two histidines coordinated to the [2Fe–2S] cluster. *Biochemistry* **30**: 11579–11584.

Han S, Ching Y, Rousseau DL. (1990) Primary intermediate in the reaction of oxygen with fully reduced cytochrome *c* oxidase. *Proc. Natl Acad. Sci. USA* **87**: 2491–2495.

Hayashi M, Unemoto T. (1984) Characterization of the Na^+-dependent respiratory chain NADH: quinone oxidoreductase of the marine bacterium, *Vibrio alginolyticus*, in relation to the primary Na^+ pump. *Biochim. Biophys. Acta* **767**: 470–478.

Hayashi M, Unemoto T. (1987) Subunit components and their roles in the sodium-transport NADH: quinone reductase of a marine bacterium, *Vibrio alginolyticus. Biochim. Biophys. Acta* **890**: 47–54.

Hayashi M, Hirai K, Unemoto T. (1995) Sequencing and the alignment of structural genes in the *nqr* operon encoding the Na^+-translocating NADH-quinone reductase from *Vibrio alginolyticus. FEBS Lett.* **363**: 75–77.

Hinkle PC, Kumar MA, Resetar A, Harris DL. (1991) Mechanistic stoichiometry of mitochondrial oxidative phosphorylation. *Biochemistry* **30**: 3576–3582.

Hirota S, Mogi T, Ogura T, Hirano T, Anraku Y, Kitagawa T. (1994) Observation of the $Fe-O_2$ and $Fe^{IV}=O$ stretching Raman bands for dioxygen reduction intermediates of cytochrome *bo* isolated from *Escherichia coli. FEBS Lett.* **352**: 67–70.

Hofhaus G, Weiss H, Leonard K. (1991) Electron microscopic analysis of the peripheral and membrane parts of mitochondrial NADH dehydrogenase (complex I). *J. Mol. Biol.* **221**: 1027–1043.

Holmberg E, Olausson T, Hultman T, Rydström J, Ahmad S, Glavas NA, Bragg PD. (1994) Prediction and site-specific mutagenesis of residues in transmembrane α-helices of proton-pumping nicotinamide nucleotide transhydrogenases from *Escherichia coli* and bovine heart mitochondria. *Biochemistry* **33**: 7691–7700.

Hope AB. (1993) The chloroplast cytochrome *bf* complex: a critical focus on function. *Biochim. Biophys. Acta* **1143**: 1–22.

Hope AB, Rich PR. (1989) Proton uptake by the chloroplast cytochrome *bf* complex. *Biochim. Biophys. Acta* **975**: 96–103.

Hosler JP, Ferguson-Miller S, Calhoun MW, Thomas JW, Hill J, Lemieux L, Ma J, Georgiou C, Fetter J, Shapleigh J, Tecklenburg MMJ, Babcock GT, Gennis RB.

(1993) Insight into the active-site structure and function of cytochrome oxidase by analysis of site-directed mutants of bacterial cytochrome aa_3 and cytochrome bo. *J. Bioenerg. Biomembr.* **25**: 121–136.

Hosokawa Y, Suzuki H, Toda H, Nishikimi M, Ozawa T. (1990) The primary structure of the precursor to core protein II, a putative member of mitochondrial processing protease family, of rat mitochondrial cytochrome bc_1 complex deduced from cDNA sequence analysis. *Biochem. Int.* **20**: 731–737.

Iwata S, Ostermeier C, Ludwig B, Michel H. (1995) Structure at 2.8 Å resolution of cytochrome c oxidase from *Paracoccus denitrificans*. *Nature* **376**: 660–669.

Jackson JB. (1991) The proton-translocating nicotinamide adenine dinucleotide transhydrogenase. *J. Bioenerg. Biomembr.* **23**: 715–742.

Karplus PA, Bruns CM. (1994) Structure–function relations for ferredoxin reductase. *J. Bioenerg. Biomembr.* **26**: 89–99.

Knaff DB. (1993) The cytochrome bc_1 complexes of photosynthetic purple bacteria. *Photosynth. Res.* **35**: 117–133.

Konstantinov AA. (1990) Vectorial electron and proton transfer steps in the cytochrome bc_1 complex. *Biochim. Biophys. Acta* **1018**: 138–141.

Krishnamoorthy G, Hinkle PC. (1988) Studies on the electron transfer pathway, topography of iron–sulfur centers, and site of coupling in NADH-Q oxidoreductase. *J. Biol. Chem.* **263**: 17566–17575.

Kuila D, Schoonover JR, Dyer RB, Batie CJ, Ballou DP, Fee JA, Woodruff WH. (1993) Resonance Raman studies of Rieske-type proteins. *Biochim. Biophys. Acta* **1140**: 175–183.

Link TA, Haase U, Brandt U, von Jagow G. (1993) What information do inhibitors provide about the structure of the hydroquinone oxidation site of ubihydroquinone: cytochrome c oxidoreductase? *J. Bioenerg. Biomembr.* **25**: 221–232.

Malkin R. (1992) Cytochrome bc_1 and b_6f complexes of photosynthetic membranes. *Photosynth. Res.* **33**: 121–136.

Marres CAM, de Vries S. (1991) Reduction of the Q-pool by duroquinol via the two quinone-binding sites of the QH_2:cytochrome c oxidoreductase. A model for the equilibrium between cytochrome b-562 and the Q-pool. *Biochim. Biophys. Acta* **1057**: 51–63.

Martinez SE, Huang D, Szczepaniak A, Cramer WA, Smith JL. (1994) Crystal structure of chloroplast cytochrome f reveals a novel cytochrome fold and unexpected haem ligation. *Structure* **2**: 95–105.

Meinhardt SW, Crofts AR. (1983) The role of cytochrome b-566 in the electron-transfer chain of *Rhodopseudomonas sphaeroides*. *Biochim. Biophys. Acta* **723**: 219–230.

Meunier B, Rodriguez-Lopez JN, Smith AT, Thorneley RNF, Rich PR. (1995) Laser photolysis behaviour of ferrous horseradish peroxidase with carbon monoxide and cyanide: effects of mutations in the distal haem pocket. *Biochemistry* **34**: 14687–14692.

Meunier B, Moody AJ, Mitchell RM, Rodgriguez-Lopez JN, Smith AT, Thorneley RNF, Rich PR. (1996) The reaction of cyanide with cytochrome c oxidase, cytochrome bo and horseradish peroxidase *Biochim. Biophys. Acta. EBEC Short Reports* **9**: (in press).

Mitchell P. (1961) Coupling of phosphorylation to electron and proton transfer by a chemiosmotic type of mechanism. *Nature* **191**: 144–148.

Mitchell P. (1966) *Chemiosmotic Coupling in Oxidative and Photosynthetic Phosphorylation*. Glynn Research Ltd, Bodmin.

Mitchell P. (1968) *Chemiosmotic Coupling and Energy Transduction*. Glynn Research Ltd, Bodmin.

Mitchell P. (1969) Chemiosmotic coupling and energy transduction. *Theor. Exp. Biophys.* **2**: 159–216.

Mitchell P. (1976) Possible molecular mechanisms of the protonmotive function of cytochrome systems. *J. Theor. Biol.* **62**: 327–367.

Mitchell P. (1987) A new redox loop formality involving metal-catalysed hydroxide-ion translocation. A hypothetical Cu loop mechanism for cytochrome oxidase. *FEBS Lett.* **222**: 235–245.

Mitchell P. (1988) Possible protonmotive osmochemistry in cytochrome oxidase. *Ann. N. Y. Acad. Sci.* **550**: 185–198.

Mitchell P. (1991) Foundations of osmochemistry. *Biosci. Rep.* **11**: 297–346.

Mitchell R, Rich PR. (1994) Proton uptake by cytochrome c oxidase on reduction and on ligand binding. *Biochim. Biophys. Acta* **1186**: 19–26.

Mitchell R, Mitchell P, Rich PR. (1992) Protonation states of the catalytic cycle inter-mediates of cytochrome c oxidase. *Biochim. Biophys. Acta* **1101**: 188–191.

Møller IM, Rasmusson AG, Fredlund KM. (1993) NAD(P)H-ubiquinone oxido-reductases in plant mitochondria. *J. Bioenerg. Biomembr.* **25**: 377–384.

Moody AJ, Rich PR. (1994) The reaction of hydrogen peroxide with pulsed cytochrome bo from *Escherichia coli. Eur. J. Biochem.* **226**: 731–737.

Moody AJ, Brandt U, Rich PR. (1991) Single electron reduction of 'slow' and 'fast' cytochrome-c oxidase. *FEBS Lett.* **293**: 101–105.

Moody AJ, Rumbley JN, Ingledew WJ, Gennis RB, Rich PR. (1993) The reaction of hydrogen peroxide with cytochrome bo from *E. coli. Biochem. Soc. Trans.* **21**: 255S.

Moore AL, Rich PR. (1985) Organisation of the respiratory chain and oxidative phosphorylation. In: *Encyclopaedia of Plant Physiology*, Vol. 18 (eds R Douce, DA Day). Springer-Verlag, pp. 134–172.

Moore AL, Bonner WD Jr, Rich PR. (1978) The determination of the protonmotive force during cyanide insensitive respiration in plant mitochondria. *Arch. Biochem. Biophys.* **186**: 298–306.

Moore GR, Pettigrew GW. (1990) *Cytochromes c: Evolutionary, Structural and Physicochemical Aspects.* Springer-Verlag, Berlin.

Moore GR, Pettigrew GW, Rogers NK. (1986) Factors influencing redox potentials of electron transfer proteins. *Proc. Natl Acad. Sci. USA* **83**: 4998–4999.

Moser CC, Keske JM, Warncke K, Farid RS, Dutton PL. (1992) Nature of biological electron transfer. *Nature* **355**: 796–802.

Ogura T, Takahashi S, Hirota S, Shinzawa-Itoh K, Yoshikawa S, Appelman EH, Kitagawa T. (1993) Time-resolved resonance Raman elucidation of the pathway for dioxygen reduction by cytochrome c oxidase. *J. Am. Chem. Soc.* **115**: 8527–8536.

Ohnishi T. (1975) Thermodynamic and EPR characterization of iron–sulfur centers in the NADH-ubiquinone segment of the mitochondrial respiratory chain in pigeon heart. *Biochim. Biophys. Acta* **387**: 475–490.

Ohnishi T, Schägger H, Meinhardt SW, LoBrutto R, Link TA, von Jagow G. (1989) Spatial organization of the redox active centers in the bovine heart ubiquinol-cytochrome c oxidoreductase. *J. Biol. Chem.* **264**: 735–744.

Olausson T, Fjellström O, Meuller J, Rydström J. (1995) Molecular biology of nicotinamide nucleotide transhydrogenase – a unique proton pump. *Biochim. Biophys. Acta* **1231**: 1–19.

Oliveberg M, Malmström BG. (1992) Reaction of dioxygen with cytochrome c oxidase reduced to different degrees: indications of a transient dioxygen complex with copper-B. *Biochemistry* **31**: 3560–3563.

Orii Y. (1984) Formation and decay of the primary oxygen compound of cytochrome oxidase at room temperature as observed by stopped-flow, laser flash photolysis and rapid scanning. *J. Biol. Chem.* **259**: 7187–7190.

Paddock ML, McPherson PH, Feher G, Okamura MY. (1990) Pathway of proton transfer in bacterial reaction centers: replacement of serine-L223 by alanine inhibits electron and proton transfers associated with reduction of quinone to dihydroquinone. *Proc. Natl Acad. Sci. USA* **87**: 6803–6807.

Ragan CI. (1987) Structure of NADH-ubiquinone reductase (complex I). *Curr. Top. Bioenerg.* **15**: 1–36.

Ragan CI. (1990) Structure and function of an archetypal respiratory chain complex: NADH-ubiquinone reductase. *Biochem. Soc. Trans.* **18**: 515–516.

Rich PR. (1984) Electron and proton transfers through quinones and cytochrome bc complexes. *Biochim. Biophys. Acta* **768**: 53–79.

Rich PR. (1988) A critical examination of the supposed variable proton stoichiometry of the chloroplast cytochrome bf complex. *Biochim. Biophys. Acta* **932**: 33–42.

Rich PR. (1991) The osmochemistry of electron-transfer complexes. *Biosci. Rep.* **11**: 539–571.

Rich PR. (1995) Towards an understanding of the chemistry of oxygen reduction and proton translocation in the iron–copper respiratory oxidases. *Aust. J. Plant. Physiol.* **22**: 479–486.

Rich PR. (1996) Quinone binding sites of membrane proteins as targets for inhibitors. *Pesticide Sci.* (in press).

Rich PR, Jeal AE, Madgwick SA, Moody AJ. (1990) Inhibitor effects on the redox-linked protonations of the b haems of the mitochondrial bc_1 complex. *Biochim. Biophys. Acta* **1018**: 29–40.

Rich PR, Meunier B, Ward FB. (1995) Predicted structure and possible ionmotive mechanism of the sodium-linked NADH-ubiquinone oxidoreductase of *Vibrio alginolyticus*. *FEBS Lett.* **375**: 5–10.

Robertson DE, Dutton PL. (1988) The nature and magnitude of the charge-separation reactions of ubiquinol cytochrome c_2 oxidoreductase. *Biochim. Biophys. Acta* **935**: 273–291.

Robertson DE, Ding H, Chelminski PR, Slaughter C, Hsu J, Moomaw C, Tokito M, Daldal F, Dutton PL. (1993) Hydroubiquinone-cytochrome c_2 oxidoreductase from *Rhodobacter capsulatus*: definition of a minimal, functional isolated preparation. *Biochemistry* **32**: 1310–1317.

Rousseau DL, Ching Y-C, Wang J. (1993) Proton translocation in cytochrome c oxidase: redox linkage through proximal ligand exchange on cytochrome a_3. *J. Bioenerg. Biomembr.* **25**: 165–177.

Salerno JC, Ohnishi T. (1980) Studies on the stabilized ubisemiquinone species in the succinate-cytochrome c reductase segment of the intact mitochondrial membrane system. *Biochem. J.* **192**: 769–781.

Saraste M. (1990) Structural features of cytochrome oxidase. *Q. Rev. Biophys.* **23**: 331–366.

Schägger H, Link TA, Engel WD, von Jagow G. (1986) Isolation of the eleven protein subunits of the bc_1 complex from beef heart. *Methods Enzymol.* **126**: 224–237.

Shinkarev VP, Drachev LA, Mamedov MD, Mulkidjanian AJ, Semenov AY, Verkhovsky MI. (1993) Effect of pH and surface potential on the rate of electric potential generation due to proton uptake by secondary quinone acceptor of reaction centers in *Rhodobacter sphaeroides* chromatophores. *Biochim. Biophys. Acta* **1144**: 285–294.

Simpkin D, Palmer G, Devlin FJ, McKenna MC, Jensen GM, Stephens PJ. (1989) The axial ligands of haem in cytochromes: a near-infrared magnetic circular dichroism study of yeast cytochromes c, c_1 and b and spinach cytochrome f. *Biochemistry* **28**: 8033–8039.

Skulachev VP. (1992) Chemiosmotic systems and the basic principles of cell energetics. In: *Molecular Mechanisms in Bioenergetics* (ed. L Ernster). Elsevier Science Publishers B.V., Amsterdam, pp. 37–73.

Sled VD, Friedrich T, Leif H, Weiss H, Meinhardt S, Fukumori Y, Ohnishi T. (1993) Bacterial NADH-quinone oxidoreductases: iron–sulfur clusters and related problems. *J. Bioenerg. Biomembr.* **25**: 347–356.

Takahashi E, Wraight CA. (1992) Proton and electron transfer in the acceptor quinone complex of *Rhodobacter sphaeroides* reaction centers: characterization of site-directed mutants of the two ionizable residues, Glu^{L212} and Asp^{L213}, in the Q_B binding site. *Biochemistry* **31**: 855–867.

Tan K, Beattie P, Leach DRF, Rich PR, Ward FB. (1996) Expression and analysis of the gene for the catalytic β subunit of the sodium-translocating NADH-ubiquinone oxidoreductase of *Vibrio alginolyticus*. *Biochem. Soc. Trans.* **24**: 12S.

Thomas JW, Puustinen A, Alben JO, Gennis RB, Wikström M. (1993) Substitution of asparagine for aspartate-135 in subunit I of the cytochrome *bo* ubiquinol oxidase of *Escherichia coli* eliminates proton-pumping activity. *Biochemistry* **32**: 10923–10928.

Tokuda H, Unemoto T. (1984) Na^+ is translocated at NADH: quinone oxidoreductase segment in the respiratory chain of *Vibrio alginolyticus*. *J. Biol. Chem.* **259**: 7785–7790.

Trumpower BL, Gennis RB. (1994) Energy transduction by cytochrome complexes in mitochondrial and bacterial respiration: the enzymology of coupling electron transfer reactions to transmembrane proton translocation. *Annu. Rev. Biochem.* **63**: 675–716.

Tsukihara T, Aoyama H, Yamashita E, Tomizaki T, Yamaguchi H, Shinzawa-Itoh K, Nakashima R, Yaono R, Yoshikawa S. (1995) Structures of metal sites of oxidized bovine heart cytochrome c oxidase at 2.8 Å. *Science* **269**: 1069–1074.

Unemoto T, Hayashi M. (1993) Na^+-translocating NADH-quinone reductase of marine and halophilic bacteria. *J. Bioenerg. Biomembr.* **25**: 385–391.

van Belzen R, van Gaalen MCM, Cuypers PA, Albracht SPJ. (1990) New evidence for the dimeric nature of NADH:Q oxidoreductase in bovine-heart submitochondrial particles. *Biochim. Biophys. Acta* **1017**: 152–159.

Varotsis C, Babcock GT. (1990) Appearance of the $v(Fe^{IV}=O)$ vibration from a ferryl-oxo intermediate in the cytochrome oxidase/dioxygen reaction. *Biochemistry* **29**: 7357–7362.

Varotsis C, Babcock GT, Lauraeus M, Wikström M. (1995) Raman detection of a peroxy intermediate in the hydroquinone-oxidizing cytochrome aa_3 of *Bacillus subtilis*. *Biochim. Biophys. Acta* **1231**: 111–116.

Vinogradov AD. (1993) Kinetics, control and mechanism of ubiquinone reduction by the mammalian respiratory chain-linked NADH-ubiquinone reductase. *J. Bioenerg. Biomembr.* **25**: 367–376.

Watmough NJ, Cheesman MR, Greenwood C, Thomson AJ. (1994) Cytochrome *bo* from *Escherichia coli*: reaction of the oxidized enzyme with hydrogen peroxide. *Biochem. J.* **300**: 469–475.

Weidner U, Geier S, Ptock A, Friedrich T, Leif H, Weiss H. (1993) The gene locus of the proton-translocating NADH:ubiquinone oxidoreductase in *Escherichia coli*. Organization of the 14 genes and relationship between the derived proteins and subunits of mitochondrial complex I. *J. Mol. Biol.* **233**: 109–122.

Weiss H, Friedrich T, Hofhaus G, Preis D. (1991) The respiratory chain NADH dehydrogenase (complex I) of mitochondria. *Eur. J. Biochem.* **197**: 563–576.

Whitmarsh J, Bowyer JR, Crofts AR. (1982) Modification of the apparent redox reaction between cytochrome *f* and the Rieske iron–sulfur protein. *Biochim. Biophys. Acta* **682**: 404–412.

Widger WR, Cramer WA. (1991)The cytochrome *b6/f* complex. In: *Cell Culture and Somatic Cell Genetics of Plants*. Academic Press Inc, pp. 149–176.

Wikström M. (1989) Identification of the electron transfers in cytochrome oxidase that are coupled to proton-pumping. *Nature* **338**: 776–778.

Wikström M, Krab K, Saraste M. (1981) *Cytochrome Oxidase A Synthesis*. Academic Press, London.

Wikström M, Bogachev A, Finel M, Morgan JE, Puustinen A, Raitio M, Verkhovskaya M, Verkhovsky MI. (1994) Mechanism of proton translocation by the respiratory oxidases. The histidine cycle. *Biochim. Biophys. Acta* **1187**: 106–111.

Woodruff WH. (1993) Coordination dynamics of heme–copper oxidases. The ligand shuttle and the control and coupling of electron transfer and proton translocation. *J. Bioenerg. Biomembr.* **25**: 177–188.

Xu X, Koyama N, Cui M, Yamagishi A, Nosoh Y, Oshima T. (1991) Nucleotide sequence of the gene encoding NADH dehydrogenase from an alkalophile, *Bacillus* sp. strain YN-1. *J. Biochemistry* **109**: 678–683.

Yagi T. (1991) Bacterial NADH-quinone oxidoreductases. *J. Bioenerg. Biomembr.* **23**: 211–226.

Yagi T. (1993) The bacterial energy-transducing NADH-quinone oxidoreductases. *Biochim. Biophys. Acta* **1141**: 1–17.

Yagi T, Yano T, Matsuno-Yagi A. (1993) Characteristics of the energy-transducing NADH-quinone oxidoreductase of *Paracoccus denitrificans* as revealed by biochemical, biophysical, and molecular biological approaches. *J. Bioenerg. Biomembr.* **25**: 339–345.

Yamazaki I, Araiso T, Hayashi Y, Yamada H, Makino R. (1978) Analysis of acid–base properties of peroxidase and myoglobin. *Adv. Biophys.* **11**: 249–281.

Yonetani T. (1970) Cytochrome *c* peroxidase. *Adv. Enzymol.* **33**: 309–335.

Young IG, Rogers BL, Campbell HD, Jaworowski A, Shaw DC. (1981) Nucleotide sequence coding for the respiratory NADH dehydrogenase of *Escherichia coli*. *Eur. J. Biochem.* **116**: 165–170.

Yun C-H, Wang Z, Crofts AR, Gennis RB. (1992) Examination of the functional roles of 5 highly conserved residues in the cytochrome *b* subunit of the bc_1 complex of *Rhodobacter sphaeroides*. *J. Biol. Chem.* **267**: 5901–5909.

Electron transfer reactions in chemistry. Theory and experiment*

Rudolph A. Marcus

10.1 Electron transfer experiments since the late 1940s

Since the late 1940s, the field of electron transfer processes has grown enormously, both in chemistry and biology. The development of the field, experimentally and theoretically, as well as its relationship to the study of other kinds of chemical reactions, represents to us an intriguing history, one in which many threads have been brought together. In this chapter, some history, recent trends and my own involvement in this research are described.

The early experiments in the electron transfer field were on 'isotopic exchange reactions' (self-exchange reactions) and, later, 'cross reactions.' These experiments reflected two principal influences. One of these was the availability after the Second World War of many radioactive isotopes, which permitted the study of a large number of isotopic exchange electron transfer reactions, such as

$$Fe^{2+} + Fe^{\star 3+} \rightarrow Fe^{3+} + Fe^{\star 2+}, \tag{10.1}$$

and

$$Ce^{3+} + Ce^{\star 4+} \rightarrow Ce^{4+} + Ce^{\star 3+}, \tag{10.2}$$

in aqueous solution, where the asterisk denotes a radioactive isotope.

There is a twofold simplicity in typical self-exchange electron transfer reactions (so-called since other methods beside isotopic exchange were later used to study some of them): (i) the reaction products are identical with the reactants, thus eliminating one factor which usually influences the rate of a chemical reaction in a major way, namely the relative thermodynamic stability of the reactants and products; and (ii) no chemical bonds are broken or formed

---

Protein Electron Transfer, Edited by D.S. Bendall
* This chapter is a slightly modified version of the text of the Nobel Lecture delivered by Professor Marcus in Stockholm on 8 December 1992. It was originally published in *Les Prix Nobel* 1992 and is reproduced with permission from The Nobel Foundation. © The Nobel Foundation 1993 or Les Prix Nobel 1992.

in *simple* electron transfer reactions. Indeed, these self-exchange reactions represent, for these combined reasons, the simplest class of reactions in chemistry. Observations stemming directly from this simplicity were to have major consequences, not only for the electron transfer field but also, to a lesser extent, for the study of other kinds of chemical reactions as well (see Lewis 1989; Lewis and Hu, 1984; Shaik *et al.*, 1992).

A second factor in the growth of the electron transfer field was the introduction of new instrumentation, which permitted the study of the rates of rapid chemical reactions. Electron transfers are frequently rather fast, compared with many reactions which undergo, instead, a breaking of chemical bonds and a forming of new ones. Accordingly, the study of a large body of fast electron transfer reactions became accessible with the introduction of this instrumentation. One example of the latter was the stopped flow apparatus, pioneered for inorganic electron transfer reactions by N. Sutin. It permitted the study of bimolecular reactions in solution on the millisecond time scale (a fast time scale at the time). Such studies led to the investigation of what have been termed electron transfer 'cross-reactions', that is electron transfer reactions between two different redox systems, as in

$$Fe^{2+} + Ce^{4+} \rightarrow Fe^{3+} + Ce^{3+}, \tag{10.3}$$

which supplemented the earlier studies of the self-exchange electron transfer reactions. A comparative study of these two types of reaction, self-exchange and cross-reactions, stimulated by theory, was also later to have major consequences for the field and, indeed, for other areas.

Again, in the field of electrochemistry, the new post-war instrumentation in chemical laboratories led to methods which permitted the study of fast electron transfer reactions at metal electrodes. Prior to the late 1940s, only relatively slow electrochemical reactions, such as the discharge of an H_3O^+ ion at an electrode to form H_2, had been investigated extensively. They involved the breaking of chemical bonds and the forming of new ones.

Numerous electron transfer studies have now also been made in other areas, some depicted in *Figure 10.1*. Some of these investigations were made possible by a newer technology, lasers particularly, and now include studies in the picosecond and subpicosecond time regimes. Just recently, (non-laser) nanometre-sized electrodes have been introduced to study electrochemical processes that are still faster than those hitherto investigated. Still other recent investigations, important for testing aspects of the electron transfer theory at electrodes, involve the new use of an intervening ordered adsorbed monolayer of long chain organic compounds on the electrode to facilitate the study of various effects, such as varying the metal–solution potential difference on the electrochemical electron transfer rate.

In some studies of electron transfer reactions in solution, there has also been a skilful blending of these measurements of chemical reaction rates with various organic or inorganic synthetic methods, as well as with site-directed mutagenesis, to obtain still further hitherto unavailable information. The use of chemically modified proteins to study the distance dependence of electron transfer, notably by Gray and co-workers, has opened a whole new field of activity.

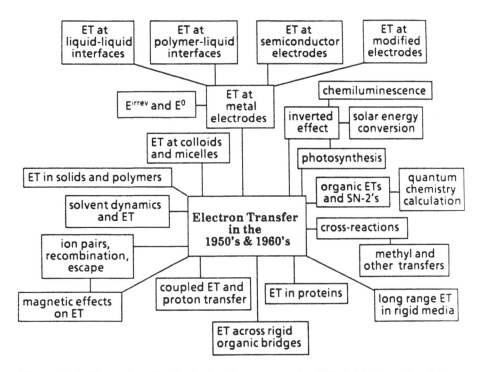

Figure 10.1. Examples of topics in the electron transfer (ET) field. Reproduced from Marcus and Siddarth (1992) *Photoprocesses in Transition Metal Complexes, Biosystems and other Molecules: Experiment and Theory* (ed. E Kochanski). NATO ASI Series C, Vol. 376, pp. 49–88, with permission from Kluwer Academic Publishers.

The interaction of theory and experiment in these many electron transfer fields has been particularly extensive and exciting, and each has stimulated the other. The present chapter addresses the underlying theory and this interaction.

10.2 The early experience

My own involvement in the electron transfer field began in a rather circuitous way. My early background was in experimental measurements of reaction rates as a chemistry graduate student at McGill University (1943–1946) and as a post-doctoral associate at the National Research Council of Canada (1946–1949). A subsequent post-doctoral study at the University of North Carolina (1949–1951) on the theory of reaction rates resulted in what is now known in the literature as RRKM theory (Rice, Ramsperger, Kassel, Marcus).

This unimolecular reaction field reflects another long and extensive interaction between theory and experiment. RRKM theory enjoys widespread use and is now usually referred to in the literature only by its acronym (or by the texts written about it (Forst, 1973; Gilbert and Smith, 1990; Robinson and Holbrook, 1972), instead of by citation of the original articles.

After the theoretical post-doctoral, I joined the faculty of the Polytechnic Institute of Brooklyn in 1951 and wondered what theoretical research to do next after writing the RRKM papers (1951–1952). I remember vividly how a friend of mine, a colleague at Brooklyn Poly, Frank Collins, came down to my office every day with a new idea on the liquid state transport theory which he was developing, while I, for theoretical research, had none. Perhaps this gap in not doing anything immediately in the field of theory was, in retrospect, fortunate. In not continuing with the study of the theory of unimolecular reactions, for which there were too few legitimate experimental data at the time to make the subject one of continued interest, I was open to investigate quite different problems in other areas. I did, however, begin a programme of experimental studies in gas phase reactions, prompted by my earlier studies at NRC and by the RRKM work.

While I was at Brooklyn Poly, a student in my statistical mechanics class in this period (Abe Kotliar) asked me about a particular problem in polyelectrolytes. It led to my writing two papers on the subject (1954–1955), one of which required a considerable expansion in my background in electrostatics, so as to analyse different methods for calculating the free energy of these systems: in polyelectrolyte molecules, it may be recalled, the ionic charges along the organic or inorganic molecular backbone interact with each other and with the solvent. In the process, I read the relevant parts of the texts that were readily available to me on electrostatics (Caltech's Mason and Weaver's was later to be particularly helpful!). When shortly thereafter I encountered some papers on electron transfer, a field entirely new to me, I was reasonably well prepared to treat the problems which lay ahead.

10.3 Developing an electron transfer theory

10.3.1 *Introduction*

My first contact with electron transfers came in 1955 as a result of chancing upon a 1952 symposium issue on the subject in the *Journal of Physical Chemistry*. An article by Bill Libby caught my eye – a use of the Franck–Condon principle to explain some experimental results, namely, why some isotopic exchange reactions which involve electron transfer between pairs of small cations in aqueous solution, such as the reaction shown in Equation 10.1, are relatively slow, whereas electron transfers involving larger ions, such as $Fe(CN)_6^{3-}$–$Fe(CN)_6^{4-}$ and MnO_4^-–MnO_4^{2-}, are relatively fast (Libby, 1952).

Libby explained this observation in terms of the Franck–Condon principle, as discussed below. The principle was used extensively in the field of spectroscopy for interpreting spectra for the excitation of the molecular electronic–vibrational quantum states. An application of that principle to chemical reaction rates was novel and caught my attention. In that paper Libby gave a 'back-of-the-envelope' calculation of the resulting solvation energy barrier which slowed the reaction. However, I felt instinctively that even though the idea – that somehow the Franck–Condon principle was involved – seemed strikingly right, the calculation itself was incorrect. The next month of study of the problem was, for me, an especially busy one. To place the topic in some perspective I first digress and describe the type of theory that was used for other types of chemical reaction rates at the time and continues to be useful today.

10.3.2 *Reaction rate theory*

Chemical reactions are often described in terms of the motion of the atoms of the reactants on a potential energy surface. This potential energy surface is really the electronic energy of the entire system, plotted versus the positions of all the atoms. A very common example is the transfer of an atom or a group B from AB to form BC

$$AB + C \rightarrow A + BC \tag{10.4}$$

An example of Equation 10.4 is the transfer of an H, such as in $IH + Br \rightarrow I + HBr$, or the transfer of a CH_3 group from one arene sulphonate to another. To aid in visualizing the motion of the atoms in this reaction, this potential energy function is frequently plotted as constant energy contours in a space whose axes are chosen to be two important relative coordinates such as, in Equation 10.4, a scaled AB bond length and a scaled distance from the centre of mass of AB to C, as in *Figure 10.2*.

A point representing this reacting system begins its trajectory in the lower right region of the figure in a valley in this plot of contours, the 'valley of the reactants.' When the system has enough energy, appropriately distributed between the various motions, it can cross the 'mountain pass' (saddle-point region) separating the initial valley from the products' valley in the upper left, and so form the reaction products. There is a line in the figure, XY, analogous to the 'continental divide' in the Rocky Mountains in the US, which separates systems which could flow spontaneously into the reactants' valley from those which could flow into the products' one. In chemists' terminology, this line represents the 'transition state' of the reaction.

In transition state theory, a quasi-equilibrium between the transition state and the reactant is frequently postulated, and the reaction rate is then calculated using equilibrium statistical mechanics. A fundamental dynamic basis, which

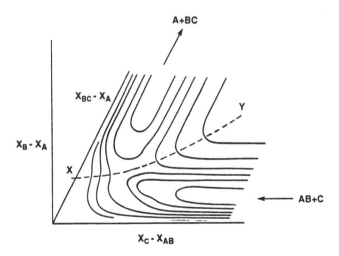

Figure 10.2. Potential energy contours for Equation 10.4, $AB + C \rightarrow A + BC$, in the co-linear case.

replaces this apparently *ad hoc* but common assumption of transition state theory and which is perhaps not as well known in the chemical literature as it deserves to be, was given many years ago by the physicist and one-time chemical engineer, Eugene Wigner (Wigner, 1938). He used a classical mechanical description of the reacting system in the many-dimensional space (of coordinates and momenta). Wigner pointed out that the quasi-equilibrium would follow as a dynamic consequence, if each trajectory of a moving point representing the reacting system in this many-dimensional space did not re-cross the transition state (and if the distribution of the reactants in the reactants' region were a Boltzmann one). In recent times, the examination of this re-crossing has been a common one in classical mechanical trajectory studies of chemical reactions. Usually, re-crossings are relatively minor, except in non-adiabatic reactions, where they are readily treated (see Section 10.3.3).

In practice, transition state theory is generalized, so as to include as many coordinates as are needed to describe the reacting system. Further, when the system can 'tunnel' quantum mechanically through the potential energy barrier (the 'pass') separating the two valleys, as for example frequently happens at low energies in H transfer reactions, the method of treating the passage across the transition state region needs, and has received, refinement. (The principal problem encountered here has been the lack of 'dynamical separability' of the various motions in the transition state region.)

10.3.3 *Electron transfer theory. Formulation*

In contrast to the above picture, we have already noted that in simple electron transfer reactions no chemical bonds are broken or formed, and so a somewhat different picture of the reaction is needed for the electron transfer reaction.

In his 1952 symposium paper, Libby noted that when an electron is transferred from one reacting ion or molecule to another, the two new molecules or ions formed are in the wrong environment of the solvent molecules, since the nuclei do not have time to move during the rapid electron jump: in Equation 10.1 a Fe^{2+} ion would be formed in some configuration of the many nearby dipolar solvent molecules that was appropriate to the original Fe^{3+} ion. Analogous remarks apply to the newly formed Fe^{3+} ion in the reaction. On the other hand, in reactions of 'complex ions,' such as those in the $Fe(CN)_6^{3-}$–$Fe(CN)_6^{4-}$ and MnO_4^-–MnO_4^{2-} self-exchange reactions, the two reactants are larger, and so the change of electric field in the vicinity of each ion, upon electron transfer, would be smaller. The original solvent environment would therefore be less foreign to the newly formed charges, and so the energy barrier to reaction would be less. In this way, Libby explained the faster self-exchange electron transfer rate for these complex ions. Further confirmation was noted in the ensuing discussion in the symposium: the self-exchange $Co(NH_3)_6^{3+}$–$Co(NH_3)_6^{2+}$ reaction is very slow, and it was pointed out that there was a large difference in the equilibrium Co–N bond lengths in the 3+ and the 2+ ions, and so each ion would be formed in a very 'foreign' configuration of the vibrational coordinates, even though the ions are 'complex ions.'

After studying Libby's paper and the symposium discussion, I realized that what troubled me in this picture for reactions occurring in the dark was that energy was not conserved: the ions would be formed in the wrong high energy environment, but the only way such a non-energy-conserving event could happen would be by the absorption of light (a 'vertical transition'), and not in the dark. Libby had perceptively introduced the Franck–Condon principle to chemical reactions, but something was missing.

In the present discussion, as well as in Libby's treatment, it was supposed that the electronic interaction of the reactants which causes the electron transfer is relatively weak. That view is still the one that seems appropriate today for most of these reactions. In this case of weak electronic interaction, the question becomes: how does the reacting system behave in the dark so as to satisfy both the Franck–Condon principle and energy conservation? I realized that fluctuations had to occur in the various nuclear coordinates, such as in the orientation coordinates of the individual solvent molecules and indeed in any other coordinates whose most probable distribution for the products differs from that of the reactants. With such fluctuations, values of the coordinates could be reached which satisfy both the Franck–Condon and energy conservation conditions and so permit the electron transfer to occur in the dark.

For a reaction such as that of Equation 10.1, an example of an initial and final configuration of the solvent molecules is depicted in *Figure 10.3*. Fluctuations from the original equilibrium ensemble of configurations were ultimately needed, prior to the electron transfer, and were followed by a relaxation to the equilibrium ensemble for the products, after electron transfer.

The theory then proceeded as follows. The potential energy U_r of the entire system, reactants plus solvent, is a function of the many hundreds of relevant coordinates of the system, coordinates which include, among others, the position and orientation of the individual solvent molecules (and hence of their dipole moments, for example), and the vibrational coordinates of the reactants, particularly those in any inner coordination shell of the reacting ions. (For example, the inner coordination shell of an ion such as Fe^{2+} or Fe^{3+} in water is known from EXAFS experiments to contain six water molecules.) No longer were there just the two or so important coordinates that were dominant in Equation 10.4.

Similarly, after the electron transfer, the reacting molecules have the ionic charges appropriate to the reaction products, and so the relevant potential energy function U_p is that for the products plus solvent. These two potential

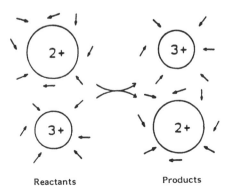

Reactants   Products

Figure 10.3. Typical nuclear configurations for reactants, products, and surrounding solvent molecules in Equation 10.1. The longer M-OH$_2$ bond length in the +2 state is indicated schematically by the larger ionic radius. Reproduced from Sutin *et al.* (1988) Nuclear reorganization barriers to electron transfer. *Pure Appl. Chem.* **60:** 1817–1830, with permission from the International Union of Pure and Applied Chemistry.

energy surfaces will intersect if the electronic coupling which leads to electron transfer is neglected. For a system with N coordinates this intersection occurs on an $(N-1)$ dimensional surface, which then constitutes in our approximation the transition state of the reaction. The neglected electronic coupling causes a well-known splitting of the two surfaces in the vicinity of their intersection. A schematic profile of the two potential energy surfaces in the N-dimensional space is given in *Figure 10.4*. (The splitting is not shown.)

Due to the effect of the previously neglected electronic coupling and the coupling to the nuclear motion near the intersection surface S, an electron transfer can occur at S. In classical terms, the transfer at S occurs at fixed positions and momenta of the atoms, and so the Franck–Condon principle is satisfied. Since U_r equals U_p at S, energy is also conserved. The details of the electron transfer depend on the extent of electronic coupling and how rapidly the point representing the system in this N-dimensional space crosses S. (It has been treated, for example, using as an approximation the well-known one-dimensional Landau–Zener expression for the transition probability at the near-intersection of two potential energy curves.)

When the splitting caused by the electronic coupling is large enough at the intersection, a system crossing S from the lower surface on the reactants' side of S continues onto the lower surface on the products' side, and so an electron transfer in the dark has then occurred. When the coupling is, instead, very weak ('non-adiabatic reactions'), the probability of successfully reaching the lower surface on the products' side is small and can be calculated using quantum mechanical perturbation theory, for example, using Fermi's golden rule, an improvement over the one-dimensional Landau–Zener treatment.

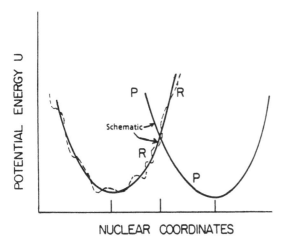

Figure 10.4. Profile of potential energy surfaces for reactants plus environment, R, and for products plus environment, P. Solid curves: schematic. Dashed curves: schematic but slightly more realistic. The typical splitting at the intersection of U_r and U_p is not shown in the figure. Reproduced from Marcus and Siddarth (1992) *Photoprocesses in Transition Metal Complexes, Biosystems and other Molecules: Experiment and Theory* (ed. E Kochanski). NATO ASI Series C, Vol. 376, pp. 49–88, with permission from Kluwer Academic Publishers.

Thus, there is some difference and some similarity with a more conventional type of reaction such as that of Equation 10.4, whose potential energy contour plots were depicted in *Figure 10.2*. In both cases, fluctuations of coordinates are needed to reach the transition state, but since so many coordinates can now play a significant role in the electron transfer reaction, because of the major and relatively abrupt change in charge distribution on passing through the transition state region, a rather different approach from the conventional one was needed to formulate the details of the theory.

10.3.4 *Electron transfer theory. Treatment*

In the initial paper (Marcus, 1956), I formulated the above picture of the mechanism of electron transfer and, to make the calculation of the reaction rate tractable, treated the solvent as a dielectric continuum. In the transition state the position-dependent dielectric polarization $\mathbf{P}_u(\mathbf{r})$ of the solvent, due to the orientation and vibrations of the solvent molecules, was not the one in equilibrium with the reactants' or the products' ionic charges. It represented instead, some macroscopic fluctuation from them. The electronic polarization for the solvent molecules, on the other hand, can respond rapidly to any such fluctuations and so is that which is dictated by the reactants' charges and by the instantaneous $\mathbf{P}_u(\mathbf{r})$.

With these ideas as a basis, what was then needed was a method of calculating the electrostatic free energy G of this system with its still unknown polarization function $\mathbf{P}_u(\mathbf{r})$. I obtained this free energy G by finding a reversible path for reaching this state of the system. Upon then minimizing G, subject to the constraint imposed by the Franck–Condon principle (reflected in the electron transfer occurring at the intersection of the two potential energy surfaces), I was able to find the unknown $\mathbf{P}_u(\mathbf{r})$ and, hence, to find the G for the transition state. That G was then introduced into transition state theory and the reaction rate calculated.

In this research, I also read and was influenced by a lovely paper by Platzmann and Franck (1952) on the optical absorption spectra of halide ions in water and later by work of physicists such as Pekar (1954) and Fröhlich (1954) on the closely related topic of polaron theory. As best as I can recall now, my first expressions for G during this month of intense activity seemed rather clumsy, but then, with some rearrangement, a simple expression emerged that had the right feel to it and that I was also able to obtain by a somewhat independent argument. The expression also reduced reassuringly to the usual one, when the constraint of arbitrary $\mathbf{P}_u(\mathbf{r})$ was removed. Obtaining the result for the mechanism and rate of electron transfer was indeed one of the most thrilling moments of my scientific life.

The expression for the rate constant k of the reaction is given by

$$k = A\exp\left(\frac{-\Delta G^\star}{k_{\mathrm{B}}T}\right) \tag{10.5a}$$

where $\Delta G^\star$, in turn, is given by

$$\Delta G^\star = \frac{\lambda}{4}\left(1 + \frac{\Delta G^\circ}{\lambda}\right)^2 \tag{10.5b}$$

The A in Equation 10.5a is a term depending on the nature of the electron transfer reaction (e.g. bimolecular or intramolecular), ΔG° is the standard free energy of reaction (and equals zero for a self-exchange reaction), λ is a 'reorganization term', composed of solvational (λ_{o}) and vibrational (λ_{i}) components.

$$\lambda = \lambda_{o} + \lambda_{i} \tag{10.6}$$

In a two-sphere model of the reactants, λ_{o} was expressed in terms of the two ionic radii a_{1} and a_{2} (including in the radius any inner coordination shell), the centre-to-centre separation distance R of the reactants, the optical (ε_{op}) and static (ε_{s}) dielectric constants of the solvent (Marcus, 1956), and the charge transferred Δe from one reactant to the other:

$$\lambda_{o} = (\Delta e)^{2} \left(\frac{1}{2a_{1}} + \frac{1}{2a_{2}} - \frac{1}{R} \right) \left(\frac{1}{\varepsilon_{op}} - \frac{1}{\varepsilon_{s}} \right) \tag{10.7}$$

For a bimolecular reaction, work terms, principally electrostatic, are involved in bringing the reactants together and in separating the reaction products, but are omitted from Equation 10.5 for notational brevity. The expression for the vibrational term λ_{i} is given by

$$\lambda_{i} = \frac{1}{2} \sum_{j} k_{j} \left(Q_{j}^{r} - Q_{j}^{p} \right)^{2} \tag{10.8}$$

where Q_{j}^{r} and Q_{j}^{p} are equilibrium values for the jth normal mode coordinate Q, k_{j} is a reduced force constant $2k_{j}^{r}k_{j}^{p} / (k_{j}^{r} + k_{j}^{p})$ associated with it, k_{j}^{r} being the force constant for the reactants and k_{j}^{p} being that for the products. (I introduced a 'symmetrization' approximation for the vibrational part of the potential energy surface, to obtain this simple form of Equations 10.5–10.8, and tested it numerically.)

In 1957, I published the results of a calculation of the λ_{i} arising from a stretching vibration in the innermost coordination shell of each reactant (Marcus, 1957) [the equation used for λ_{i} was given in the 1960 paper (Marcus, 1960)]. An early paper on the purely vibrational contribution using chemical bond length coordinates and neglecting bond–bond correlation had already been published for self-exchange reactions by George and Griffith (1959).

I also extended the theory to treat electron transfers at electrodes, and distributed it as an Office of Naval Research Report in 1957 (Marcus, 1977a,b), the equations being published later in a journal paper in 1959 (Marcus, 1959). I had little prior knowledge of the subject, and my work on electrochemical electron transfers was facilitated considerably by reading a beautiful and logically written survey article of Roger Parsons on the equilibrium electrostatic properties of electrified metal–solution interfaces (Parsons, 1954).

In the 1957 and 1965 work I showed that the electrochemical rate constant was again given by Equations 10.5–10.7, but with A now having a value appropriate to the different 'geometry' of the encounter of the participants in the reaction. The $1/2a_{2}$ in Equation 10.7 was now absent (there is only one reacting ion) and R now denotes twice the distance from the centre of the reactant's charge to the electrode (it equals the ion-image distance). A term $e\eta$ replaced the ΔG° in Equation 10.5b, where e is the charge transferred between

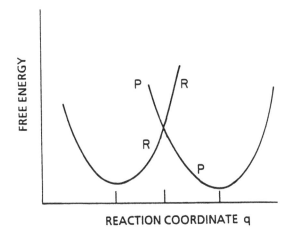

Figure 10.5. Free energy of reactants plus environment versus the reaction coordinate *q* (R curve), and free energy of products plus environment versus reaction coordinate *q* (P curve). The three vertical lines on the abscissa denote, from left to right, the value for the reactants, for the transition state, and for the products. Reproduced from Marcus and Siddarth (1992) *Photoprocesses in Transition Metal Complexes, Biosystems and other Molecules: Experiment and Theory* (ed. E Kochanski). NATO ASI Series C, Vol. 376, pp. 49–88, with permission from Kluwer Academic Publishers.

the ion and the electrode, and η is the activation overpotential, namely the metal–solution potential difference, relative to the value it would have if the rate constants for the forward and reverse reactions were equal. These rate constants are equal when the minima of the two G curves in *Figure 10.5* have the same height.

When $|e\eta| < \lambda$, most electrons go into or out of quantum states in the metal that are near the Fermi level. However, because of the continuum of states in the metal, the inverted effect was now predicted to be absent for this process, that is the counterpart of Equation 10.5 is applicable only in the region $|e\eta| < \lambda$: in the case of an intrinsically highly exothermic electron transfer reaction at an electrode, the electron can remove the immediate 'exothermicity' by (if entering) going into a high unoccupied quantum state of the metal, or (if leaving) departing from a low occupied quantum state, each far removed from the Fermi level. (The inverted region effect should, however, occur for the electron transfer when the electrode is a narrow band semiconductor.)

After these initial electron transfer studies, which were based on a dielectric continuum approximation for the solvent outside the first coordination shell of each reactant, I introduced a purely molecular treatment of the reacting system. Using statistical mechanics, the solvent was treated as a collection of dipoles in the 1960 paper (Marcus, 1960), and later in 1965 a general charge distribution was used for the solvent molecules and for the reactants (Marcus, 1965). At the same time I found a way in this 1960 paper of introducing rigorously a global reaction coordinate in this many-dimensional (N) coordinate space of the reacting system. The globally defined coordinate so introduced was equivalent to using $U_p - U_r$, the potential energy difference between the products plus solvent (U_p) and the reactants plus solvent (U_r) (see Warshel, 1982). It was, thereby, a coordinate defined everywhere in this N-dimensional space.

The free energy G_r of a system containing the solvent and the reactants, and that of the corresponding system for the products, G_p, could now be defined along this globally defined reaction coordinate. (In contrast, in reactions such as that depicted by *Figure 10.2*, it is customary, instead, to define a reaction coordinate locally, namely, in the vicinity of a path leading from the valley of the reactants through the saddle-point region and into the valley of the products.)

The potential energies U_r and U_p in the many-dimensional coordinate space are simple functions of the vibrational coordinates but are complicated functions of the hundreds of relevant solvent coordinates: there are many local minima corresponding to locally stable arrangements of the solvent molecules. However, I introduced a 'linear response approximation', in which any hypothetical change in charge of the reactants produces a proportional change in the dielectric polarization of the solvent. (Recently, I utilized a central limit theorem to understand this approximation better – beyond simple perturbation theory – and plan to submit the results for publication.) With this linear approximation, the free energies G_r and G_p became simple quadratic functions of the reaction coordinate.

Such an approach had major consequences. This picture permitted a depiction of the reaction in terms of parabolic free energy plots in simple and readily visualized terms, as in *Figure 10.5*. With them the trends predicted from the equations were readily understood. It was also important to use the free energy curves, instead of oversimplified potential energy profiles, because of the large entropy changes which occur in many electron transfer cross-reactions, due to changes in strong ion–polar solvent interactions. (The free energy plot is legitimately a one-coordinate plot while the potential energy plot is at most a profile of the complicated U_r and U_p in N-dimensional space.)

With the new statistical mechanical treatment of 1960 and 1965, one could also see how certain relationships between rate constants initially derivable from the dielectric continuum-based equations in the 1956 paper could also be valid more generally. The relations were based, in part, on Equations 10.4, 10.5 and (initially via Equations 10.7 and 10.8) on the approximate relationship

$$\lambda_{12} \cong \frac{1}{2}\left(\lambda_{11} + \lambda_{22}\right) \tag{10.9}$$

where λ_{12} is the λ for the cross-reaction and the λ_{11} and λ_{22} are those of the self-exchange reactions.

10.3.5 *Predictions*

In the 1960 paper, I had listed a number of theoretical predictions resulting from these equations (Marcus, 1960), in part to stimulate discussion with experimentalists in the field at a Faraday Society meeting on oxidation–reduction reactions, where this paper was to be presented. At the time, I certainly did not anticipate the subsequent involvement of the many experimentalists in testing these predictions. Among the latter was one which became one of the most widely tested aspects of the theory, namely, the 'cross-relation'. This expression, which follows from Equations 10.5 and 10.9, relates the rate constant k_{12} of a cross-reaction to the two self-exchange rate constants,

k_{11} and k_{22}, and to the equilibrium constant K_{12} of the reaction:

$$k_{12} \cong (k_{11}k_{22}K_{12}f_{12})^{1/2} \qquad (10.10)$$

where f_{12} is a known function of k_{11}, k_{22} and K_{12} and is usually close to unity.

Another prediction in the 1960 paper concerned what I termed there the inverted region: in a series of related reactions, similar in λ but differing in ΔG°, a plot of the activation free energy $\Delta G^\star$ versus ΔG° is seen from Equation 10.5 to first decrease as ΔG° is varied from 0 to some negative value, vanish at $\Delta G^\circ = -\lambda$ and then increase when ΔG° is made still more negative. This initial decrease of $\Delta G^\star$ with increasingly negative ΔG° is the expected trend in chemical reactions and is similar to the usual trend in 'Bronsted plots' of acid- or base-catalysed reactions and in 'Tafel plots' of electrochemical reactions. I termed that region of ΔG° the 'normal' region. However, the prediction for the region where $-\Delta G^\circ > 1$, the 'inverted region', was the unexpected behaviour, or

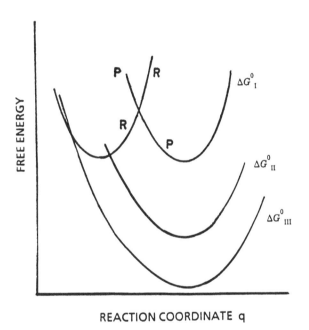

REACTION COORDINATE q

Figure 10.6. Plot of the free energy G versus the reaction coordinate q, for reactants' (R) and products' (P), for three different values of ΔG°, the cases I–III indicated in *Figure 10.7*. Reproduced from Marcus and Siddarth (1992) *Photoprocesses in Transition Metal Complexes, Biosystems and other Molecules: Experiment and Theory* (ed. E Kochanski). NATO ASI Series C, Vol. 376, pp. 49–88, with permission from Kluwer Academic Publishers.

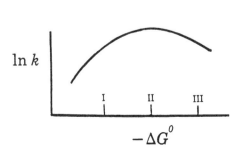

Figure 10.7. Plot of ln k_r versus $-\Delta G^\circ$ Points I and III are in the normal and inverted regions, respectively, while point II, where ln k_r is a maximum, occurs at $-\Delta G^\circ = \lambda$. Reproduced from Marcus and Siddarth (1992) *Photoprocesses in Transition Metal Complexes, Biosystems and other Molecules: Experiment and Theory* (ed. E Kochanski). NATO ASI Series C, Vol. 376, pp. 49–88, with permission from Kluwer Academic Publishers.

at least unexpected until the present theory was introduced.

This inverted region is also easily visualized using *Figures 10.6* and *10.7*: successively making ΔG° more negative, by lowering the products' G curve vertically relative to the reactant curve, decreases the free energy barrier $\Delta G^\star$ (given by the intersection of the reactants' and products' curves): that barrier is seen in *Figure 10.6* to vanish at some ΔG° and then to increase again.

Other predictions dealt with the relationship between the electrochemical and the corresponding self-exchange electron transfer rates, the numerical estimate of the reaction rate constant k and, in the case of non-specific solvent effects, the dependence of the reaction rate on solvent dielectric properties. The testing of some of the predictions was delayed by an extended sabbatical in 1960–1961, which I spent auditing courses and attending seminars at the nearby Courant Mathematical Institute.

10.3.6 *Comparisons of experiment and theory*

Around 1962, during one of my visits to Brookhaven National Laboratory, I showed Norman Sutin the 1960 predictions. Norman had either measured via his stopped flow apparatus or otherwise knew rate constants and equilibrium constants, which permitted the cross-relation Equation 10.10 to be tested. There were about six such sets of data which he had available. I remember vividly the growing sense of excitement we both felt as, one by one, the observed k_{12}s more or less agreed with the predictions of the relationship. I later collected the results of this and of various other tests of the 1960 predictions and published them in 1963 (Marcus, 1963). Perhaps by showing that the previously published expressions were not mere abstract formulae, but rather had concrete applications, this 1963 paper, and many tests by Sutin and others, appear to have stimulated numerous subsequent tests of the cross-relation and of the other predictions. A few examples of the cross-relation test are given in *Table 10.1*.

The encouraging success of the experimental tests given in the 1963 paper suggested that the theory itself was more general than the approximations (e.g. solvent dipoles, unchanged force constants) used in 1960 and stimulated me to give a more general formulation (Marcus, 1965). The latter paper also contains a unified treatment of electron transfers in solution and at metal electrodes, and served, thereby, to generalize my earlier (1957) treatment of the electrochemical electron transfers (Marcus, 1977a,b).

The best experimental evidence for the inverted region was provided by Miller, Calcaterra and Closs (Miller *et al.*, 1984), almost 25 years after it was predicted. This successful experimental test, which was later obtained for other electron transfer reactions in other laboratories, is reproduced in *Figure 10.8*. Possible reasons for not observing it in the earlier tests are several-fold and have been discussed elsewhere.

Previously, indirect evidence for the inverted region had been obtained by observing that electron transfer reactions with a very negative $\Delta G^\star$ may result in chemiluminescence: when the G_r and G_p curves intersect at a high $\Delta G^\star$ because of the inverted region effect, there may be an electron transfer to a more easily accessible G_p curve, one in which one of the products is electronically excited and which intersects the G_r curve in the normal region at

Table 10.1. Comparison of calculated and experimental k_{12} values

Reaction	k_{12} (M^{-1} s^{-1})	
	Observed	Calculated
$IrCl_6^{2-} + W(CN)_8^{4-}$	6.1×10^7	6.1×10^7
$IrCl_6^{2-} + Fe(CN)_6^{4-}$	3.8×10^5	7×10^5
$IrCl_6^{2-} + Mo(CN)_8^{4-}$	1.9×10^6	9×10^5
$Mo(CN)_8^{3-} + W(CN)_8^{4-}$	5.0×10^6	4.8×10^6
$Mo(CN)_8^{3-} + Fe(CN)_6^{4-}$	3.0×10^4	2.9×10^4
$Fe(CN)_6^{3-} + W(CN)_8^{4-}$	4.3×10^4	6.3×10^4
$Ce^{IV} + W(CN)_8^{4-}$	$>10^8$	4×10^8
$Ce^{IV} + Fe(CN)_6^{4-}$	1.9×10^6	8×10^6
$Ce^{IV} + Mo(CN)_6^{4-}$	1.4×10^7	1.3×10^7
L-$Co[(-)PDTA]^{2-} + Fe(bipy)_3^{3+}$	8.1×10^4	$\geq 10^5$
L-$Fe[(-)PDTA]^{2-} + Co(EDTA)^-$	1.3×10^1	1.3×10^1
L-$Fe[(-)PDTA]^{2-} + Co(ox)_3^{3-}$	2.2×10^2	1.0×10^3
$Cr(EDTA)^{2-} + Fe(EDTA)^-$	$\geq 10^6$	10^9
$Cr(EDTA)^{2-} + Co(EDTA)^-$	$\approx 3 \times 10^5$	4×10^7
$Fe(EDTA)^{2-} + Mn(CyDTA)^-$	$\approx 4 \times 10^5$	6×10^6
$Co(EDTA)^{2-} + Mn(CyDTA)$	9×10^{-1}	2.1
$Fe(PDTA)^{2-} + Co(CyDTA)^-$	1.2×10^1	1.8×10^1
$Co(terpy)_2^{2+} + Co(bipy)_3^{3+}$	6.4×10	3.2×10
$Co(terpy)_2^{2+} + Co(phen)_3^{3+}$	2.8×10^2	1.1×10^2
$Co(terpy)_2^{2+} + Co(bipy)(H_2O)_4^{3+}$	6.8×10^2	6.4×10^4
$Co(terpy)_2^{2+} + Co(phen)(H_2O)_4^{3+}$	1.4×10^3	6.4×10^4
$Co(terpy)_2^{2+} + Co(H_2O)_6^{3+}$	7.4×10^4	2×10^{10}
$Fe(phen)_3^{2+} + MnO_4^-$	6×10^3	4×10^3
$Fe(CN)_6^{4-} + MnO_4^-$	1.3×10^4	5×10^3
$V(H_2O)_6^{2+} + Ru(NH_3)_6^{3+}$	1.5×10^3	4.2×10^3
$Ru(en)_3^{2+} + Fe(H_2O)_6^{3+}$	8.4×10^4	4.2×10^5
$Ru(NH_3)_6^{2+} + Fe(H_2O)_6^{3+}$	3.4×10^5	7.5×10^6
$Fe(H_2O)_6^{2+} + Mn(H_2O)_6^{3+}$	1.5×10^4	3×10^4

Reproduced from Bennett (1973) Metalloprotein redox reactions. *Prog. Inorg. Chem.* **18**: 1–176, with permission from John Wiley & Sons.

a low $\Delta G^\star$, as in *Figure 10.9*. Indeed, experimentally in some reactions 100% formation of an electronically excited state of a reaction product has been observed by Bard and co-workers (Wallace and Bard, 1973), and results in chemiluminescence.

Another consequence of Equation 10.5 is the linear dependence of $k_B T \ln k$ on $-\Delta G^\circ$ with a slope of 1/2, when $|\Delta G^\circ/\lambda|$ is small, and a similar behaviour at electrodes, with ΔG° replaced by $e\eta$, the product of the charge transferred and the activation overpotential. Extensive verification of both these results has been obtained. More recently, the curvature of plots of $\ln k$ versus $e\eta$, expected from these equations, has been demonstrated in several experiments. The very recent use of ordered organic molecular monolayers on electrodes, either to slow down the electron transfer rate or to bind a redox-active agent to the electrode, but in either case to avoid or minimize diffusion control of the fast electron transfer processes, has considerably facilitated this study of the curvature in the $\ln k$ versus $e\eta$ plot (Chidsey, 1991).

Comparison of experiment and theory has also included that of the absolute

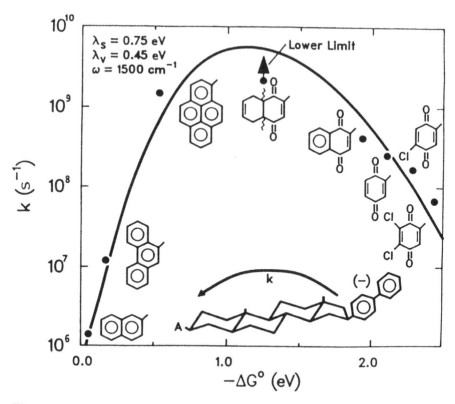

Figure 10.8. Inverted region effect in chemical electron transfer reactions. Reprinted with permission from Miller *et al.* (1984) Intramolecular long-distance electron transfer in radical anions. The effects of free energy and solvent on the reaction rates. *J. Am. Chem. Soc.* **106**: 3047–3049. Copyright 1984 American Chemical Society.

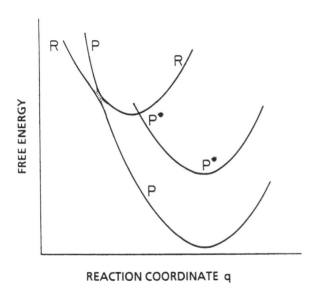

REACTION COORDINATE q

Figure 10.9. A favoured formation of an electronically excited state of the products. Reproduced from Marcus and Siddarth (1992) *Photoprocesses in Transition Metal Complexes, Biosystems and other Molecules: Experiment and Theory* (ed. E Kochanski). NATO ASI Series C, Vol. 376, pp. 49–88, with permission from Kluwer Academic Publishers.

reaction rates of the self-exchange reactions, the effect on the rate of varying the solvent, an effect sometimes complicated by ion pairing in the low dielectric constant media involved, and studies of the related problem of charge transfer spectra, such as

$$DA + h\nu \rightarrow D^+A^- \tag{10.11}$$

Here, the frequency of the spectral absorption maximum is given by ν_{max}

$$h\nu_{max} = \lambda + \Delta G^o \tag{10.12}$$

Comparisons with Equation 10.12, using Equation 10.7 for λ, have included those of the effects of separation distance and of the solvent dielectric constant.

Comparisons have also been made of the self-exchange reaction rates in solution with the rates of the corresponding electron transfer reactions at electrodes. An example of the latter is the plot given in *Figure 10.10*, where the self-exchange rates are seen to vary by some 20 orders of magnitude. The discrepancy at high ks is currently the subject of some reinvestigation of the fast electrode reaction rates, using the new nanotechnology. Most recently, a new type of interfacial electron transfer rate has also been measured, electron transfer at liquid–liquid interfaces (Geblewicz and Schriffren, 1988). In treating the latter, I extended the 'cross-relation' to this two-phase system. It is clear that much is to be learned from this new area of investigation. (The study of the transfer of ions across such an interface, on the other hand, goes back to the time of Nernst and of Planck, around the turn of the century.)

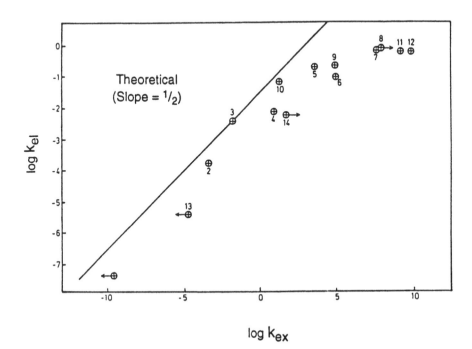

$$\log k_{ex}$$

Figure 10.10. Comparison of isotopic exchange electron transfer rates in solution, covering 20 orders of magnitude, with rates of correponding electron transfers at metal electrodes. Reproduced from Cannon (1980) *Electron Transfer Reactions*, with permission from Butterworth-Heinemann and the author.

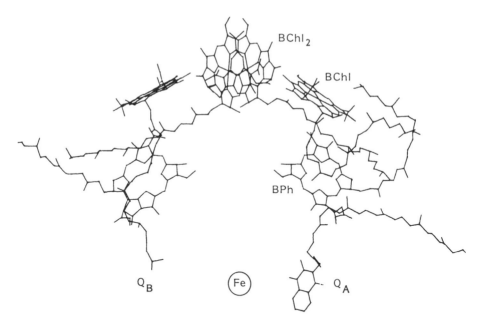

Figure 10.11. Redox-active species involved in the initial charge separation for a photosynthetic bacterium (Deisenhofer and Michel, 1989; Deisenhofer *et al.*, 1984; Yeates *et al.*, 1987), with labels added, to conform to the present text; they include a missing Q_B. Adapted from Deisenhofer *et al.* (1984) X-ray structure analysis of a membrane protein complex. Electron density map at 3 Å resolution and a model of the chromophores of the photosynthetic reaction centre from *Rhodopseudomonas viridis. J. Mol. Biol.* **180:** 385–398, with permission from Academic Press Ltd.

10.3.7 *Other applications and extensions*

As noted in *Figure 10.1*, one aspect of the electron transfer field has been its continued and, indeed, ever-expanding growth in so many directions. One of these is in the biological field, as is evident from the detailed theoretical and experimental studies discussed in the earlier chapters of this book. The three-dimensional structure of a photosynthetic reaction centre, the first membrane protein to be so characterized, was obtained by Deisenhofer, Michel and Huber, who received the Nobel Prize in Chemistry in 1988 for this work (Deisenhofer and Michel, 1989; Deisenhofer *et al.*, 1984) A bacterial photosynthetic system is depicted in *Figure 10.11*, where the protein framework holding fast the constituents in this reaction centre is not shown.

In the photosynthetic system there is a transfer of electronic excitation from 'antenna' chlorophylls (not shown in *Figure 10.11*) to a special pair $BChl_2$. The latter then transfers an electron to a bacteriopheophytin (BPh) within a very short time (~3 ps) and from it to a quinone Q_A in 200 ps and thence to the other quinone Q_B. (Other chemical reactions then occur with these separated charges at each side of the membrane, bridged by this photosynthetic reaction centre.)

To avoid wasting the excitation energy of the $BChl_2^*$ unduly, it is necessary that the $-\Delta G^0$ of this first electron transfer to BPh be small. (It is only ~0.25 eV out of an overall excitation energy of $BChl_2^*$ of 1.38 eV.) In order that this electron transfer also be successful in competing with two wasteful processes, the fluorescence and the radiationless transition of $BChl_2^*$, it is also necessary that $\Delta G^\star$ for that first electron transfer step be small and hence, by Equation 10.5, that the λ be small. The size of the reactants is large, and the immediate protein environment is largely non-polar, so leading to a small λ (see Equation 10.7). Nature appears, indeed, to have constructed a system with this desirable property.

Furthermore, to avoid another form of wasting the energy, it is also important that an unwanted back electron transfer reaction from the BPh^- to the $BChl_2^+$ not compete successfully with a second forward electron transfer step from BPh^- to Q_A. That is, it is necessary that the back transfer, a 'hole-electron recombination' step, be slow, even though it is a very highly exothermic process (~1.1 eV). It has been suggested that the small λ (~0.25 eV) and the resulting inverted region effect play a role in providing this essential condition for the effectiveness of the photosynthetic reaction centre.

There is now a widespread interest in synthesizing systems which can mimic the behaviour of nature's photosynthetic systems, and so offer other routes for the harnessing of solar energy. The current understanding of how nature works has served to provide some guidelines. In this context, as well as that of electron transfer in other proteins, there are also relevant experiments in long-range electron transfer. Originally the studies were of electron transfer in rigid glasses and were due to Miller and co-workers. (Miller, 1975) More recently, the studies have involved a donor and receptor held together by synthetically made rigid molecular bridges (Closs and Miller, 1988). The effect of varying the bridge length has been studied in the various systems. A theoretical estimate of the distance dependence of electron transfers in a photosynthetic system was first made by Hopfield (1974), who used a square barrier model and an approximate molecular estimate of the barrier height.

Recently, in their studies of long-range electron transfer in chemically modified proteins, Gray and co-workers have studied systematically the distance or site dependence of the electronic factor, by attaching an appropriate electron donor or acceptor to a desired site (Karpishin et al., 1994; Wuttke et al., 1992). For each such site, the reactant chosen should be such that $-\Delta G^\circ \approx \lambda$, that is which has a k at the maximum of the $\ln k$ versus $-\Delta G^\circ$ curve (see Equations 10.4 and 10.5). The value of k then no longer depends on a $\Delta G^\star$. Since $\Delta G^\star$ is distance dependent (see Equation 10.7), it is particularly desirable to make $\Delta G^\star \approx 0$, so that the relative ks at the various sites now reflect only the electronic factor. Dutton and co-workers have treated data similarly for a number of reactions by using, where possible, the k at the maximum of each $\ln k$ versus ΔG° curve (Moser et al., 1992). Of particular interest in such studies is whether there is a simple exponential decrease of the electronic factor on the separation distance between donor and acceptor, or whether there are deviations from this monotonic behaviour, due to local structural factors.

In a different development, the mechanism of various organic reactions has been explored by several investigators, notably by Eberson (1987), in the light of current electron transfer theory. Other organic reactions have been explored by Shaik and Pross (1989), in their analysis of a possible electron transfer mechanism versus a conventional mechanism, and by Shaik et al. (1992).

Theoretical calculations of the donor–acceptor electronic interactions, initially by McConnell (1961) and by Larsson (1981, 1983), and later by others (Beratan and Hopfield, 1984), our group among them, have been used to treat long-range electron transfer. The methods recently have been adapted to large protein systems. In our studies with Siddarth (Siddarth and Marcus, 1993), we used an 'artificial intelligence' searching technique to limit the number of amino acids used in the latter type of study (see also Stuchebrukhov and Marcus, 1995).

Another area of much current activity in electron transfers is that of solvent dynamics, following the pioneering treatment for general reactions by Kramers (1940). Important later developments for electron transfer were made by many contributors. Solvent dynamics affect the electron transfer reaction rate when the solvent is sufficiently sluggish. As we showed recently with Sumi and Nadler, the solvent dynamics effect can also be modified significantly, when there are vibrational (λ_i) contributions to λ.

Computational studies, such as the insightful one of David Chandler and co-workers (Bader et al., 1990) on the $Fe^{2+} + Fe^{3+}$ self-exchange reaction, have also been employed recently. Using computer simulations, they obtained a verification of the parabolic G curves, even for surprisingly high values of the fluctuation in G. They also extended their studies to dynamic and quantum mechanical effects of the nuclear motion. Studies of the quantum mechanical effects on the nuclear motion on electron transfer reactions were initiated in 1959 by Levich and Dogonadze (1959, 1960), who assumed a harmonic oscillator model for the polar solvent medium and employed perturbation theory. Their method was related to that used for other problems by Huang and Rhys (1950) and Kubo and Toyozawa (1955).

There were important subsequent developments by various authors on these quantum effects, including the first discussion of quantum effects for the vibrations of the reactants by Sutin in 1962 and the important work of Jortner and co-workers in 1974–1975 (Kestner et al., 1974; Ulstrup and Jortner, 1975), who combined a Levich and Dogonadze type approach to treat the high frequency vibrations of the reactants with the classical expression which I described earlier for the polar medium (see also Ovchinnikov and Ovchinnikova, 1969; Ulstrup, 1979). These quantum effects have implications for the temperature dependence of k, among other effects. Proceeding in a different (classical) direction, Savéant showed recently how to extend Equation 10.5 to reactions which involved the rupture of a chemical bond by electron transfer and which he had previously studied experimentally: $M(e) + RX \rightarrow M + R + X^-$, where R is an alkyl group, X a halide and M a metal electrode (Andrieux et al., 1992; Savéant, 1993).

A particularly important early development was that by Taube in the 1950s (see Taube, 1975), who received the Nobel Prize for his work in 1983. Taube introduced the idea of different mechanisms for electron transfer – outer sphere and inner sphere electron transfers, which he had investigated experimentally. His experimental work on charge transfer spectra of strongly interacting systems ['Creutz–Taube' ion, 1959, (Creutz and Taube, (1973)] and of weakly interacting ones has been similarly influential. Also notable has been Hush's theoretical work on charge transfer spectra, both of intensities and absorption maxima (Hush, 1967), which supplemented his earlier theoretical study of electron transfer rates (Hush, 1961).

There has been a 'spin-off' of the original electron transfer theory to other types of chemical reactions as well. In particular, the $\Delta G^\star$ versus ΔG° relationship and the cross-relation have been extended to these other reactions, such as the transfer of atoms, protons or methyl groups. (Even an analogue of Equations 10.5 and 10.9, but for binding energies instead of energy barriers, has been introduced to relate the stability of isolated proton-bound dimers AHB^+ to those of AHA^+ and BHB^+!)

Since the transfer of these nuclei involves strong electronic interactions, it is not well represented by intersecting parabolic free energy curves, and so a different theoretical approach was needed. For this purpose I adapted a 'bond-energy-bond-order' model of H. Johnston, in order to treat the problem for a reaction of the type given by Equation 10.4 (Marcus, 1968). The resulting simple expression for $\Delta G^\star$ is similar to Equation 10.5, when $|\Delta G^\circ/\lambda|$ is not large (<1/2), but differs from it in not having any inverted region. It has the same λ property as that given by Equation 10.9, and has resulted in a cross-relation analogous to Equation 10.10. The cross-relation has been tested experimentally by Lewis (Lewis, 1989; Lewis and Hu, 1984) for the transfer of methyl groups, and the $\Delta G^\star$ versus ΔG° relationship has been used or tested for other transfers by Albery and by Kreevoy and their co-workers, among others.

It is naturally gratifying to see one's theories used. A recent article, which showed the considerable growth in the use of papers such as the 1956 and 1964 articles [*Science Watch* 3(9), November 1992, p. 8], points up the impressive and continued vitality of the field itself. The remarks above on many areas of electron transfer and on the spin-off of such work on the study of other types of reactions represent a necessarily brief picture of these broad-based investigations.

Acknowledgements

My acknowledgments are to my many fellow researchers in the electron transfer field, notably Norman Sutin, with whom I have discussed so many of these matters for the past 30 or more years. I also thank my students and post-doctorals, whose presence was a constant source of stimulation to me, both in the electron transfer field and in the other fields of research which we have explored. In its earliest stage and for much of this period, this research was supported by the Office of Naval Research and also later by the National Science Foundation. The support of both agencies continues to this day and I am very pleased to acknowledge its value and timeliness here. In my Nobel lecture, I concluded on a personal note with a slide of my great-uncle, Henrik Steen (né Markus), who came to Sweden in 1892. He received his doctorate in theology from the University of Uppsala in 1915, and was an educator and a prolific writer of pedagogic books. As I noted in the biographical sketch in *Les Prix Nobel*, he was one of my childhood idols. Coming here, visiting with my Swedish relatives – some 30 or so of his descendants – has been an especially heartwarming experience for me and for my family. In a sense I feel that I owed him a debt, and that it is most fitting to acknowledge that debt here.

References

Andrieux CP, Le Gorande A, Savéant J-M. (1992) Electron transfer and bond breaking. Examples of passage from a sequential to a concerted mechanism in the electrochemical reductive cleavage of arylmethyl halides. *J. Am. Chem. Soc.* **114:** 6892–6904.

Bader JS, Kuharski RA, Chandler D. (1990) Role of nuclear tunneling in aqueous ferrous–ferric electron transfer. *J. Chem. Phys.* **93:** 230–236.

Bennett LE. (1973) Metalloprotein redox reactions. *Prog. Inorg. Chem.* **18:** 1–176.

Beratan DN, Hopfield JJ. (1984) Failure of the Born–Oppenheimer and Franck–Condon approximations for long distance electron transfer rate calculations. *J. Chem. Phys.* **81:** 5753–5759.

Cannon RD. (1980) *Electron Transfer Reactions.* Butterworths, London.

Chidsey CED. (1991) Free energy and temperature dependence of electron transfer at metal-electrolyte interfaces. *Science* **251:** 919–922.

Closs GL, Miller JR. (1988) Intramolecular long-distance electron transfer in organic molecules. *Science* **240:** 440–447.

Creutz C, Taube H. (1973) Binuclear complexes of ruthenium ammines. *J. Am. Chem. Soc.* **95:** 1086–1094.

Deisenhofer J, Michel M. (1989) The photosynthetic reaction centre from the purple bacterium *Rhodopseudomonas viridis* (Nobel lecture). *Angew. Chem. Engl. Ed.* **28:** 829–968.

Deisenhofer J, Epp O, Miki K, Huber R, Michel H. (1984) X-ray structure analysis of a membrane protein complex. Electron density map at 3 Å resolution and a model of the chromophores of the photosynthetic reaction centre from *Rhodopseudomonas viridis*. *J. Mol. Biol.* **180:** 385–398.

Eberson L. (1987) *Electron Transfer Reactions in Organic Chemistry.* Springer, New York.

Forst W. (1973) *Theory of Unimolecular Reactions.* Academic Press, New York.

Fröhlich H. (1954) Electron in lattice fields. *Adv. Phys.* **3:** 325–361.

Geblewicz G, Schriffren DI. (1988) Electron transfer between immiscible solutions. The hexacyanoferrate–lutecium biphthalocyanine system. *J. Electroanal. Chem.* **244:** 27–37.

George P, Griffith JS. (1959) Electron transfer and enzyme catalysis. In: *The Enzymes* (eds PD Boyer, H Lardy, K Myrbäck). Academic Press, New York, pp. 347–389.

Gilbert RG, Smith SC. (1990) *Theory of Unimolecular and Recombination Reactions.* Blackwells, Oxford.

Hopfield JJ. (1974) Electron transfer between biological molecules by thermally activated tunnelling. *Proc. Natl Acad. Sci. USA* **71:** 3640–3644.

Huang K, Rhys A. (1950) Theory of light absorption and non-radiative transitions in F-centres. *Proc. R. Soc. Lond. A* **204:** 406–423.

Hush NS. (1961) Adiabatic theory of outer sphere electron-transfer reactions in solution. *Trans. Faraday Soc.* **57:** 557–580.

Hush NS. (1967) Intervalence-transfer absorption. 2. Theoretical considerations and spectroscopic data. *Prog. Inorg. Chem.* **8:** 391–444.

Karpishin TB, Grinstaff MW, Komar-Panicucci S, McLendon G, Gray HB. (1994) Electron transfer in cytochrome *c* depends upon the structure of the intervening medium. *Structure* **2:** 415–422.

Kestner NR, Logan J, Jortner J. (1974) Thermal electron transfer reactions in polar solvents. *J. Phys. Chem.* **21:** 2148–2166.

Kramers HA. (1940) Brownian motion in a field of force and the diffusion model of chemical reactions. *Physica* **7:** 284–304.

Kubo R, Toyozawa Y. (1955) Application of the method of generating function to radiative and non-radiative transitions of a trapped electron in a crystal. *Prog. Theor. Phys.* **13:** 160–182.

Larsson S. (1981) Electron transfer in chemical and biological systems. Orbital rules for nonadiabatic transfer. *J. Am. Chem. Soc.* **103:** 4034–4040.

Larsson S. (1983) Electron transfer in proteins. *J. Chem. Soc., Faraday Trans. 2* **79:** 1375–1388.

Levich VG, Dogonadze RR. (1959) Theory of radiationless electronic transitions between ions in solutions. *Proc. Acad. Sci. USSR Phys. Chem. Sect.* **124:** 9–13.

Levich VG, Dogonadze RR. (1960) Adiabatic theory for electron transfers in solution. *Proc. Acad. Sci. USSR Phys. Chem. Sect.* **133:** 591–594.

Lewis ES. (1989) S_N2 and single-electron-transfer mechanisms. The distinction and relationship. *J. Am. Chem. Soc.* **111:** 7576–7578.

Lewis ES, Hu DD. (1984) Methyl transfers 8. The Marcus equation. *J. Am. Chem. Soc.* **106:** 3292–3296.

Libby WF. (1952) Theory of electron exchange reactions in aqueous solution. *J. Phys. Chem.* **56:** 863–868.

Marcus RA. (1956) On the theory of oxidation–reduction reactions involving electron transfer. I. *J. Chem. Phys.* **24:** 966–978.

Marcus RA. (1957) On the theory of oxidation–reduction reactions and of related processes. *Trans. N. Y. Acad. Sci.* **19:** 423–431.

Marcus RA. (1959) On the theory of electrochemical and chemical electron transfer processes. *Can. J. Chem.* **37:** 155–163.

Marcus RA. (1960) Theory of oxidation–reduction reactions involving electron transfer. Part 4. A statistical–mechanical basis for treating contributions from solvent, ligands and inert salt. *Disc. Faraday Soc.* **29:** 21–31.

Marcus RA. (1963) On the theory of oxidation–reduction reactions involving electron transfer. V. Comparison and properties of electrochemical and chemical rate constants. *J. Phys. Chem.* **67:** 853–857.

Marcus RA. (1965) On the theory of electron-transfer reactions. VI. Unified treatment for homogeneous and electrode reactions. *J. Chem. Phys.* **43:** 679–701.

Marcus RA. (1968) Theoretical relations among rate constants, barriers, and Brønsted slopes of chemical reactions. *J. Chem. Phys.* **72:** 891–899.

Marcus RA. (1977a) On the theory of overvoltage for electrode processes possessing electron transfer mechanism. I. In: *Special Topics in Electrochemistry* (ed. PA Rock). Elsevier, New York, pp. 180–209.

Marcus RA. (1977b) Electrostatic free energy and other properties of states having nonequilibrium polarization. II. Electrode systems. In: *Special Topics in Electrochemistry* (ed. PA Rock). Elsevier, New York, pp. 161–179.

Marcus RA, Siddarth P. (1992) Theory of electron transfer reactions and comparison with experiments. In: *Photoprocesses in Transition Metal Complexes, Biosystems and other Molecules: Experiment and Theory* (ed. E Kochanski). NATO ASI Series C, Vol. 376, Kluwer Academic Publishers, Dordrecht, pp. 49–88.

McConnell HM. (1961) Intramolecular charge transfer in aromatic free radicals. *J. Chem. Phys.* **35:** 508–515.

Miller JR. (1975) Intermolecular electron transfer by quantum mechanical tunneling. *Science* **189:** 221–222.

Miller JR, Calcaterra LT, Closs GL. (1984) Intramolecular long-distance electron transfer in radical anions. The effects of free energy and solvent on the reaction rates. *J. Am. Chem. Soc.* **106:** 3047–3049.

Moser CC, Keske JM, Warncke K, Farid RS, Dutton PL. (1992) Nature of biological electron transfer. *Nature* **355:** 796–802.

Ovchinnikov AA, Ovchinnikova MY. (1969) Contribution to theory of elementary electron-transfer reactions in polar liquids. *Soviet Phys. JETP* **29:** 688.

Parsons R. (1954) Equilibrium properties of electrified interphases. In: *Modern Aspects of Electrochemistry* (ed. JO'M Bockris). Academic Press, New York, pp. 103–179.

Pekar SI. (1954) *Untersuchungen uber die Electronentheorie der Kristalle.* Akademie Verlag, Berlin.

Platzman R, Franck J. (1952) Theory of the absorption spectra of halide ions in solution. In: *L. Farkas Memorial Volume* (eds A Farkus, EP Wigner). Research Council of Israel, Jerusalem, pp. 21–36.

Robinson PJ, Holbrook HA. (1972) *Unimolecular Reactions.* Wiley, New York.

Savéant J-M. (1993) Electron transfer, bond breaking, and bond formation. *Acc. Chem. Res.* **26:** 455–461.

Shaik SS, Pross A. (1989) Nucleophilic attack on cation radicals and cations. A theoretical analysis. *J. Am. Chem. Soc.* **111:** 4306–4312.

Shaik SS, Schlegel HB, Wolfe S. (1992) *Theoretical Aspects of Physical Organic Chemistry.* Wiley, New York.

Siddarth P, Marcus RA. (1993) Electron-transfer reactions in proteins: an artificial intelligence approach to electronic coupling. *J. Phys. Chem.* **97:** 2400–2405.

Stuchebrukhov AA, Marcus RA. (1995) Theoretical study of electron transfer in ferrocytochromes. *J. Phys. Chem.* **99:** 7581–7590.

Sutin N, Brunschwig BS, Creutz C, Winkler JR. (1988) Nuclear reorganization barriers to electron transfer. *Pure Appl. Chem.* **60:** 1817–1830.

Taube H. (1975) Intramolecular electron transfer. *Pure Appl. Chem.* **44:** 25–42.

Ulstrup J. (1979) *Charge Transfer Processes in Condensed Media.* Springer-Verlag, New York.

Ulstrup J, Jortner J. (1975) The effect of intramolecular quantum modes on free energy relationships for electron transfer reactions. *J. Chem. Phys.* **63:** 4358–4368.

Wallace WL, Bard AJ. (1973) Electrogenerated chemiluminescence. 35. Temperature dependence of the ECL efficiency of Ru(bpy)$_3^{2+}$ in acetonitrile and evidence for very high excited state yields from electron transfer reactions. *J. Phys. Chem.* **83:** 1350–1357.

Warshel A. (1982) Dynamics of reactions in polar solvents – semi-classical trajectory studies of electron-transfer and proton-transfer reactions. *J. Phys. Chem.* **86:** 2218–2224.

Wigner E. (1938) The transition state method. *Trans. Faraday Soc.* **34:** 29–41.

Wuttke DS, Bjerrum MJ, Chang I-J, Winkler JR, Gray HB. (1992) Electron tunneling in ruthenium-modified cytochrome *c. Biochim. Biophys. Acta* **1101:** 168–170.

Yeates TO, Komiya H, Rees DC, Allen JP, Feher G. (1987) Structure of the reaction centre from *Rhodobacter sphaeroides* R-26: membrane–protein interactions. *Proc. Natl Acad. Sci. USA* **84:** 6438–6442.

Further reading

Bolton JR, Mataga N, McLendon G, eds. (1991) *Electron Transport in Inorganic, Organic and Biological Systems, Adv. Chem. Ser. 228.* American Chemical Society, Washington.

Endicott JF, Schwarzenkamp JW, Watzky MA. (1994) Electron transfer: general and theoretical. In: *Mechanisms of Inorganic and Organometallic Reactions*, Vol. 8 (ed. MV Twigg). Plenum Press, New York, pp. 3–6.

Fox MA, Chanon M, eds. (1988) *Photoinduced Electron Transfer.* Elsevier, Amsterdam.

Macartney DH. (1994) Redox reactions between two metal complexes. In: *Mechanisms of Inorganic and Organometallic Reactions*, Vol. 8 (ed. MV Twigg). Plenum Press, New York, pp. 7–29.

Marcus RA. (1956) Electrostatic free energy and other properties of states having nonequilibrium polarization. I. *J. Chem. Phys.* **24:** 979–989.

Marcus RA. (1957) On the theory of oxidation–reduction reactions involving electron transfer. II. Applications to data on the rates of isotopic exchange reactions. *J. Chem. Phys.* **26:** 867–871.

Marcus RA. (1957) On the theory of oxidation–reduction reactions involving electron transfer. III. Applications to data on the rates of organic redox reactions. *J. Chem. Phys.* **26:** 872–877.

Marcus RA. (1963) Free energy of nonequilibrium polarization systems. II. Homogeneous and electrode systems. *J. Chem. Phys.* **38:** 1858–1862.

Marcus RA. (1963) Free energy of nonequilibrium polarization systems. III. Statistical mechanics of homogeneous and electrode systems. *J. Chem. Phys.* **39:** 1734–1740.

Marcus RA. (1964) Chemical and electrochemical electron-transfer theory. *Annu. Rev. Phys. Chem.* **15:** 155–196.

Marcus RA. (1965) On the theory of chemiluminescent electron-transfer reactions. *J. Chem. Phys.* **43:** 2654–2657.

Marcus RA. (1965) On the theory of shifts and broadening of electronic spectra of polar solutes in polar media. *J. Chem. Phys.* **43:** 1261–1274.

Marcus RA, Sutin N. (1985) Electron transfers in chemistry and biology. *Biochim. Biophys. Acta* **811:** 265–322.

Newton MD, Sutin N. (1984) Electron transfer reactions in condensed phases. *Annu. Rev. Phys. Chem.* **35:** 437–498.

Newton MD. (1991) Quantum chemical probes of electron-transfer kinetics: the nature of donor–acceptor interactions. *Chem. Rev.* **91:** 767–792.

Sutin N. (1983) Theory of electron transfer reactions: insights and hindsights. *Prog. Inorg. Chem.* **30:** 441–498.

Sutin N, Brunschwig BS, Creutz C, Winkler JR. (1988) Nuclear reorganization barriers to electron transfer. *Pure Appl. Chem.* **60:** 1817–1830.

Appendix A
Electron transfer rate calculations

José Nelson Onuchic and David N. Beratan

The goal of this appendix is to provide the reader with a connection between the common expressions for electron transfer rates and some of the deeper theoretical underpinnings. We hope to present a somewhat unified view of the popular rate expressions that are valid in limiting cases. We take this strategy because a number of excellent conventional approaches for deriving the rate expressions exist (DeVault, 1984; Jortner, 1980; Marcus and Sutin, 1985; Mikkelsen and Ratner, 1987; Newton and Sutin, 1984; Onuchic *et al.*, 1986; Ulstrup, 1979), but a unified view utilizing modern perspectives and notation seems needed.

We will introduce the basic concepts associated with the calculation of unimolecular electron transfer (ET) rates. Special attention is given to non-adiabatic ET, since this is the limit most relevant to long-distance biological ET processes. In ET reactions, an electronic state is 'created' in some spatial region (by a prior ET reaction, photoexcitation, etc.), and one wants to calculate the rate at which this electron transfers to a different site localized elsewhere in space. Following creation of the initial state, the ET reaction is generally a radiationless process (although in some cases electron-hole recombination is associated with luminescence). The task before us is how to describes this ET event.

A.1 The ET Hamiltonian

Clearly, we need to define two electronic states, the donor (D) and acceptor (A) localized states. Our goal is to show how to obtain the effective Hamiltonian that has been used extensively to study the generic ET problem (Garg *et al.*, 1985; Onuchic, 1987; Onuchic *et al.*, 1992; Skourtis and Onuchic, 1993; Wolynes, 1987):

$$H_{et} = T_{DA}(\vec{Q})\sigma_x + \frac{1}{2}\left(\alpha_D^{eff}(\vec{Q}) + \alpha_A^{eff}(\vec{Q})\right)$$

$$+ \frac{1}{2}\left(\alpha_D^{eff}(\vec{Q}) - \alpha_A^{eff}(\vec{Q})\right)\sigma_z + H_Q \tag{A.1}$$

Protein Electron Transfer, Edited by D.S. Bendall
© 1996 BIOS Scientific Publishers Ltd, Oxford

T_{DA} is the tunnelling matrix element between donor and acceptor. σ_x and σ_z are the Pauli matrices. 'Spin-states' of $+1$ or -1 correspond to donor or acceptor localized states respectively. We are not dealing with an actual spin flip, but rather with another two-state (D and A) system. The spin-up/spin-down language is borrowed simply so that we can utilize a standard notation.

$$<\text{up}|\sigma_z|\text{up}> = 1, \quad <\text{down}|\sigma_z|\text{down}> = -1 \quad \text{and} \quad <\text{up}|\sigma_z|\text{down}> = 0$$

and

$$<\text{up}|\sigma_x|\text{up}> = 0, \quad <\text{down}|\sigma_x|\text{down}> = 0 \quad \text{and} \quad <\text{up}|\sigma_x|\text{down}> = 1$$

Terms that appear in brackets, $|>$ and $<|$, represent spin wave functions and the complex conjugate of a spin wave function respectively. The 'up' state refers to the electron on D and the 'down' state to the electron on A.

H_Q supplies the dynamics for the nuclear coordinates $(\vec{Q})$. $\alpha_D^{\text{eff}}(\vec{Q})(\alpha_A^{\text{eff}}(\vec{Q}))$ is the instantaneous (electronic) energy for the reactant (product) state. This equation assumes that the D and A electronic states are computed at frozen nuclear configurations, that is the Born–Oppenheimer approximation is assumed.

A.2 Two-level systems

For a given nuclear configuration, the strategy for constructing the electronic part of the effective Hamiltonian is Löwdin partitioning (Löwdin, 1962). This approach was introduced to ET by Larsson (1981) and has been used extensively to investigate the effect of bridge structure on T_{DA}. There has been a significant effort aimed at computing T_{DA} in different ET systems; this is the topic of Chapter 2. As discussed there, the reduction of a many-orbital system to two states requires that the tunnelling energy of a transferring electron (E_{tun}) be specified. This tunnelling energy reflects the binding energy of an electron associated with the donor in the geometry of the activated complex. This energy (and its relative value with respect to the orbital energies of the bridge) and the structure of the bridge determine the donor–acceptor coupling. E_{tun} appears as a parameter in the effective matrix element T_{DA}. In actual computations, E_{tun} is usually set equal to the average of the D and A energies and its value can be improved iteratively. Skourtis and co-workers have examined how to perform such a reduction for single- many-electron Hamiltonians (Skourtis and Onuchic, 1993; Skourtis et al., 1993). They also described how to obtain the optimal value for E_{tun} and the limits of validity for this approach. As an example, we will show how this reduction is done for a general one-electron Hamiltonian. The electronic Hamiltonian can be written in matrix notation as:

$$\begin{pmatrix} H_{DA} & H_{DA,B} \\ H_{B,DA} & H_{Bridge} \end{pmatrix}$$

$$\text{(A.2)}$$

where H_{DA} is the Hamiltonian matrix that includes only the relevant donor and acceptor orbitals. The direct coupling between D and A in the case of long distance transfer is negligible. H_{Bridge} is the Hamiltonian matrix for the bridge and $H_{\text{B,DA}}$ is the matrix that couples the donor and acceptor to the bridge. Löwdin diagonalization yields a reduced 2 x 2 matrix.

$$\overline{H}_{\text{DA}} = H_{\text{DA}} - H_{\text{DA,B}}\left(E - H_{\text{Bridge}}\right)^{-1} H_{\text{B,DA}} \qquad (\text{A.3})$$

This reduced matrix is:

$$\overline{H}_{\text{DA}} = \begin{pmatrix} \alpha_D^{\text{eff}}(E) & T_{\text{DA}}(E) \\ T_{\text{AD}}(E) & \alpha_A^{\text{eff}}(E) \end{pmatrix} \qquad (\text{A.4a})$$

where

$$\alpha_{\text{D(A)}}^{\text{eff}}(E) = \alpha_{\text{D(A)}} + \Delta_{\text{D(A)}}(E) \qquad (\text{A.4b})$$

$$\Delta_{\text{D(A)}} = \sum_{i,j} \upsilon_{\text{D(A)}\,i}\, G_{ij}(E)\, \upsilon_{j\text{D(A)}} \qquad (\text{A.4c})$$

and

$$T_{\text{DA}} = \sum_{i,j} \upsilon_{\text{D}i}\, G_{ij}(E)\, \upsilon_{j\text{A}} \qquad (\text{A.4d})$$

The is and js in the sums run over the bridge orbitals. G is the Green function for the bridge, that is the Green's function associated with H_{el} in the absence of the donor and acceptor. $G = (E - H_{\text{Bridge}})^{-1}$.

A.2.1 E_{tun} and the Condon approximation

Equation A.4a is equivalent to Equation A.3; the eigenvalues of the two equations are the same. Of the thousands of molecular orbitals present, we are only interested in two states that define the two-level system. The first step in analysing the two states of interest is to compute the tunnelling energy. The effective donor energy can be obtained from Equation A.4 by solving

$$\overline{\alpha}_{\text{D}} = \alpha_{\text{D}}^{\text{eff}}\left(\overline{\alpha}_{\text{D}}\right) \qquad (\text{A.5})$$

The roots of this equation closest to α_{D} is the effective donor energy. A similar calculation can be performed for the acceptor. The tunnelling energy, E_{tun}, is obtained for the nuclear configuration $\overline{Q}$ that brings donor and acceptor energies into coincidence:

$$E_{\text{T}} = \overline{\alpha}_{\text{D}} = \overline{\alpha}_{\text{A}} \qquad (\text{A.6})$$

This configurational constraint is required because the process is radiationless, and is known as the Condon approximation. After calculating the tunnelling energy, we finally extract the two-level system by fixing the value of E in Equation A.4 equal to E_{tun}. How good is this approximation? It is reasonable for most of the ET problems discussed in this book, and limitations have been considered (Skourtis and Onuchic, 1993; Skourtis et al., 1993). Calculations of the matrix element T_{DA} given by Equation A.4d is the focus of Chapter 2. In most cases (and this is assumed in the following discussion), ET occurs only when the donor and acceptor energy levels match, so T_{DA} is assumed constant with a value equal to the one obtained for the nuclear configuration in which these two energies are the same. Problems may arise for the Condon approximation when T_{DA} itself is strongly E_{tun} dependent.

The electronic coupling computed at an appropriate tunnelling energy can be rewritten in a molecular orbital language as

$$T_{DA} = \sum_n {}^{(MO)} \sum_{i,j} {}^{(AO)} \frac{V_{Di} C_i^{(n)} C_j^{(n)} V_{Aj}}{E_{tun} - E^{(n)}} \qquad (A.7)$$

As noted in Chapter 2: (i) the summation involves an energy denominator so, as E_{tun} approaches the bridge states, coupling is enhanced. If E_{tun} falls below the bridge HOMO or above the bridge LUMO, bridge-localized intermediates can form and this is not the case of interest in most protein-mediated ET reactions. (ii) The bridge wave function coefficients $C_i^{(n)}$s are oscillatory, and only upon summation of the product of coefficients does roughly exponential decay of coupling with distance emerge. (iii) The nature of the bridge interaction with D and A enters through pre-factors (Vs) that are expected to be large near D and A, but small elsewhere.

A.2.2 Born–Oppenheimer approximation

The prior discussion is framed in the context of the Born–Oppenheimer separation of electronic and nuclear motion. For the Born–Oppenheimer approximation to hold, electronic energy separations must be large compared with relevant nuclear excitation energies. In addition, as the electron tunnels from the donor to the acceptor, it spends a 'time' in the classically forbidden region (Beratan and Hopfield, 1984; Bialek et al., 1989; Onuchic et al., 1986). If this time is much shorter than the period of the vibrational motion coupled to ET, the tunnelling electron sees nuclei at fixed positions, that is the Born–Oppenheimer approximation works. These approximations are reasonably good for ET at moderate distances in proteins. In extremely long-range ET, for example between quinone B of the photosynthetic reaction centre and cytochrome c (Feher et al., 1989) or between D and A at random distances in a frozen glass (at long times) (Miller et al., 1989), the Born–Oppenheimer approximation itself is called into question.

A.3   Rate expressions

Now that the electronic part of the effective Hamiltonian has been described, we turn to the nuclear coordinates $\bar{Q}$ that cause the D and A electronic levels $[\alpha_D^{eff}(\bar{Q})$ and $\alpha_A^{eff}(\bar{Q})$ of Equation A.1] to fluctuate.

Simple models of vibronic (electron–nuclear) coupling lead to electronic energies that vary linearly with nuclear coordinate, $\bar{Q}$ (Onuchic et al., 1986). If the nuclear coordinate is a single harmonic (collective) coordinate and the donor and acceptor energies (αs) vary linearly with $\bar{Q}$, one can combine this linear term with the quadratic nuclear term by completing the square when writing the full nuclear potential energy and recover two parabolic nuclear potential energy surfaces, or 'Marcus parabolas' for $\sigma_z = \pm 1$ (see *Figure A.1*).

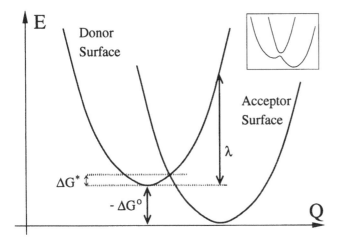

Figure A.1. Schematic representation of the 'Marcus parabolas.' Note that the donor surface is obtained by $\sigma_z = 1$ in Equation A.1 and the acceptor surface is obtained from $\sigma_z = -1$. A graphical representation of the reorganization energy (λ), that is the energy to move from the donor equilibrium position to the acceptor equilibrium position on the donor surface is shown. The driving force, that is the difference in energy between the donor and acceptor states ($-\Delta G^{\circ}$) is also called ε in the text (Equation A.9). $\Delta G^{\star}$ is the classical activation barrier and is equal to $(\varepsilon-\lambda)^2/4\lambda$ in Equation A.13.

The linear terms in the nuclear coordinate are different, depending upon whether the state is D⁻A or DA⁻. H_Q is harmonic with reorganization energy $\lambda = M\Omega^2(Q_D - Q_A)^2/2$ and ε is the driving force. $(Q_D - Q_A)$ represents the shift in the equilibrium geometry upon ET.

An important part of Equation A.1 not discussed so far represents the remaining nuclear degrees of freedom in H_Q that do not belong to the reaction coordinate. Without the coupling to this 'bath of modes', there is no way that friction and temperature can be introduced to the problem. Energy flows between the reaction coordinate and this bath. Since the bath is 'large' compared with the reaction coordinate, its average kinetic energy sets the temperature of the system and it is also responsible for damping (Caldeira and Leggett, 1983; Garg et al., 1985; Onuchic and Wolynes, 1988). For details about this formalism, the reader is referred to the two previous references. However, it is important to note that, without the existence of this thermal bath, a rate would not exist. In most simpler treatments, the existence of the bath is included by artificially broadening the energy levels of the reaction coordinate. The strength of the coupling to these 'environment modes' determines whether the reaction rate is adiabatic or non-adiabatic. A brief discussion of this topic is presented in Chapter 2 and at the end of this appendix. The interested reader is referred to the three references above.

The tremendous success of the Marcus theory leaves us with the question of why such a simple representation based upon a harmonic mode, in most cases overdamped by solvent, is such a good starting point for complex environments? As suggested by Equation A.6, electron tunnelling occurs only when the donor and acceptor energies are resonant. Therefore, the important nuclear factor that indicates resonance is the energy gap (energy difference) between the donor and acceptor states. In the harmonic case, this energy gap is linearly related to the reaction coordinate. (This relationship appears later in this section, see Equation A.9.) For this reason, the ideal reaction coordinate for the problem is the energy gap, and the question becomes how complex is its dynamics. This gap may have a very complicated dependence on the conventional nuclear coordinates.

In the case of polar solvents, a single collective reaction coordinate, normally treated as an effective overdamped harmonic mode, has been successful for describing the coupling of the mode's polarization to the ET reaction. The dynamics are described by the overdamped time τ_D/ε_s [τ_D is the Debye time for the solvent and ε_s is its static dielectric constant (Barbara and Jarzeba, 1990; Maroncelli et $al.$, 1989; Simon, 1988)]. Recent theoretical advances have shown how a many-dimensional complex solvent can be reduced to this simple representation (Bacgchi, 1989; Leite and Onuchic, 1996; Onuchic and Wolynes, 1993).

Now that we have computed the D–A interaction, we need to compute the ET rate itself. There are numerous approaches to this problem, the most familiar one begins with transition state theory and incorporates corrections as needed. We discuss this strategy at the end of this section (in the context of adiabatic versus non-adiabatic ET rates). The non-adiabatic limit is the ET rate limit of most relevance to biological ET. If the tunnelling matrix element is very small (see discussion at the end of the appendix), the ET rate can be written as

$$k_{et} = \frac{2\pi}{\hbar} T_{DA}^2 \sum_{i,f} P_i^T \delta(E_i - E_f) \qquad (A.8)$$

P_i^T is the thermal distribution of the donor–acceptor state configurations. For the harmonic case described above and assuming Q to be classical,

$$\Delta E = E_i - E_f = M\Omega^2(Q_A - Q_D)Q + \varepsilon - \lambda \qquad (A.9)$$

and $P_i^T(\Delta E)$ can be obtained from the change of variable $Q \rightarrow \Delta E$ since

$$P_i^T(Q) = \frac{1}{\sqrt{2\pi k_B T / M\Omega^2}} \exp{-\frac{M\Omega^2(Q-Q_D)^2}{2k_B T}} \qquad (A.10)$$

leading to

$$P_i^T(\Delta E) = \frac{1}{\sqrt{4\pi\lambda k_B T}} \exp{-\frac{[\Delta E - (\varepsilon - \lambda)]^2}{4\lambda k_B T}} \qquad (A.11)$$

The electron transfer rate becomes

$$k_{et} = \frac{2\pi}{\hbar} T_{DA}^2 P_i^T(\Delta E = 0) = \frac{2\pi}{\hbar} T_{DA}^2 < \delta(\Delta E) > \qquad (A.12)$$

This is the well known expression for non-adiabatic ET in the classical limit, that is

$$k_{et} = \frac{2\pi}{\hbar} T_{DA}^2 \frac{1}{\sqrt{4\pi\lambda k_B T}} \exp\left[-\frac{(\varepsilon-\lambda)^2}{4\lambda k_B T}\right] \tag{A.13}$$

A full quantum expression can be obtained by replacing

$$<\delta(\Delta E)> \;\rightarrow\; <\delta(\Delta\hat{E})> \;=\; \int dt \left\langle T\exp\left[-i\int_0^t d\tau \frac{\Delta\hat{E}(\tau)}{\hbar}\right]\right\rangle \tag{A.14}$$

For the case of a single harmonic reaction coordinate, the full quantum expression can be written as (Bialek and Goldstein, 1986; Garg et al., 1985):

$$k_{et} = \frac{T_{DA}^2}{\hbar^2} \int_{-\infty}^{\infty}$$

$$d\tau \exp\left\{ \frac{i\varepsilon\tau}{\hbar} - i\frac{\delta Q^2}{\hbar}\int_0^{\infty}\frac{d\omega}{2\pi}\sin(\omega\tau)\frac{\gamma(\omega)}{\omega}\times\frac{\Omega^4}{\mid\Omega^2-\omega^2-i\omega\gamma(\omega)\mid^2} \right.$$
$$\left. -\frac{\delta Q^2}{\hbar}\int_0^{\infty}\frac{d\omega}{2\pi}[1-\cos(\omega\tau)]\frac{\gamma(\omega)}{\omega}\times\frac{\Omega^4\coth(\hbar\omega/2k_B T)}{\mid\Omega^2-\omega^2-i\omega\gamma(\omega)\mid^2}\right\} \tag{A.15}$$

where $\gamma(\omega)$ is the damping constant. If γ is assumed constant we are in the 'ohmic' limit, that is the damping force is proportional to the velocity. An approximate solution of Equation A.15 may be obtained by applying saddle-points or steepest descent methods (Bialek and Goldstein, 1986; Onuchic et al., 1990). For underdamped modes there are multiple saddle points, and their contributions must be summed. One obtains

$$k_{et} = A\exp[-G]\sum_{n=0}^{\infty}\cos(2\pi n\varepsilon/\hbar\Omega)\exp[-n\gamma/\Gamma] \tag{A.16a}$$

where

$$\Gamma^{-1}(T=0) = \frac{2\pi\varepsilon}{\hbar\Omega^2} \tag{A.16b}$$

$$\Gamma^{-1}(T\gg\hbar\Omega/k_B T) \approx \frac{2\pi\varepsilon}{\hbar\Omega^2}\times\frac{k_B T}{\hbar\Omega} \tag{A.16c}$$

and

$$A \approx \frac{2\sqrt{\pi}T_{DA}^2}{\hbar}\left\{2(\hbar\Omega)^2 S(2\bar{n}+1)+...\right\}^{-1/2} \tag{A.16d}$$

$$G \approx \frac{(\varepsilon \pm S\hbar\Omega)^2}{2S(\hbar\Omega)^2(2\bar{n}+1)} - \frac{(\varepsilon \pm S\hbar\Omega)^3}{6S(\hbar\Omega)^3(2\bar{n}+1)^5}\left[3+(2\bar{n}+1)^2\right]+... \tag{A.16e}$$

where $\bar{n}= [\exp(\hbar\Omega/k_{\mathrm{B}}T) - 1]^{-1}$ and $\lambda = S\hbar\Omega$. We have expanded G around the term that dominates when $\hbar \to 0$ in order to emphasize that we are looking for quantum corrections to the classical and semi-classical results. For example, if we retain only the $n = 0$ saddle-point and only the first term in G, the result reduces to the expression obtained by Hopfield (1974). The so-called quantum limit (Jortner, 1980) is obtained by setting $\gamma \to 0$. In this limit, Equation A.16b would reduce to a delta function that is inhomogeneously broadened.

The discussion to this point has been based on the existence of a single reaction coordinate. Often, in addition to a classical or semi-classical reaction coordinate, quantum modes are also important. These are the inner sphere modes discussed by Marcus. The existence of quantum modes complicates the parabolic dependence of the activation barrier on the driving force ε. For completeness, we present the rate expression for the case of a transferring electron coupled to two reaction coordinates, a quantum one, y, and a classical one, z. λ_y and λ_z are their respective reorganization energies. y is a high frequency fully quantum mode, so $\hbar\Omega_y >> k_{\mathrm{B}}T$. Quantum modes coupled to ET are mostly localized vibrations of the ET chromophores (at room temperature, $k_{\mathrm{B}}T = 200$ cm^{-1}, covalent chain skeletal vibrations with $\hbar\Omega_y \gg k_{\mathrm{B}}T$ are quantized). In this case (Onuchic, 1987), the electron transfer rate is

$$k_{\mathrm{et}} = \frac{2\pi}{\hbar}\sum_{m_{\mathrm{A}}}\left\{T_{\mathrm{DA}}^{\mathrm{eff}}(m_{\mathrm{A}})\right\}^2 \frac{1}{\sqrt{4\pi\lambda_x k_{\mathrm{B}}T}}\exp\left[-\frac{\left(\varepsilon_{m_{\mathrm{A}}}^{\mathrm{eff}} - \lambda_x\right)^2}{4\lambda_x k_{\mathrm{B}}T}\right] \tag{A.17a}$$

and

$$T_{\mathrm{DA}}^{\mathrm{eff}}(m_{\mathrm{A}}) = T_{\mathrm{DA}} < n_{\mathrm{D}} = 0\,|\,m_{\mathrm{A}} > \text{ and } \varepsilon_{m_{\mathrm{A}}}^{\mathrm{eff}} = \varepsilon - m_{\mathrm{A}}\hbar\Omega_y \tag{A.17b}$$

where $|n_{\mathrm{D}}>$ is the n_{D}th level for y when the electron is on the donor and $|m_{\mathrm{A}}>$ is the m_{A}th level for y when the electron is on the acceptor.

We conclude this section by relating the theoretical approaches presented here to transition state theory, the most common strategy for computing chemical reaction rates. We use this method to discuss the question of adiabaticity versus non-adiabaticity introduced in Chapter 2. For simplicity, we present a brief discussion for classical barrier crossing. For a full description of the quantum and classical case, the reader is referred to Onuchic and Wolynes (1988). The transition state theory of classical barrier crossing replaces the question of how many reactive events occur by the question of how many crossings from reactant to product occur. The problem therefore is much simpler, we only have to calculate the flux across the top of the barrier. (Eyring, 1935; Wigner, 1937). Corrections to this value are normally necessary because multiple crossings can occur for each reactive event. Thus, the transition state rate has to be corrected by a transmission coefficient $\kappa = k/k_{TST}$ that can be estimated by the inverse of the typical number of forward crossings per reactive event.

We now present two examples of κ estimates. First, consider an adiabatic reaction with a diffusive reaction coordinate [Kramer's rate (Kramers, 1940)]. In this case, a typical trajectory resides mostly around the reactant or product well, and a fluctuation rarely causes a barrier crossing. Following the crossing event, once the trajectory has an energy lower than the barrier height by about $k_B T$, it will typically settle into either side of the well. The distance from either well to the transition state is l_{TST}. Once the system diffuses this distance, it is trapped in one of the two wells. Using a well-known relationship for random walks, the number of steps taken in the transition state region is of the order of $(l_{TST}/l)^2$ where l is the corresponding mean free path and is inversely proportional to the damping coefficient, η; η measures the coupling between the reaction coordinate and the bath. In the overdamped limit, the diffusion coefficient is $k_B T/\eta$. The number of re-crossings is the fraction of these steps that actually go across the saddle-point. Since a step is of length l, this fraction is l/l_{TST}. The transmission coefficient is hence proportional to the mean free path and it is $\kappa = l/l_{TST}$. When the appropriate algebra is performed, we recover exactly the adiabatic limit associated with Equation 2.3 of Chapter 2.

The second case we discuss is the question of adiabaticity versus non-adiabaticity. A different but equivalent discussion to the one presented in Chapter 2 is the following. Using the idea of re-crossings, we can ask the following question: what is the probability of going from reactant (donor) to product (acceptor) in each crossing? This can be answered using the standard Landau–Zener approach to ballistic crossings (Landau, 1932; Zener, 1932), and it can be extended to non-ballistic ones. When these corrections are included, we find exactly the result given by Equation 2.3 in Chapter 2 for the overdamped regime. The question of adiabatic versus non-adiabatic rates is a very subtle one. Single crossings may each have a small probability to switch between donor and acceptor states, but if one has multiple crossings per event, these probabilities add up. This interplay between coupling to the environment and Landau–Zener factors is important in determining the nature of the reaction rate.

We hope that this appendix has provided you with a few new ideas about strategies for computing ET reaction rates as well as some clues about the origins of standard rate expressions. We have not provided all the details, but this would itself require a separate volume. However, we hope we have provided the key references (and not an overwhelming number of them) that will allow you to deepen your understanding on this subject if you so desire.

Acknowledgements

We are grateful to our collaborators for their thoughtful discussion of these ideas with us. This work is supported by the National Science Foundation (MCB-9316186 and CHE-9257093), the National Institutes of Health (GM48043) and the Department of Energy (DE-FG36-94G010051).

References

Barbara PF, Jarzeba, W. (1990) Ultrafast photochemical intramolecular charge and excited state solvation. *Adv. Photochem* **15**: 1–68.

Bacgchi B. (1989) Dynamics of solvation and charge transfer reactions in dipolar liquids. *Annu. Rev. Phys. Chem.* **40**: 115–141.

Beratan DN, Hopfield JJ. (1984) Failure of the Born–Oppenheimer and Franck–Condon approximations for long distance electron transfer rate calculations. *J. Chem. Phys.* **81**: 5753–5759.

Bialek W, Goldstein RF. (1986) Protein dynamics, tunneling, and all that. *Phys. Scripta* **34**: 273–282.

Bialek W, Bruno WJ, Joseph J, Onuchic JN. (1989) Quantum and classical dynamics in biochemical reactions. *Photosynth. Res.* **22**: 15–27.

Caldeira AO, Leggett AJ. (1983) Quantum tunnelling in a dissipative system. *Ann. Phys.* **149**: 374–456.

DeVault D. (1984) *Quantum Mechanical Tunneling in Biological Systems.* 2nd edn. Cambridge University Press, New York.

Eyring HJ. (1935) The activated complex in chemical reactions. *J. Chem. Phys.* **3**: 107–115.

Feher G, Allen J, Okamura M, Rees D. (1989) Structure and function of bacterial photosynthetic reaction centres. *Nature* **339**: 111–116.

Garg A, Onuchic JN, Ambegaokar V. (1985) Effect of friction on electron transfer in biomolecules. *J. Chem. Phys.* **83**: 4491–4503.

Hopfield JJ. (1974) Electron transfer between biological molecules by thermally activated tunneling. *Proc. Natl Acad. Sci. USA* **71**: 3640–3644.

Jortner J. (1980) Dynamics of electron transfer in bacterial photosynthesis. *Biochim. Biophys. Acta* **594**: 193–230.

Kramers HA. (1940) Brownian motion in a field of force and the diffusion model of chemical reactions. *Physica* **7**: 284–304.

Landau L. (1932) Zur Theorie der Energieübertragung bei Stössen. *Phys. Z. SowjU* **1**: 88–98.

Larsson S. (1981) Electron transfer reactions in chemical and biological systems. Orbital rules for nonadiabatic transfer. *J. Am. Chem. Soc.* **103**: 4034–4040.

Leite VBP, Onuchic JN. (1996) Structure and dynamics of solvent landscapes in charge transfer reactions. *J. Phys. Chem.* (in press).

Löwdin PO. (1962) Studies in perturbation theory. IV. Solution of eigenvalue problem by projection operator formalism. *J. Math. Phys.* **3**: 969–982.

Marcus RA, Sutin N. (1985) Electron transfers in chemistry and biology. *Biochim. Biophys. Acta* **811**: 265–322.

Maroncelli M, MacInnis J, Fleming GR. (1989) Polar solvent dynamics and electron-transfer reactions. *Science* **243**: 1674–1681.

Miller JR, Beitz JV, Huddleston RK. (1984) Effect of free energy on rates of electron transfer between molecules. *J. Am. Chem. Soc.* **106**: 5057–5068.

Mikkelsen KV, Ratner MA. (1987) Electron tunneling in solid-state electron-transfer reactions. *Chem. Rev.* **87**: 113–153.

Newton MD, Sutin N. (1984) Electron transfer reactions in condensed phases. *Annu. Rev. Phys. Chem.* **35**: 437–480.

Onuchic, JN. (1987) Effect of friction on electron transfer – the two reaction coordinate case. *J. Chem. Phys.* **86**: 3925–3943.

Onuchic JN, Wolynes, PG. (1988) Classical and quantum pictures of reaction dynamics in condensed matter: resonances, dephasing and all that. *J. Phys. Chem.* **92**: 6495–6503.

Onuchic JN, Wolynes, PG. (1993) Energy landscapes, glass transitions, and chemical reaction dynamics in biomolecular or solvent environment. *J. Chem. Phys.* **98**: 2218–2224.

Onuchic JN, Beratan DN, Hopfield JJ. (1986) Some aspects of electron transfer reaction dynamics. *J. Phys. Chem.* **90**: 3707–3721.

Onuchic JN, Goldstein RF, Bialek, W. (1990) Biomolecular dynamics – quantum or classical? Results for photosynthetic electron transfer. In: *Perspectives in Photosynthesis, Proceedings of the 22nd Jerusalem Symposium on Quantum Chemistry and Biochemistry,* (eds J Jortner, B Pullman). D. Reidel Publishing Co., Dordrecht, pp. 211–226.

Onuchic JN, Beratan DN, Winkler JR, Gray HB. (1992) Pathway analysis of protein electron-transfer reactions. *Annu. Rev. Biophys. Biomol. Struct.* **21:** 349–377.

Simon JD. (1988) Time-resolved studies of solvation in polar medium. *Acc. Chem. Res.* **21:** 128–134.

Skourtis SS, Onuchic JN. (1993) Effective two-state systems in bridge-mediated electron transfer: a Green function analysis. *Chem. Phys. Lett.* **209:** 171–177.

Skourtis SS, Beratan DN, Onuchic JN. (1993) The two-state reduction for electron and hole transfer in bridge-mediated electron-transfer reactions. *Chem. Phys.* **176:** 501–520.

Ulstrup J. (1979) *Charge Transfer Process in Condensed Matter.* Springer Verlag, New York.

Wigner EJ. (1937) Calculation of the rate of elementary association reactions. *J. Chem. Phys.* **5:** 720–725.

Wolynes, PG. (1987) Dissipation, tunneling, and adiabicity criteria for curve crossing problems in the condensed phase. *J. Chem. Phys.* **86:** 1957–1966.

Zener C. (1932) Non-adiabatic crossing of energy levels. *Proc. R. Soc. Lond. A* **137:** 696–702.

Appendix B
Kinetic analysis

D.S. Bendall

B.1 Introduction

An understanding of electron transfer reactions depends on the measurement of rate constants and their evaluation in terms of assumed kinetic mechanisms. This appendix gives a fuller discussion of the variety of mechanisms observed amongst proteins than was possible in Chapter 3, and of the types of kinetic analysis that have been employed. In the more complex cases, however, the reader is referred to the original papers.

The variety of analytical equations that have been used to describe ionic strength effects is also discussed. The limitations of them all are shown by the fact that they depend on simple models of the distribution of charges in order to be amenable to analysis, and yet the equations derived involve more parameters than can be evaluated confidently.

B.2 Observed rate constants

B.2.1 *First order and pseudo-first order reactions*

Most electron transfer reactions are intrinsically first order, as in Scheme B.1:

$$(AB) \overset{k_{et}}{\to} (A^-B^+)$$

Scheme B.1

(AB) may refer either to two redox centres within a protein or to the active form of the complex between two proteins interacting by diffusion (see Scheme 3.1). In the latter case, the overall reaction is bimolecular and second order. Nevertheless, the reaction conditions are usually arranged so that the observed time course is exponential and the reaction may then be described as pseudo-first order. This is the case if one reactant, say A, is in considerable excess so that its concentration does not change significantly during the course of the reaction.

Protein Electron Transfer, Edited by D.S. Bendall
© 1996 BIOS Scientific Publishers Ltd, Oxford

The rate equation for Scheme B.1 is

$$\frac{d[AB]}{dt} = -k_{et}[AB] \tag{B.1}$$

Rearrangement and integration between $t = 0$ and $t = t$ gives

$$\ln[AB]_t = \ln[AB]_0 - k_{et}t \tag{B.2}$$

or in exponential form

$$[AB]_t = [AB]_0 \, e^{-k_{et}t} \tag{B.3}$$

Although Equation B.2 is linear and convenient for manual plotting, the rate constant, k_{et}, is usually extracted from the experimental data by fitting them to Equation B.3 with a computer program employing a non-linear least squares procedure. The Marquardt (1963) algorithm is commonly used. A reaction may also be represented in terms of its time constant, τ, which is $1/k_{et}$ and represents the time required for [AB] to fall to $1/e$ (37%) of its initial value. Non-linear least squares programs can also be used to fit results to two, or occasionally more, exponential processes.

In the pseudo-first order case, the observed rate constant, k_1, should be plotted as a function of the concentration of A, the reactant present in excess. In the simplest case, the experimental points are best fitted to a straight line, the slope of which gives the second order rate constant (k_2) directly. In general, however, a curve is obtained which is usually a rectangular hyperbola. The initial slope of this curve gives k_2, and the saturating rate extrapolated to infinite concentration of A gives further kinetic information depending on the appropriate kinetic model (see Section B.3).

Two or more exponentials occur not infrequently in electron transfer reactions, for example when a single protein contains more than one redox centre (Tollin, 1995). The case of a bimolecular flash-induced reaction [(photooxidation of plastocyanin by photosystem (PSI)] has been discussed in several recent papers (Drepper et al., 1996; Nordling et al., 1991; Sigfridsson et al., 1995, 1996). The reaction is described in Scheme B.2, in which A is plastocyanin, present in excess, and B^+ is $P700^+$ generated by flash activation:

$$A + B^+ \underset{k_{-a}}{\overset{k_a}{\rightleftharpoons}} (AB^+) \overset{k_{et}}{\rightarrow} (A^+B)$$

Scheme B.2

Although this is superficially similar to Scheme 3.1, it differs significantly in that B^+ and (AB^+) are generated instantaneously by the flash, and the concentration of (AB^+) is established by the binding equilibrium during the previous dark period. The steady-state approximation is inappropriate in this case, but the differential equations can be solved using Laplace transforms. The time course of re-reduction of B^+ can be resolved into two exponentials with rate constants k_f and k_s and relative amplitudes which vary with [A]. The original papers should be consulted for details.

B.2.2 *Second order reactions*

Electron transfer between two proteins in solution is second order in terms of the concentrations of *free* protein, and becomes so in terms of total concentrations when these are low. In general, the integrated rate equation for a second order reaction involves measurement of the instantaneous concentrations of both reactants, which is often difficult to achieve. A relatively simple situation applies, however, when the initial concentrations of the reactants are equal. In this case ($[A]_0 = [B]_0$), the differential rate equation

$$\frac{d[A]}{dt} = k_2[A][B] \tag{B.4}$$

yields a simple integrated equation which can be fitted to experimental data to yield a value for k_2 directly:

$$[B]_t = \frac{[B]_0}{1 + k_2[B]_0 t} \tag{B.5}$$

This method of analysis is sometimes used in experiments involving stopped-flow spectrophotometry because the ability to use low concentrations of *both* proteins renders very fast reactions easier to measure (Northrup *et al.*, 1993).

B.3 Kinetic models

The differential equations that describe simple collision processes become more complex and difficult to solve when intermediates are introduced into the reaction scheme, as is almost always necessary with the type of reaction we are considering. A useful technique is the steady-state method, which reduces the mathematical problem to simple algebra. The simplifying approximation that makes this possible is the assumption that the rate of change of the concentration of a reaction intermediate is small compared with those of the reactants and products. Under these conditions, it is a fair approximation that the rate of formation of the complex equals its rate of decay. Complex mechanisms can be handled by this technique, the limitation being the large number of terms that may be generated.

An alternative to the production of an analytical equation is numerical integration of the appropriate differential rate equation. A computer program for this purpose is KINSIM which allows one to simulate the time course associated with any mechanism (Frieden, 1993a,b, 1995); success depends on the initial choice of rate constants. Another program called FITSIM uses KINSIM to fit a given mechanism to experimental data. Such a technique is useful not only to simulate a mechanism for which no analytical equation is available, but also to test the applicability of an assumed mechanism to any particular set of experimental results.

B.3.1 *Reversible reactions*

When the equilibrium constant for a reaction is small, the back reaction has a significant effect on observed pseudo-first order rate constants and the relatively simple kinetic analysis discussed in Section 3.2 is no longer valid.

$$A + B \underset{k_{-a}}{\overset{k_a}{\rightleftharpoons}} (AB) \underset{k_{-et}}{\overset{k_{et}}{\rightleftharpoons}} (A^-B^+) \underset{k_{-d}}{\overset{k_d}{\rightleftharpoons}} A^- + B^+$$

Scheme B.3

The steady-state approximation may be applied to the above scheme and an analytical equation derived, but the result is complex. However, a simulation of the reaction using KINSIM shows that certain terms are too small to be significant if reasonable values for individual rate constants are assumed. Elimination of these yields the following approximate equation (E.J. McLaughlin, personal communication) for the observed pseudo-first order rate constant:

$$k_f = \frac{k_a k_f [A]}{k_{-a} + k_f + k_a [A]} \tag{B.6}$$

Equation B.6 is identical to Equation 3.1, which describes the irreversible reaction, except that k_{et} has been substituted by k_f where

$$k_1 = \frac{k_{et} k_d K_{et}}{k_{et} + k_d K_{et}} \tag{B.7}$$

K_{et} being the equilibrium constant of the electron transfer step. The interpretation of Equation B.6 is more straightforward if the rates of association and dissociation are independent of oxidation state ($k_d = k_{-a}$), when K_{et} equals the overall equilibrium constant K_{eq}. In this case, Equation B.7 can be written as

$$k_f = \frac{k_{et} k_{-a} K_{eq}}{k_{et} + k_{-a} K_{eq}} \tag{B.8}$$

and it becomes clear that the equation appropriate for a reversible reaction reverts to that for the 'irreversible' case when $k_{-a} K_{eq} \gg k_{et}$. For rapid reactions k_{-a} tends to be of comparable magnitude to k_{et} and, unless $K_{eq} \gg 1$, equating k_f and k_{et} will not be justified.

An example of a thorough analysis of a flash-activated reaction in which the reverse electron transfer should be taken into account is given by Drepper et al. (1996).

B.3.2 *Two binding sites and rearrangement*

The type of model discussed above in which a single binding and reaction site is postulated may be too simple, and evidence for at least two binding sites has been described for several systems (see Section 3.5.2). A general equation for this type of mechanism would be complex, and some simplification is necessary to yield a manageable solution. Zhou and Kostic (1992, 1993) have treated a case in which two interconvertible binding sites seem to be associated with slow and fast electron transfer respectively, with the equilibrium constants for rearrangement ($K_R = k_r/k_{-r}$) and binding strongly favouring the slowly reacting site. Under these circumstances, flash activation of the pre-formed complex (which avoids dealing with the rates of diffusional encounter and dissociation) gave a monophasic response interpreted with the help of the following equation for the observed first order rate constant:

$$k_1 = k_s + \frac{k_r k_f}{k_{-r} + k_f}$$ (B.9)

where k_f and k_s are the fast and slow electron transfer rate constants. Rearrangement was regarded as a type of surface diffusion, the rate of which would be decreased by increasing the viscosity of the medium, which would not be expected to affect k_s and k_f. Simple theory suggests that k_r should vary with η^{-1} (Kramers, 1940), but Qin and Kostic (1994) have discussed how this relationship might be modified to take account of non-solvent friction.

The simplest case of a bimolecular reaction in which rearrangement takes place is one with two forms of the reactant complex with a compulsory order of formation and a negligible back reaction, as in Scheme B.4:

$$A + B \underset{k_{-a}}{\overset{k_a}{\rightleftharpoons}} (AB)_1 \underset{k_{-r}}{\overset{k_r}{\rightleftharpoons}} (AB)_2 \overset{k_{et}}{\rightarrow} (A^-B^+) \rightarrow A^- + B^+$$

Scheme B.4

Under pseudo-first order conditions, with A in excess, the following expression can be derived for the observed rate constant:

$$k_1 = \frac{k_a k_r k_{et}[A]}{k_{-a}(k_{et} + k_{-r}) + k_r k_{et} + k_a(k_{et} + k_r + k_{-r})[A]}$$ (B.10)

Two simplifying conditions can be envisaged, depending on whether the electron transfer step is fast or relatively slow. On the one hand, when $k_{et} \gg k_r \approx k_{-r}$ Equation B.10 reverts to Equation 3.1, except that k_{et} becomes k_r. This is a warning that an apparent k_{et} not obtained by direct measurement may not be the true k_{et} but the rate constant for some other rate-limiting step in formation of the reactive form of the complex. On the other hand, when $k_r \gg k_{-r} > k_{et}$, Equation B.10 also reverts to Equation 3.1, except that now K_A has to be interpreted as $K_A K_R$, K_R being the equilibrium constant for the rearrangement step, k_r/k_{-r}.

Matthis et al. (1995) have derived equations to describe the behaviour of an irreversible reaction under pseudo-first order conditions involving two non-overlapping binding sites both of which are active in electron transfer. The case of a rate-limiting rearrangement in a flash-activated system has been discussed by Sigfridsson et al. (1995, 1996).

B.4 Ionic strength effects

This section extends the discussion of Section 3.3.4 to give the analytical equations that have been employed in simple macroscopic electrostatic models of the effects of ionic strength. In earlier work, electrostatic interactions were treated in terms of the total charge of the reactants, and equations were developed based on either the Brønsted–Bjerrum theory of ionic reactions or Marcus theory (Marcus and Sutin, 1985).

Substitution of the Debye–Hückel expression for activity coefficients into Equation 3.9 yields the following expression for the rate constant:

$$\ln k(I) = \ln k(I_0) - \frac{e^2 \kappa}{8\pi\varepsilon_0 \varepsilon k_B T} \left[\frac{Z_A^2}{1 + \kappa r_A} + \frac{Z_B^2}{1 + \kappa r_B} + \frac{(Z_A + Z_B)^2}{1 + \kappa r_{AB\ddagger}} \right] \quad \text{(B.11)}$$

which gives the effect of ionic strength on the rate constant in terms of the rate constant at infinite dilution (I_0), the ionic strength expressed as the Debye–Hückel factor κ (see Equation 3.4), the total charges on the proteins and the radii of the separate proteins and of the transition state. It applies strictly only to diffusion-controlled reactions for which the back reaction can be neglected (Koppenol, 1980).

Another approach (Wherland and Gray, 1976) uses as its starting point the introduction of a work term into the Marcus equation in the form of the potential energy U of interaction, which is calculated from the Debye–Hückel theory (Debye, 1942):

$$U = \frac{Z_A Z_B N_A e^2}{8\pi\varepsilon_0 \varepsilon} \left[\frac{e^{\kappa r_A}}{1 + \kappa r_A} + \frac{e^{\kappa r_B}}{1 + \kappa r_B} \right] \cdot \frac{e^{-\kappa r_{AB}}}{r_{AB}} \quad \text{(B.12)}$$

Substitution of Equation B.12 into Equation 3.10 results in the Debye–Marcus equation:

$$\ln k(I) = \ln k(I_\infty) - \frac{Z_A Z_B e^2}{8\pi\varepsilon_0 \varepsilon k_B T} \left[\frac{e^{-\kappa r_B}}{1 + \kappa r_A} + \frac{e^{-\kappa r_A}}{1 + \kappa r_B} \right] \cdot \frac{1}{r_{AB}} \quad \text{(B.13)}$$

which gives the effect of ionic strength in terms of the rate constant extrapolated to infinite ionic strength when the effects of protein charges have been shielded completely by mobile ions. An equivalent expression related to I_0 may also be derived.

The Brønsted and Debye–Marcus equations give similar results when the ionic strength is low, but diverge at higher values. The main difference between the two treatments is that Brønsted, but not Debye–Marcus, treats the two reactants as coalescing to form a sphere in the activated complex. They have been used with reasonable success with some systems, but break down completely in other cases, primarily because the importance of local charge at the reaction site, rather than overall charge, is not recognized.

In the above treatments, protein molecules are regarded as uniformly charged spheres. The first step in a more realistic treatment was the introduction of an overall dipole moment, which could be calculated once the three-dimensional structure of the protein was known. Van Leeuwen (1983) derived the following equation (which differs slightly from Debye's Equation B.12) for the interaction energy between two protein molecules treated as uniformly charged spheres (monopoles) when they are in contact:

$$U = \frac{Z_A Z_B e^2}{8\pi\varepsilon_o\varepsilon} \cdot \frac{\left(1-e^{-2\kappa r_B}\right)}{\kappa r_B\left(1+\kappa r_A\right)} \cdot \frac{1}{r_{AB}} \tag{B.14}$$

The effect of additional dipolar interactions was shown to modify the effective value of Z by addition of a term dependent on ionic strength and $\cos\theta$, where θ is the angle between the direction of the dipole moment and a line from the centre of the molecule to the centre of the active site. Equation B.14 for the interaction energy was substituted into expressions for the binding constant and the observed pseudo-first order rate constant, which were given in terms of U. In the case of the binding constant:

$$K_A(I) = K_A(I_\infty)e^{-U/RT} \tag{B.15}$$

The expression for the rate constant became relatively simple only in the case of certain limiting conditions. When the reaction is activation controlled ($k_{-a} \gg k_{et} + k_a[A]$), $k_1 = K_A k_{et}[A]$ and the full expression for the effect of ionic strength on k_1 becomes:

$$\ln k_1(I) = \ln k_1(I_\infty) - \left[Z_A Z_B + (ZP)(1+\kappa r_{AB}) + (PP)(1+\kappa r_{AB})^2\right]$$
$$\times \frac{Z_A Z_B e^2}{8\pi\varepsilon_o\varepsilon k_B T}\left[\frac{\left(1-e^{-2\kappa r_B}\right)}{\kappa r_B\left(1+\kappa r_A\right)}\right] \cdot \frac{1}{r_{AB}} \tag{B.16}$$

where ZP and PP refer to the monopole–dipole and dipole–dipole interactions respectively:

$$(ZP) = \frac{\left(Z_A P_B\cos\theta_B + Z_B P_A\cos\theta_A\right)}{e r_{AB}}$$
$$(PP) = \frac{\left(P_A P_B\cos\theta_A\cos\theta_B\right)}{\left(e r_{AB}\right)^2} \tag{B.17}$$

In van Leeuwen's analysis, the dipolar terms, ZP and PP, are significant at high ionic strength ($I > 100$ mM) when $1/\kappa$ is small and the proteins behave as if possessing a collection of individual point charges, but their effect disappears at low ionic strength when the influence of individual charges tends to become merged into the overall charge of the protein as a whole. The method has had considerable success in describing the behaviour of some systems and gives information about θ and hence the structure of the reaction complexes. It fails in other cases, particularly when it is evident that the charge at the reaction site of a protein is of opposite sign to the net charge of the molecule as a whole. The 'parallel plate' model of Watkins et al. (1994) was introduced to model in a simple way the case of interactions involving large proteins that have a marked concentration of charged groups of the same sign in the neighbourhood of the reaction site.

The difference between the van Leeuwen model and that of Watkins et al. lies in the geometry of the charge distribution. The latter authors treated the significant charges as being evenly distributed over a plate of a radius of a few angstroms at the reaction site and, in addition, calculated a dipole moment and allowed for an angle between the direction of the dipole moment and the line from the centre of the molecule to the centre of the plate. The interaction energy was then calculated as a function of ionic strength, but in contrast to van Leeuwen this was introduced into a simple equation for the second order rate constant of a diffusion-controlled reaction. The overall equation derived was similar in form to that of van Leeuwen:

$$\ln k_2(I) = \ln k_2(I_\infty) - V_{ii}X(I) - V_{id}X(I)Y(I) - V_{dd}Y(I)^2 Z(I) \quad \text{(B.18)}$$

In Equation B.18, the V_{xx} are numbers representing the monopole–monopole, monopole–dipole and dipole–dipole interactions respectively, and can be converted into the corresponding potential energies by multiplying by $k_B T$. The total interaction energy is given by

$$U = k_B T(V_{ii} + V_{id} + V_{dd}) \quad \text{(B.19)}$$

The V terms are similar to the corresponding ones of van Leeuwen except for the nature of the distance terms introduced. The main differences lie in $X(I)$, $Y(I)$ and $Z(I)$ which are increasingly complex functions of κ. In most of the cases examined by Watkins et al., however, the experimental results could be fitted with the monopole–monopole term only, and inclusion of dipolar terms did not improve the fit. Equation B.18 then reduces to:

$$\ln k_2(I) = \ln k_2(I_\infty) - V_{ii} \frac{e^{-\kappa\rho}}{(1 + \kappa\rho)} \quad \text{(B.20)}$$

where ρ represents the radius of the charged plate on each protein. A plot of $\ln k_2(I)$ against $I^{1/2}$ can be fitted to Equation B.20 to give three parameters, $k_2(I_\infty)$, ρ and V_{ii}. The last of these is the intercept on the $\ln k_2$ axis and represents the interaction energy (after multiplying by $k_B T$) extrapolated to zero ionic strength.

References

Debye P. (1942) Reaction rates in ionic solutions. *Trans. Electrochem. Soc.* **82**: 265–272.
Drepper F, Hippler M, Nitschke W, Haehnel W. (1996) Binding dynamics and electron transfer between plastocyanin and photosystem I. *Biochemistry* **35**: 1282–1295.

Frieden C. (1993a) Numerical integration of rate equations by computer. *Trends Biochem. Sci.* **18:** 58–60.

Frieden C. (1993b) Numerical integration of rate equations by computer: an update. *Trends Biochem. Sci.* **19:** 181–182.

Frieden C. (1995) Analysis of kinetic data: practical applications of computer simulation and fitting programs. *Methods Enzymol.* **240:** 311–322.

Koppenol WH. (1980) Effect of a molecular dipole on the ionic strength dependence of a bimolecular rate constant. Identification of the site of reaction. *Biophys. J.* **29:** 493–507.

Kramers HA. (1940) Brownian motion in a field of force and the diffusion model of chemical reactions. *Physica* **7:** 284–304.

Marcus RA, Sutin N. (1985) Electron transfers in chemistry and biology. *Biochim. Biophys. Acta* **811:** 265–322.

Marquardt DW. (1963) An algorithm for least-squares estimation of nonlinear parameters. *J. Soc. Ind. Appl. Maths* **11:** 431–441.

Matthis AL, Vitello LB, Erman JE. (1995) Oxidation of yeast iso-1 ferrocytochrome c by yeast cytochrome c peroxidase compounds I and II. Dependence upon ionic strength. *Biochemistry* **34:** 9991–9999.

Nordling M, Sigfridsson K, Young S, Lundberg L, Hansson Ö. (1991) Flash-photolysis studies of the electron transfer from genetically modified spinach plastocyanin to photosystem I. *FEBS Lett.* **291:** 327–330.

Northrup SH, Thomasson KA, Miller CM, Barker PD, Eltis LD, Guillemette JG, Inglis SC, Mauk AG. (1993) Effects of charged amino acid mutations on the biomolecular kinetics of reduction of yeast iso-1-ferricytochrome c by bovine ferrocytochrome b_5. *Biochemistry* **32:** 6613–6623.

Qin L, Kostic NM. (1994) Photoinduced electron transfer from the triplet state of zinc cytochrome c to ferricytochrome b_5 is gated by configurational fluctuations of the diprotein complex. *Biochemistry* **33:** 12592–12599.

Sigfridsson K, Hansson Ö, Karlsson BG, Baltzer L, Nordling M, Lundberg LG. (1995) Spectroscopic and kinetic characterization of the spinach plastocyanin mutant Tyr83-His: a histidine residue with a high pK value. *Biochim. Biophys. Acta* **1228:** 28–36.

Sigfridsson K, Young S, Hansson Ö. (1996) Structural dynamics in the plastocyanin–photosystem I electron-transfer complex as revealed by mutant studies. *Biochemistry* **35:** 1249–1257.

Tollin G. (1995) Use of flavin photochemistry to probe intraprotein and interprotein electron transfer mechanisms. *J. Bioenerg. Biomembr.* **27:** 303–309.

Van Leeuwen JW. (1983) The ionic strength dependence of the rate of a reaction between two large proteins with a dipole moment. *Biochim. Biophys. Acta* **743:** 408–421.

Watkins JA, Cusanovich MA, Meyer TE, Tollin G. (1994) A "parallel plate" electrostatic model for bimolecular rate constants applied to electron transfer proteins. *Protein Sci.* **3:** 2104–2114.

Wherland S, Gray HB. (1976) Metalloprotein electron transfer reactions: analysis of reactivity of horse heart cytochrome c with inorganic complexes. *Proc. Natl Acad. Sci. USA* **73:** 2950–2954.

Zhou JS, Kostic NM. (1992) Photoinduced electron transfer from zinc cytochrome c to plastocyanin is gated by surface diffusion within the metalloprotein complex. *J. Am. Chem. Soc.* **114:** 3562–3563.

Zhou JS, Kostic NM. (1993) Gating of photoinduced electron transfer from zinc cytochrome c and tin cytochrome c to plastocyanin. Effects of solution viscosity on rearrangement of the metalloprotein complex. *J. Am. Chem. Soc.* **115:** 10796–10804.

Index